About Island Press

Since 1984, the nonprofit organization Island Press has been stimulating, shaping, and communicating ideas that are essential for solving environmental problems worldwide. With more than 1,000 titles in print and some 30 new releases each year, we are the nation's leading publisher on environmental issues. We identify innovative thinkers and emerging trends in the environmental field. We work with world-renowned experts and authors to develop cross-disciplinary solutions to environmental challenges.

Island Press designs and executes educational campaigns, in conjunction with our authors, to communicate their critical messages in print, in person, and online using the latest technologies, innovative programs, and the media. Our goal is to reach targeted audiences—scientists, policy makers, environmental advocates, urban planners, the media, and concerned citizens—with information that can be used to create the framework for long-term ecological health and human well-being.

Island Press gratefully acknowledges major support from The Bobolink Foundation, Caldera Foundation, The Curtis and Edith Munson Foundation, The Forrest C. and Frances H. Lattner Foundation, The JPB Foundation, The Kresge Foundation, The Summit Charitable Foundation, Inc., and many other generous organizations and individuals.

The opinions expressed in this book are those of the author(s) and do not necessarily reflect the views of our supporters.

A New Coast

A New Coast

Strategies for Responding to Devastating Storms and Rising Seas

Jeffrey Peterson

ISLANDPRESS

Washington | Covelo | London

ISLAND PRESS is a trademark of the Center for Resource Economics.

Library of Congress Control Number: 2019937117

All Island Press books are printed on environmentally responsible materials.

Manufactured in the United States of America
10 9 8 7 6 5 4 3 2 1

Keywords: adaptation, appropriations, beach nourishment, climate change, coastal mapping, coastal planning, coastal zone management, Community Rating System, Congress, disaster recovery, displacement, federal policy, FEMA, financing, flooding, flood insurance, hazard mitigation planning, hurricane, inundation, infrastructure, IPCC, litigation, National Climate Assessment, National Flood Insurance Program, NOAA, Paris Climate Agreement, planning, policy, real estate, relocation, resilient communities, risk assessment, sea level rise, social justice, Stafford Act, storm surge, tax incentives, wetlands

For Ida and James

Strategy requires a sense of the whole that reveals the significance of respective parts.

John Lewis Gaddis

Contents

Preface

Writing this book has been a pleasure because I have been able to combine a professional interest in sorting out how government can help address pressing problems with a long fascination with the sea.

As a college student in Maine I learned to sail and to look at the sea with what Rachel Carson called "wonder and curiosity." I always thought of the sea in terms of reliable rhythms of tides and the fixed points of land meeting ocean. It has been hard to accept that, in less than a hundred years, the entire intersection between land and sea will be radically altered by a rising sea level.

As an employee of the Environmental Protection Agency, my job was to use the tools of government to adapt to a changing climate. I worked on helping coastal communities prepare for storms and rising seas and with other federal agencies on adapting to new challenges posed by climate change. Following the election of Donald Trump as president, working within the federal government to help the country adapt to climate change became much more difficult.

After retiring from the Environmental Protection Agency in the summer of 2017, I watched in sadness and disbelief as three major hurricanes devastated the American coast in the space of three weeks, exacting a staggering toll in lives, lost homes, and damaged communities. From my work at EPA, I knew that a warmer climate would make future coastal storms more severe and cause sea levels to rise, making flooding worse. In the months that followed the storms, Congress appropriated tens of billions of dollars for disaster relief but showed little interest in coming to grips with this problem and getting the country better prepared for the more damaging storms and coastal inundation that are coming. Why not?

This book explores the country's failure to recognize the growing peril that storms and rising seas pose for coastal communities. It is a story of evolving scientific assessments, ignorance of the likely scale of losses, abdication of political leadership, and the perplexing novelty of the challenges these problems pose. Yet there are some rays of hope. The scientific understanding of coastal storms and sea level rise has improved significantly, and some state and local initiatives show promise. Still, the country is not prepared for more severe coastal storms and rising seas and not even making much progress in deciding what needs to be done.

In the process of untangling all the reasons for the failure to prepare for coastal storms and rising seas, I concluded that the country needs a plan to get ahead of this problem. Relying on market forces, existing programs, or the improvised preparation efforts of individual coastal communities, is not enough. I looked at some of the local successes, consulted experts, and framed some steps that offer a starting point for preparing the coast for more damaging storms and incremental inundation as seas rise.

It will be a fight to put in place the measures needed to prepare for the disruptions that more severe storms and rising seas will bring to the coast. The present willful ignorance of coastal risks is profitable for some. Preparing for storms and rising seas will cost money, although much less than repeatedly rebuilding after a disaster. There is a bureaucratic tendency to defend existing programs, even when they are inadequate. And, the coastal inundation coming to many places will pose hard choices, often leading people to move from homes and communities they know and love.

If there is to be a new national program to prepare for more severe storms and rising seas, there will first need to be a campaign to make that happen. Who needs to be part of a campaign? How should it be organized? What should it try to accomplish first? And, what can citizens do to understand the risk to storms and rising seas and help their community be better prepared? These are some of the questions addressed at the close of the book.

Understanding why the country is not better prepared for coastal storms and rising seas first requires reviewing the science of how oceans respond to a warming planet. That provides a foundation for discussion of impacts of storms and sea level rise on communities, ecosystems, and infrastructure. Having a picture of these impacts opens questions involving existing government programs, law, economics, sociology, psychology, demographics, corporate responsibility, and ethics. Considering strategies to respond to these risks requires exploring the political landscape and the capacity of civil society to put this issue on the national agenda. Readers may not share my conclusions, but I hope the book demonstrates the multiple facets of the problem and informs future debates on how to improve preparedness.

More damaging storms and rising waters will redraw the coastline in the decades to come, giving us a new coast. We are fortunate to have the resources, talent, and still some time to decide how to step back from the old coast and fashion a new one that meets our needs. It is up to us to meet the challenge.

Jeffrey Peterson
Falls Church, Virginia
2019

Introduction

Children of Americans living today will be coping with a warmer world dramatically altered by a changing climate. One of the most damaging impacts of warmer temperatures will be the widespread and permanent inundation of coastal communities and ecosystems as a result of higher sea level. To make matters worse, surges of ocean water generated by more severe coastal storms will ride these higher sea levels, bringing flooding farther inland than ever before. The population living along the American coast, which will increase dramatically by the year 2100, will be dealing with the steady destruction of homes, businesses, communities, and ecosystems.

Some of our descendants may be inclined to ask, "How the heck did this happen?" They would do well to look back at 2017 as a year that saw pivotal changes in understanding of the risks posed by more severe storms and rising seas. Three major hurricanes—Harvey, Irma, and Maria—crashed into the US coast in a period of several months in late 2017, with devastating damage. In a far less widely reported event earlier that year, federal government scientists released new, more geographically precise and significantly increased estimates of the extent of sea level rise along the coast. And, in January of 2017, Donald Trump was inaugurated as the 45th president of the United States and promptly began dismantling and defunding the scientific and programmatic capacity to respond to these problems. What a year.

With the events of 2017 fresh in my mind, I set out to take a hard look at the American coast, including communities, ecosystems, and economic enterprises, and to understand how more severe storms and rising seas are likely to change the coast in the decades ahead. I wondered if the country is prepared to manage the changes that more severe storms and rising seas will bring. Coastal storms in the summer of 2018 highlighted the urgency of this question. Although these 2018 storms did not match the destruction of the previous year, Hurricane Florence brought widespread flooding, both coastal and inland, to the Carolinas, and Hurricane Michael strengthened rapidly before barreling into the Panhandle of Florida, virtually wiping some coastal communities from the map.

It is fair to say that more severe storms and rising seas were something of a sleeper issue until recently. There have always been major hurricanes and the scientific evidence of increasing severity of these storms has emerged slowly. In 2013, the Intergovernmental Panel on Climate Change (IPCC) reported that global mean sea level was likely to rise somewhere in the range of 1.3 to 2.0 feet by the year 2100.[1] Such a modest rise in sea level, occurring so far in the future, seemed to some observers a sufficient reason to put actions to address the impacts of rising seas on the back burner.

Alarm bells started in 2016 when veteran climate scientist James Hansen and coauthors published a paper describing new research on sea level concluding, "Our analysis paints a very different picture than IPCC (2013)...if GHG [greenhouse gas] emissions continue to grow. In that case we conclude that multi-meter sea level rise would become practically unavoidable, probably within 50–150 years."[2] The authors took the unusual step of going beyond the dispassionate reporting of methods and findings: "Social disruption and economic consequences of such large sea level rise, and the attendant increases in storms and climate extremes, could be devastating. It is not difficult to imagine that conflicts arising from forced migrations and economic collapse might make the planet ungovernable, threatening the fabric of civilization."[3]

In early 2017, a team of scientists from federal agencies fired off another warning in a new assessment of changes to global mean sea level (GMSL) by the year 2100, reporting that "a physically plausible GMSL rise in the range of 2.0 meters (m) to 2.7 m [6.5 to 8.8 feet] . . ."[4] and "recent results regarding Antarctic ice-sheet instability indicate that such outcomes may be more likely than previously thought."[5] This report also included geographically specific estimates of future sea level rise in different regions of the United States to the year 2100, finding that, for many parts of the American coastline, sea level rise is "projected to be greater than the global average for almost all future GMSL rise scenarios."[6] And, unlike some earlier studies, this paper projected sea level rise beyond the year 2100, looking out to 2200 and finding that the upper range of possible increases in sea level by 2200 included seas rising by up to 30 feet in some regions.[7]

Rising sea level is something of an abstract concept, but new online mapping tools are now widely available to visualize how vertical sea level rise intersects the coastline causing it to shift inland to differing extents in different places depending on elevation and geography. This inland shifting of the coastline translates to permanent inundation of existing land and loss of property, built infrastructure, and natural features, such as beaches and wetlands. Hundreds of communities and millions of people are at risk of inundation by rising seas.

That's on a nice day. New understanding of how a warmer climate is likely to result in more severe coastal storms suggests that sea level rise is just half the story of what coastal communities should be expecting in the decades ahead. Temporary coastal flooding from the intense rainfall and ocean surges from major storms will increasingly occur along with higher sea level. As luck would have it, the parts of the American coast expected to see the greatest increases in sea level—the southeast Atlantic and Gulf of Mexico coasts—are also most vulnerable to major coastal storms.

At about the same time that new estimates of future sea level rise emerged, researchers were publishing new studies of the economic damage sea level rise is likely to cause along the American coast. A federal government study released in 2017, evaluating the potential cost of property damage from rising seas and storms, found that "cumulative discounted damages to coastal property in the contiguous U.S. are estimated at $3.6 trillion through 2100."[8] Significantly, the authors concluded that timely implementation of adaptation measures could dramatically slash these costs, reducing the estimated impacts to about $800 billion.

These estimates of impacts of future coastal storms and sea level rise sound bad, but the reality that future generations experience may be much more disruptive. Over twenty million Americans lived in coastal areas with less than thirty-three feet of elevation in 2000,[9] and this population is projected to roughly double by 2060,[10] resulting in a clash between rising seas and demand for housing, roads, and related infrastructure needed to serve new residents.

Given what we now know about the serious risks of more severe storms and rising seas, you might think that governments, nonprofit organizations, and the private sector must have made tackling the problem a top priority. Sadly, this is not the case. Why is a country so richly endowed with scientific, legal, and economic resources—and with so much to lose by fumbling its response to coastal storms and rising seas—struggling to prepare for these risks?

One key reason is that the science in this area was, until recently, uncertain enough to support a decision to defer any action. And, without high confidence in the science, estimates of costs and impacts were hard to pin down and more alarming estimates could be dismissed.

Longstanding federal government programs offering subsidized flood insurance and generous funding for storm recovery provide people with confidence to stay in risky coastal areas. These programs, developed long before the threats of more severe coastal storms and permanent inundation from rising seas emerged, are deeply woven into the coastal property financial system and have yet to be revised with the new risks in mind.

In addition, there is no playbook for the best way to respond to more severe storms and rising seas. Never before have communities that are home to millions of people faced gradual and permanent inundation. Novel issues arise regarding land ownership, financial impacts on coastal property owners, and the economic viability of coastal communities. Without proven solutions to these challenges, many governments, businesses, and individuals are taking a wait and see approach.

Another important reason for inaction on the challenges of more severe coastal storms and rising seas is abdication of leadership from the president and Congress. The Trump administration took office in January 2017 and dynamited efforts to address climate change generally, including several initiatives intended to come to grips with coastal storms and sea level rise, and proposed to cut funding for existing programs that might help. As a result, federal agencies have been able to do little to respond to the most recent sea level rise science. Congress has also been slow to hold hearings on the topic of coastal storm preparedness and sea level rise and to develop legislation defining response strategies.

The premise of this book is that the country needs to devise and implement a national program to better cope with the challenges of more severe coastal storms and rising seas. Why is a national program needed? The new science defining the risks is strong (see part 1). The scale and costs of the problem—measured in terms of impacts on communities, critical infrastructure, ecosystems, and private assets—are significant and occur around the entire coast (see part 2). The existing programs for flood insurance, disaster assistance, and coastal management are struggling to keep up with the costs of major coastal storms and are not set up to deal with the permanent inundation that comes with rising seas (see part 3). States, local governments, and businesses are making some headway in responding to these challenges, but the legal, social, and economic issues they face are new and complex, and they need the support and financial assistance that a national program can provide (see part 4).

A national program to address more severe coastal storms and rising seas needs clear goals such as reducing loss of life and property and reducing costs to the government for disaster relief. A national program working toward these goals should have several attributes. First, it should be complete (i.e., it addresses risks to coastal communities but also ecosystems, critical infrastructure, and military assets). The program should be proportional in the sense that proposed responses can be paid for with the financial resources reasonably available. And, the program should be fair in the sense that it protects everyone's interests on the spectrum from disadvantaged to wealthy. A national program with these attributes can best be accomplished through a cooperative effort among state and local governments, the private sector, and citizens, led by the federal government (see part 5).

An intangible but important function of a national program is helping people in coastal areas come to terms with the risks they are facing and the changes that are coming to the American coast. The hard reality is that, although coastal protection structures, such as a seawall, can buy time for some communities, sooner or later the sea will inundate most low-lying coastal areas. A national program that is complete, proportional, and fair is likely to steer new development away from risky areas and lean toward plans to step back from the coast that we know today. People need to be reassured that, despite the pain and cost of relocation, the new coast that will emerge in the decades ahead can have much the same character as the old coast, albeit with greatly reduced risks of devastating losses of life, property, and ecosystems. So, a final important attribute of a national program is integrity in the sense that the planning process is smart enough to convey the opportunities that come with a new coast and solid enough not to wither in the face of diverse objections to making hard choices related to financing and relocation.

A second premise of the book is that a new national program to prepare for more severe storms and rising seas is not likely to happen without a campaign to put the issue on the national agenda. To advance the national debate on how to respond to these challenges, a campaign needs a platform of policies, programs, and funding mechanisms. A campaign can also organize interested parties, build public awareness, and advocate for adoption of key actions (see part 5). Importantly, a national campaign can deliver the political energy needed to overcome opposition to this expensive and sometimes controversial work. In effect, a campaign can help lead a transition to a new coast.

PART I

A Warming Climate Drives Coastal Storms and Rising Seas

Reason and free inquiry are the only effectual agents against error.

Thomas Jefferson

1

Coastal Storms,
Coastal Nightmare

Major coastal storms are killers. Looking back on the human costs of Hurricane Harvey, which hit the coast of Texas in late August 2017, the *Houston Chronicle* memorialized the seventy-five people killed in the resulting flooding: "A beloved pastor and his wife swept away by a raging creek in Fort Bend County. An elderly man who died alone, trapped by rising waters in his west Houston home. Six members of the Saldivar family trying to escape the torrential rains. A dedicated police officer who could not ignore his duty. Those are among the many whom this storm took from us, and many others whose names we don't yet know."[1]

Just a week later, Hurricane Irma, the strongest Atlantic basin hurricane ever recorded, devastated the Caribbean, the Florida Keys, and the western coast of Florida, resulting in the death of ninety-two people in the United States, including seventy-seven in Florida.[2] The *Florida Sun Sentinel* interviewed a resident of Cudjoe Key: "It's been a nightmare. . . . You live here in a resort, everything's nice and pretty, and the next day it's all gone. . . . Death. That's what it sounded like to me."[3] Winds exceeded 130 miles per hour[4] and sea water surged five to eight feet above ground level in the Keys.[5]

Then, less than a month later, on September 20, Hurricane Maria struck the island of Puerto Rico with winds of 155 miles per hour[6] driving a surge of water six to nine feet high.[7] The initial official death toll was sixty-four, but several organizations argued it was much higher. The *New York Times* reviewed differing estimates

Figure 1-1. Port Arthur, Texas, was among many Texas coastal communities that suffered extensive flooding of homes, businesses, and transportation systems as a result of Hurricane Harvey, August 2017. Photo by Staff Sgt. Daniel Martinez, South Carolina National Guard.

and found that 1,052 more people than usual died across the island in the forty-two days after the storm.[8] A May 2018 report by researchers at Harvard University came to a much higher estimate, finding that 4,645 people died as a result of the hurricane.[9] Many of these deaths are associated with lack of access to medical services or facilities or electric power. In August of 2018, the government of Puerto Rico settled on an estimate of 1,427 deaths directly due to storm damage while noting that estimates from other studies range between 800 and 8,000 deaths due to delayed health care.[10]

The loss of life in these storms was tragic, but not record setting by American standards. The Galveston Hurricane in 1900 is thought to have killed between 6,000 and 12,000, with winds of over 140 miles per hour and a storm surge of fifteen feet.[11] Hurricane Katrina killed at least 1,833 people in late August 2005, with wind speeds over 175 miles per hour and a storm surge of twenty-four to twenty-eight feet along the northern Gulf of Mexico.[12] Hurricane Sandy brought high winds and a storm surge over nine feet at the lower end of Manhattan Island and along the New Jersey shore, claiming 106 lives, mostly from drowning,[13] in October 2012. Roughly a dozen hurricanes have each resulted in over 100 deaths in the United States in the past century.[14]

Figure 1-2. The US Coast Guard conducts a water rescue in Jacksonville, Florida, during Hurricane Irma, September 2017. Photo by US Coast Guard.

Figure 1-3. Coast Guard Lt. Lucas Taylor provides food and water to a girl in Moca, Puerto Rico, October 2017, following Hurricane Maria. Photo by Petty Officer 3rd Class David Micallef, US Coast Guard.

One factor behind the significant loss of life and damage costs of the 2017 storms is the growth of population and the value of assets along the coast. Another consideration is that the long-standing phenomena of major coastal storms is playing out against a backdrop of a warming planet. Climate models suggest that a warmer climate will result in more intense, and perhaps more frequent, coastal storms. In addition, warming temperatures are driving a gradual rise in sea levels globally and along the American coast. Rising seas will not make coastal storms more frequent or more severe but will push storm surges farther inland.

With these concerns in mind, it is worth looking more closely at the problem that connects coastal storms and rising seas: storm surge. It is also important to understand past trends in costs of major storms and how storms may change as the planet warms.

The Role of Storm Surge in Coastal Storm Deaths and Damage

A storm surge is a wave of ocean water, over and above the predicted astronomical tide, generated by high winds and low barometric pressure associated with a coastal storm. Smaller storms combine with high tides to generate nuisance flooding or, occasionally, flooding on the "coastal flood warning" scale of several feet.

A key thing about storm surges is that the bigger the storm—the greater the winds and lower the barometric pressure—the bigger the storm surge (see fig. 1–4). Exceptionally high storm surges, such as the twenty-four to twenty-eight feet delivered by Hurricane Katrina, have occurred, but surges of five to ten feet are more common, and damages can vary widely based on the elevation of the coast. Hurricane Michael, with the third lowest barometric pressure recorded at landfall, came ashore on the Florida Panhandle with storm surges in excess of ten feet east of Panama City and fourteen feet in Mexico Beach.[15]

The other key thing to know about storm surges is that they are by far the deadliest element of a coastal storm. In 2014, Edward Rappaport of the National Hurricane Center published a paper looking at deaths from major storms over the past fifty years finding that, "roughly 90% of the deaths occurred in water-related incidents, most by drowning . . . "[16] and "storm surge was responsible for about half of the fatalities (49%)."[17] In contrast, high winds were estimated to have caused less than 10 percent of deaths.

Given the deadly effect of storm surges, it is very useful to know what land areas are at risk of flooding by a surge in the event of a storm. Fortunately, understanding of coastal areas at risk of storm surges has improved significantly in recent years. The National Hurricane Center within the National Oceanic and Atmospheric Administration (NOAA) uses a model to "estimate storm surge heights resulting from historical, hypothetical, or predicted hurricanes."[18] The "Sea, Lake and Overland Surges from Hurricanes" model, or SLOSH for short, predicts for each basin along the Atlantic and Gulf of Mexico coasts the geographic extent and depth of storm surge in the event of a given storm size (e.g., hurricane Categories 1–5).

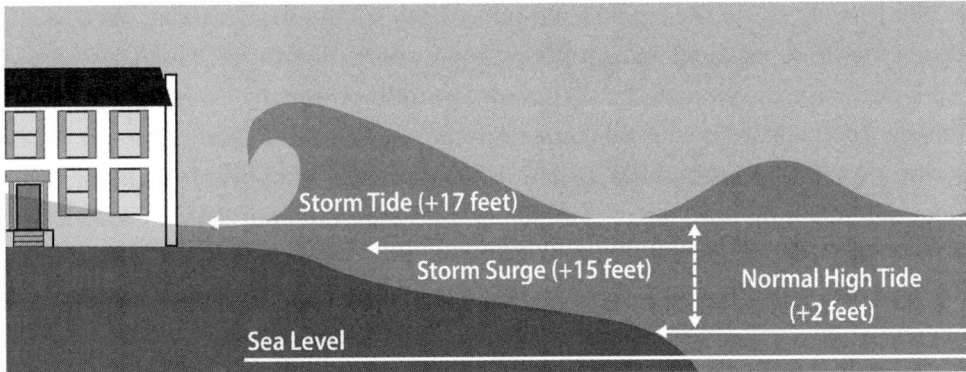

Figure 1-4. Storm surge and tide impacts. Illustration of water level differences for storm surge, storm tide, and a normal high tide, as compared to sea level. Storm surge is the rise in sea level caused solely by a storm. Storm tide is the total observed sea level during a storm, which is the combination of storm surge and normal high tide. National Ocean Service, National Oceanic and Atmospheric Administration.

Figure 1-5. Hurricane Michael caused widespread destruction along the Florida Panhandle, including in Mexico Beach, Florida, October 2018. Photo by Petty Officer 1st Class Colin Hunt, US Coast Guard.

Drawing on the SLOSH model and other data, NOAA estimates that, "in a worst-case scenario, approximately 24 million people along the East and Gulf coasts are at risk from storm surge flooding."[19] The risk consulting company CoreLogic came to a roughly comparable conclusion in a 2018 report finding 6.9 million homes at risk of storm surge.[20] NOAA found that by far the greatest number of people at risk of storm surge are in Florida, with significant populations at risk in Louisiana, New York, and New Jersey. These estimates of populations at risk of storm surge, however, are based on the land areas at risk with current sea level and do not reflect additional land area or population at risk based on future sea level, or more severe coastal storms, or growing coastal populations.

Knowing the coastal land areas most at risk of storm surge flooding in the event of a storm is a big step forward, but it would be even better to also have a sense of the risk of a major storm occurring at a specific place along the coast. The National Hurricane Center has data on that as well. This data is framed to provide a return period for major storms (i.e., hurricanes on the scale of Category 1–5) in a given coastal county, based on past experience, for the Atlantic Coast and the Gulf of Mexico.

For example, looking at NOAA's Hurricane Strike Frequency Map, it is possible to find that Miami–Dade County, Florida, has experienced six Category 1 hurricanes, and that a storm of that scale can be expected about eighteen times over an extended period (e.g., 1900–2009) (see table 1-1).[21] Because these data are drawn from past experience, it does not reflect projected increases in storm intensity due to a warming planet.

Even with these impressive statistics, however, it is impossible to know from year to year when or where a major coastal storm will form or strike. Predictions of the path and intensity of hurricanes already formed, however, are getting better, thanks to NOAA's Hurricane Forecast Improvement Project. Started in 2009, the ten-year effort is intended to reduce errors in storm track and intensity estimates by 50 percent and extend forecasts from five to seven days.[22]

Past Trends in Coastal Storms

Every four years, the United States Global Change Research Program, made up of scientists from federal agencies and other organizations, publishes a national assessment of changes in the climate. The 2014 *National Climate Assessment*, speaks to the subject of past coastal storms: "The intensity, frequency, and duration of North Atlantic hurricanes, as well as the frequency of the strongest (Category 4 and 5) hurricanes, have all increased since the early 1980s."[23] The 2017 Climate Science Special

Table 1-1. Hurricane strike frequency data for Miami–Dade County, Florida

Storm category	# of strikes	Return period, years
1	6	18
2	5	22
3	8	14
4	4	28
5	2	55

Note: Return period is defined as the average recurrence interval of a hurricane of similar in magnitude over an extended period of time (e.g., 1900–2009).

Source: Data from "Hurricane Frequency," Storm Surge Inundation Map; Miami–Dade County (click on Miami–Dade County) Environmental Protection Agency website.

Report, which is part 1 of the 2018 *National Climate Assessment*, linked these storm changes to human activity: "Human activities have contributed substantially to observed ocean—atmosphere variability in the Atlantic Ocean (medium confidence), and these changes have contributed to the observed upward trend in North Atlantic hurricane activity since the 1970s (medium confidence)."[24]

Exploring the question of trends in the costs of past major coastal storms, NOAA looked at a subset of all disasters costing over a billion dollars and found steady cost increases. In 2019, NOAA concluded that, since 1980, the United States has sustained 241 weather and climate disasters where overall damages and costs reached or exceeded $1 billion (including Consumer Price Index adjustment to 2018). The total cost of these 241 events exceeds $1.6 trillion.[25] The trend, however, is upward. From 1980 to 2013, about a half a dozen billion-dollar disasters occurred each year while in the last five years the number has increased to about a dozen.

More important, NOAA evaluated costs of different types of disasters and found that hurricanes and related storms were the single largest type of disaster event: "In short, tropical cyclones are the most costly of the weather and climate disasters . . . 40 tropical cyclones have caused a combined $862 billion in total damages—with an average of $21.6 billion per event. Accounting for just under a fifth (17 %) of the total number of events, tropical cyclones have caused almost half (55%) of the total damages attributed to billion-dollar weather and climate disasters since 1980."[26] The average costs of other types of events NOAA evaluated include drought ($9.4 billion per event), flooding ($4.3 billion per event), and wildfires ($2.5 billion per event).[27]

Average costs, however, can hide the fact that a single storm can cost over $100 billion. In January 2018, NOAA announced the final total cost of Hurricanes Harvey, Irma, and Maria to be a staggering $265 billion (Harvey, $125 billion; Irma, $50 billion; and Maria, $90 billion).[28] As a point of reference, NOAA cites the cost of Hurricane Katrina as $161 billion (adjusted to 2017 dollars), and the cost of Hurricane Sandy as $71 billion (adjusted to 2017 dollars).[29] Costs of the 2018 storms were more modest but still above average (i.e., $25 billion for Hurricane Michael and $24 billion for Hurricane Florence).[30] The social and psychological costs of lost homes, disrupted lives, and broken communities, although significant, are not monetized.

In early 2019, the Congressional Budget Office (CBO) estimated that the country should expect annual economic losses of $54 billion due to hurricanes and tropical storms under current conditions and policies (i.e., not accounting for more severe storms or rising seas). Storm surge flooding of residential property generated the largest category of losses, followed by residential wind damage and flood and wind damage to commercial and public property.[31]

As crushing as these costs of storm damages are for the communities and coastal property owners hit by a storm, much of the cost of recovering from major storms is now paid by federal taxpayers across the country. Costs to the federal government come in the form of losses by the National Flood Insurance Program and costs for disaster assistance, including billions of dollars in supplemental appropriations for the most damaging storms. CBO looked at expected annual costs to the federal government under current conditions and policies and estimated costs to be $17 billion,[32] but these costs can be dramatically higher in the event of multiple major storms.

In addition to direct costs, major coastal storms also have more general economic impacts. Following Hurricane Harvey, the *Wall Street Journal* reported that "gasoline prices surged to a two-year high at the pump Thursday after the owner of the largest pipeline in the U.S. reported that shipments are being sharply curtailed, spreading the economic pain from Hurricane Harvey throughout the nation."[33]

Anecdotal reports of broad economic impacts are generally confirmed by academic research. In a 2014 paper looking at the global record of economic impacts of tropical cyclones, Solomon M. Hsiang and Amir S. Jina came to a troubling conclusion: "We find robust evidence that national incomes decline, relative to their pre-disaster trend, and do not recover within twenty years. Both rich and poor countries exhibit this response. . . . Income losses arise from a small but persistent suppression of annual growth rates spread across the fifteen years following disaster, generating large and significant cumulative effects."[34]

Future Coastal Storms on a Warming Planet

Looking for long-term trends in the record of past coastal storms is one way to try to understand future storm patterns. Another approach is to develop models of changing climate conditions, such as air temperature and ocean condition, and estimate the frequency and severity of future storms. The science of "geophysical fluid dynamics" addresses the question of whether a warming planet will result in coastal storms that are more frequent or severe than those experienced today or in the past.

Use of models to project future coastal storms is a tricky business. The 2014 National Climate Assessment evaluated multiple studies and concluded that, "by late this century, models, on average, project a slight decrease in the annual number of tropical cyclones, but an increase in the number of the strongest (Category 4 and 5) hurricanes. . . . There is some uncertainty in this as the individual models do not always agree on the amount of projected change."[35]

The 2017 Climate Science Special Report generally backed the earlier assessment: "Both theory and numerical modeling simulations generally indicate an increase in tropical cyclone (TC) intensity in a warmer world, and the models generally show an increase in the number of very intense TCs."[36] Focusing in on the Atlantic Basin, NOAA concluded that "it is likely that climate warming will cause Atlantic hurricanes in the coming century to have higher rainfall rates than present-day hurricanes, and medium confidence that they will be more intense (higher peak winds and lower central pressures) on average,"[37] and "it's likely the number of major hurricanes (Category 3 and higher) would increase by two in a similar active year at the end of century."[38] Of course, more intense storms generate larger storm surges that reach higher elevations and farther inland than surges from less intense storms.

To make matters worse, more intense storms can increase rainfall, and the extent of coastal inundation, beyond the flooding that would result from just storm surge. The 2014 *National Climate Assessment* concluded that warmer air results in greater rainfall from hurricanes, finding that "almost all existing studies project greater rainfall rates in hurricanes in a warmer climate, with projected increases of about 20% averaged near the center of hurricanes."[39] In September 2018, researchers at Stony Brook University evaluated the impact of Hurricane Florence, which deluged the Carolinas in 2018, reporting that "rainfall amounts over the Carolinas are increased by over 50% due to climate change."[40]

This more intense rainfall might be manageable if greater storm intensity also resulted in storms passing over a given place faster. Unfortunately, these more intense storms are slowing down, raining on a given place longer. In a study that evaluated hurricanes between 1949 and 2016, James Kossin reported that hurricanes around the globe are moving more slowly; up to 10 percent slower globally, and as much as 20 percent slower when over land in the Atlantic region.[41]

The stalling of Hurricane Harvey over Houston, which produced over four feet of rain in some locations, is an example of this storm-slowing effect. The extreme rainfall from Harvey is also considered a preview of future storm rainfall. Although some observers dismissed any role for climate change, an international team of scientists looked at data from Harvey and prior storms finding that, "global warming made the precipitation about 15% (8%–19%) more intense, or equivalently made such an event three (1.5–5) times more likely. This analysis makes clear that extreme rainfall events along the Gulf Coast are on the rise."[42]

How might this slowing of storms, combined with greater storm intensity, translate to changes in rainfall in the future as the climate warms? In 2018, Ethan Gutmann and colleagues at the National Center for Atmospheric Research looked at more than twenty Atlantic storms and modeled how they would have been different if they had

Figure 1-6. Hurricane Florence damaged coastal communities in North and South Carolina, September 2018, and caused major flooding inland, including flooding of poultry and swine operations. Photo by Larry Baldwin, Crystal Coast Waterkeeper, Creative Commons Attribution 2.0. https://creativecommons.org/licenses/by-nc-nd/2.0/.

occurred in the warmer conditions expected later this century.[43] These simulations, which took over a year to run on a supercomputer in Wyoming, supported the conclusion that, "while each storm's transformation would be unique, on balance, the hurricanes would become a little stronger, a little slower-moving, and a lot wetter. . . . The rainfall rate of simulated future storms would increase by an average of 24 percent."[44]

Although elevation is the critical geographic factor in storm surge inundation risk, more rainfall might contribute a lot or a little to coastal inundation based on other aspects of geography, including the size of the watershed draining to coastal waters. For example, the flat landscape of eastern North and South Carolina played a part in spreading the flood damages from Hurricane Florence beyond the coast to inland communities.

Still another confounding impact of a warmer climate on coastal storms is that, even as storms move more slowly, they are intensifying more rapidly. Researchers led by Kieran Bhatia at NOAA found in 2019 that, for storms in the Atlantic basin, the number of times that storm intensity increased by more than 30 knots of wind speed over 24 hours tripled over the period 1982 to 2009.[45] Hurricane Michael, for example, went from a Category 1 to a Category 4 storm in just 24 hours. Rapidly intensifying hurricanes are more difficult to forecast and more likely to result in loss of life and property. The NOAA researchers point to a changing climate as a key cause of rapid intensification, noting "natural variability cannot explain the magnitude of the observed upward trend in the Atlantic basin."[46]

Scientists will continue to try to reduce the uncertainty surrounding coastal storms but it seems safe to say that more extensive flooding due to increased rainfall and higher storm surges, perhaps occurring more frequently, will need to be taken into account in coastal disaster planning. More important, coastal disaster plans also need to account for rising sea levels. Rising seas will push the occasional storm surge farther inland as well as gradually and permanently inundate low-lying coastal areas. Unfortunately, as the next chapter indicates, the parts of the American coast with long histories of severe coastal storms also happen to be the areas expected to see the greatest rise in sea level.

2

Sea Level Rise Projections: Trending Upward

The science related to future global sea level is changing fast, and new research consistently points to greater increases in sea level than previously predicted. This upward trend suggests a need for greater urgency in preparing for inundation of coastal areas as well as greater urgency in international efforts to reduce the release of greenhouse gases causing climate change and rising seas.

Global averages are helpful, but sea levels are projected to increase to different degrees in different places, and a simple, global average sea level projection is not that useful when trying to prepare the United States for the impacts of rising seas. Fortunately, new research now makes understanding likely future sea level changes in specific locations around the American coast much easier. Unfortunately, sea levels along the American coast are expected to increase more than the global average.

It is worth taking a moment to sketch the big picture of climate change. The consensus view of the scientific community is that global atmospheric temperatures have warmed by about 1.0°C (1.8° Fahrenheit (F)) over the past 115 years and "it is extremely likely that human activities, especially emissions of greenhouse gases, are the dominant cause of the observed warming since the mid-20th century."[1] Greenhouse gases, including carbon dioxide, methane, and nitrous oxide, trap heat in the atmosphere causing the planet to slowly warm. Although warming air temperatures are the primary consequence of the release of greenhouse gases, the warmer air temperatures contribute to a range of other environmental changes including more severe storms and rising sea level. International efforts to limit release of greenhouse gases under the Paris Climate Agreement seek to keep warming from increasing more than 2°C (3.6° F) above preindustrial levels or 1°C above current levels.

Past Sea Level Rise

The International Panel on Climate Change (IPCC) is the preeminent international body evaluating climate change science. In its fifth assessment report in 2013, the IPCC evaluated tide gauge data from around the world and concluded that between 1901 and 2010 the rise in global sea level was roughly 7.5 inches or only about .07 inches per year, which seems small upon first consideration.[2]

Since 1993, however, a network of satellites has circled the globe equipped with laser altimeters that precisely measure sea level. Drawing on this data, the IPCC reported that global sea level rose almost twice as fast from 1993 to 2012 as over the 1901–2010 period as a whole—an annual rate of .12 inches.[3] Figure 2-1 illustrates annual global average sea level change as measured by both tide gauges and satellites.

In 2017, the United States Global Change Research Program reported that more recent satellite data showed a slight upward increase in the annual rate of sea level rise for the period 1993–2015 to .13 inches per year.[4] This gradual rise in global sea levels is the result of a warmer planet. About one-third of the rise to date is due to thermal expansion of the ocean and the remainder is the result of the melting of ice, such as glaciers, around the world and ice sheets at the polar caps.

Projections of Global Sea Level Rise

Drawing on what is known about the past increases and causes of rising sea level, scientists are refining their predictive models. In general, research over the past dozen years resulted in steady increases in projected future global sea level.

A good place to start in looking at global sea level projections is the most recent major assessment report of the IPCC. In 2013 the IPCC reported that global mean sea level was expected to rise somewhere in the range of 1.3 to 2.0 feet by the year 2100.[5] This 2013 estimate was a small upward adjustment from estimates provided in the 2007 IPCC report, when the lower range was about seven inches and the upper range for all scenarios was just shy of 2 feet (23.2 inches).[6] Like the 2007 estimates, however, the 2013 IPCC estimates were quickly challenged by new research.

In 2014, the United States Global Change Research Program concluded that "on the high end, recent work suggests that 4 feet [by 2100] is plausible."[7] (See fig. 2-2.) The report went on to note that, "in the context of risk-based analysis, some

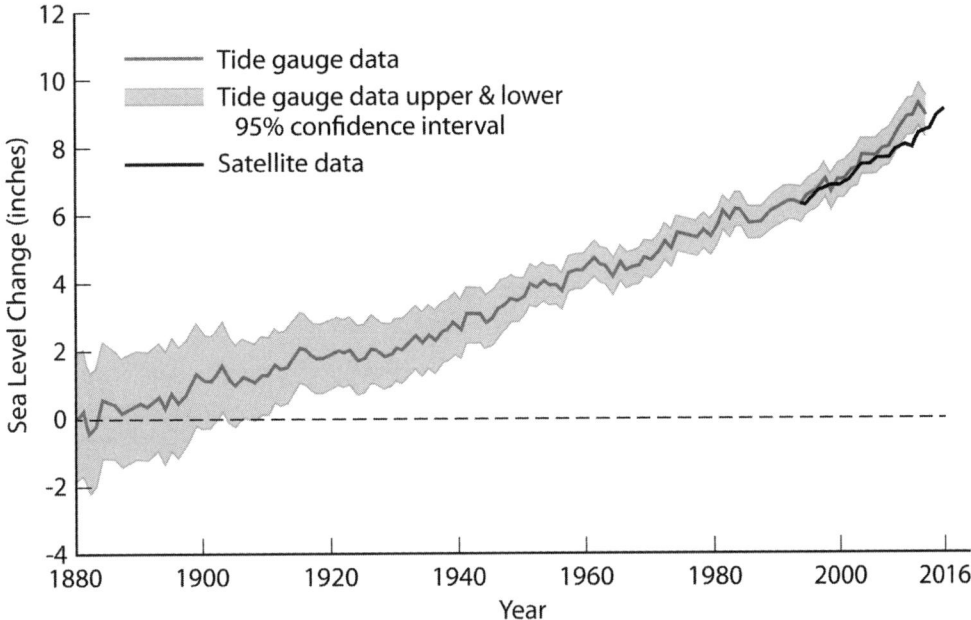

Figure 2-1. Global average absolute sea level change, 1880–2015. Cumulative changes in sea level for the world's oceans since 1880, based on a combination of long-term tide gauge measurements and recent satellite measurements. US Environmental Protection Agency, *Climate Change Indicators in the United States, 2016,* 4th ed. (Washington, DC: Environmental Protection Agency, 2016), 34.

decision-makers may wish to use a wider range of scenarios, from 8 inches to 6.6 feet by 2100. In particular, the high end of these scenarios may be useful for decision-makers with a low tolerance for risk."[8]

Then, in March 2016, veteran climate scientist James Hansen and coauthors published a paper describing new research including sea level rise predictions that "differ fundamentally from existing climate change assessments."[9] They concluded that "continued high fossil fuel emissions this century are predicted to yield . . . nonlinearly growing sea level rise, reaching several meters over a timescale of 50–150 years."[10] The authors posit that melting of Antarctic ice sheets will inject freshwater into the oceans, disrupting ocean currents and resulting in cascading impacts—including acceleration of additional loss of ice sheets.

Antarctic ice sheets also are at the center of research by Robert DeConto and David Pollard, published shortly after the Hansen paper. DeConto and Pollard reported the following: "Antarctica has the potential to contribute more than a meter of sea level rise by 2100"[11] on top of sea level rise caused by other factors such as thermal expansion and ice melting in other places. This research "suggests that the most recent

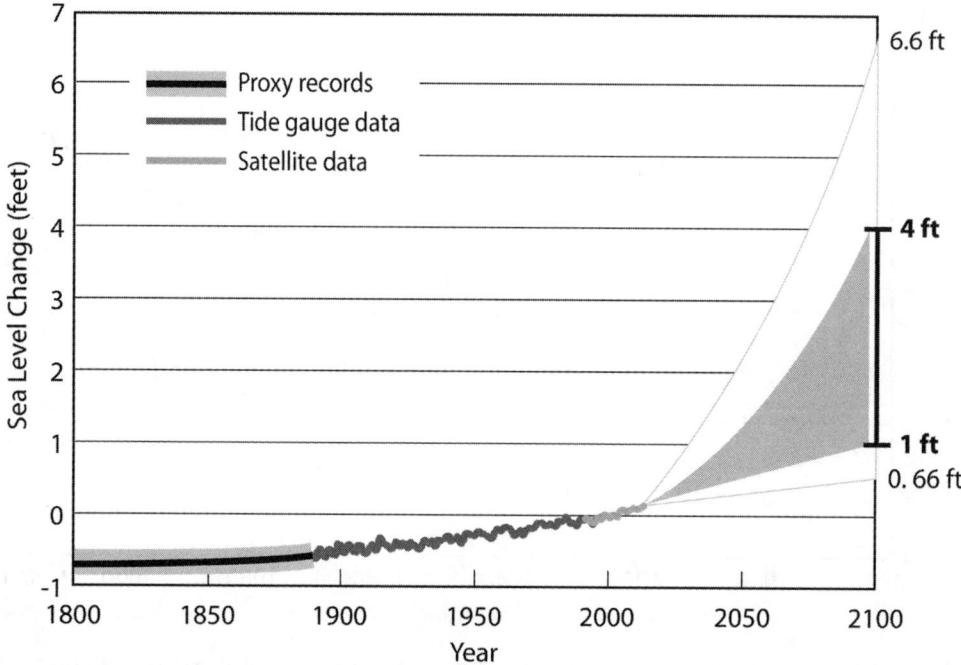

Figure 2-2. Past and projected changes in global sea level. Graph from the *2014 National Climate Assessment* illustrates the consensus view of the scientific community on future sea levels. Source: J. D. Walsh et al., "Our Changing Climate," ch. 2, *The Third National Climate Assessment* (Washington, DC: US Global Change Research Program, 2014), 45.

estimates by the Intergovernmental Panel on Climate Change for future sea-level rise over the next 100 years could be too low by almost a factor of two."[12] A key idea of the DeConto and Pollard paper is that warming air and ocean water temperatures will significantly reduce the fringes of the existing West Antarctic ice sheet resulting in sheer cliffs that are unstable and vulnerable to rapid breakup and melting.

In early 2017, scientists from federal agencies in the United States, led by William Sweet at the National Oceanic and Atmospheric Administration (NOAA), published a paper with new assessments of changes to global mean sea level concluding that "the projections and results presented in several peer-reviewed publications provide evidence to support a physically plausible GMSL [global mean sea level] rise in the range of 2.0 meters (m) to 2.7 m [6.5–8.8 feet]."[13] The authors defined the full range of "plausible" sea level rise and divided that into six scenarios and provided estimated sea level (see fig. 2-3 and app. 1). Although the probability of the "physically plausible" higher estimates occurring was judged to be very small, the NOAA paper authors concluded that "new evidence regarding the Antarctic ice

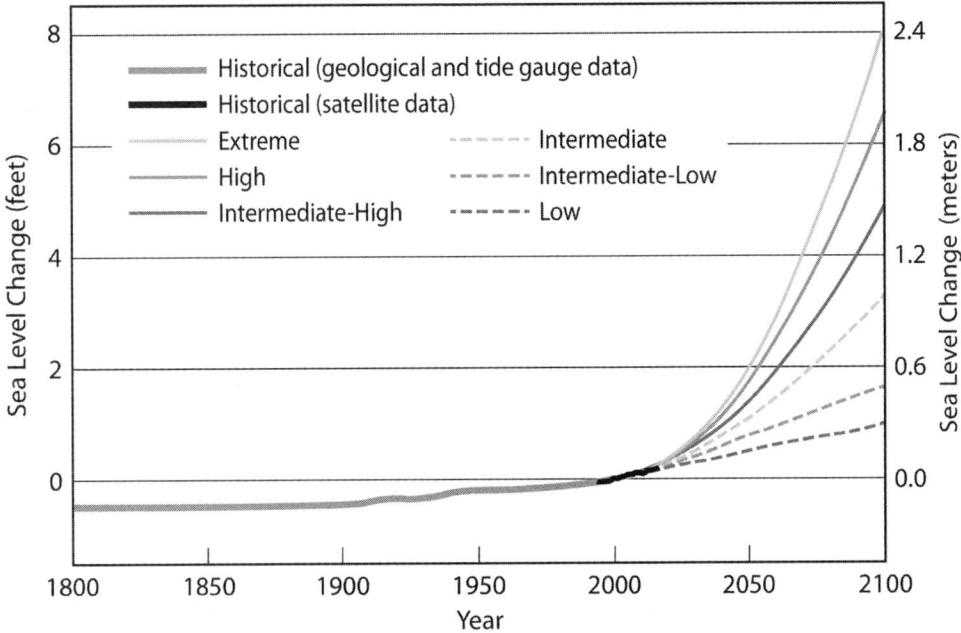

Figure 2-3. Historical and projected global average sea level rise. Six global average sea level rise scenarios developed by the US Federal Interagency Sea Level Rise Taskforce to describe the range of future possible rise this century. Adapted from David R. Easterling et al., "Our Changing Climate," ch. 2, *The Fourth National Climate Assessment* (Washington, DC: US Global Change Research Program, 2018), fig. 2.3.

sheet, if sustained, may significantly increase the probability"[14] of the physically plausible higher sea levels.

Many observers considered the 2016 Hansen paper controversial at the time, partly for its conclusion that Antarctic ice sheet loss rate was doubling every ten years. But in June 2018, a new assessment of the rate of melting of Antarctic ice sheets by a team of eighty scientists reported that ice sheet melting is occurring at an accelerating rate and that the rate has not doubled in the past ten years; it has tripled.[15] In late 2018, researchers analyzing ice cores from Greenland concluded that the rate of melting of the ice sheet is "off the charts."[16] In January 2019, the National Aeronautics and Space Administration reported a "gigantic cavity"[17] in the Thwaites Glacier in West Antarctica that "holds enough ice to raise the world ocean a little over 2 feet (65 centimeters) and backstops neighboring glaciers that would raise sea levels an additional 8 feet (2.4 meters) if all the ice were lost."[18] In April, 2019, an international team of scientists reported that loss of mass from glaciers, distinct from Greenland and Antarctic ice sheets, "may be larger than previously reported."[19]

Figure 2-4. Iceberg B-46 (*left*), estimated to be about 87 square miles in size, broke from Pine Island Glacier in West Antarctica in late October 2018. Photo by Brooke Medley, National Aeronautics and Space Administration.

Sea Level Rise on the United States Coast

In thinking about how to prepare for rising sea level in the United States, predictions of change in the global mean sea level are less useful than predictions of sea level that apply to the specific location of interest. It turns out that, for many of the places along the American coastline, sea level rise is expected to be higher than the average around the globe.

The more site-specific assessment of sea level is commonly called Relative Sea Level (RSL) and includes variables such as regional ocean currents, winds, upstream flood control, and the earth's gravity field. The most important variable, however, is vertical land movement that could include land subsidence (which has the effect of increasing the relative rise in sea level) or rebounding of land from the compressive weight of Ice Age glaciers (which has the effect of offsetting increases in sea level relative to the land).

How significant is the geographic variation within the global mean estimate of sea level rise? The IPCC addressed this question in 2013 finding that local deviation from the global mean projection can be between 10 percent and 25 percent in some ocean areas but that, "regional changes in sea level reach values of up to 30% above the global mean value in the Southern Ocean and around North America."[20]

Not only is much of the American coast expected to experience sea level increases well above the global mean, it is also subject to considerable vertical land movement most of which is land subsidence. In another stroke of bad luck, this subsidence is occurring in places, such as the central Gulf of Mexico and the Chesapeake Bay, which are also at risk of flooding during coastal storms.

For example, at Eugene Island, Louisiana, southwest of New Orleans, the projected relative sea level rise by 2100 is 6.32 feet, assuming the Intermediate sea level rise increment established in the 2017 NOAA report. Sea level rise accounts for about 3.7 feet of this change and land subsidence accounts for most of the difference.[21] In other places along the coast, however, land subsidence is less dramatic, and there is some upward land movement in Alaska and the Pacific Northwest due to rebounding of land from the compressive weight of former glaciers.

Coastal states and communities interested in preparing for sea level rise face a difficult challenge of translating global sea level rise estimates to more meaningful local estimates (i.e., they need to find the data to localize the variability in projected changes to global mean sea level to their specific part of the coast and then account for vertical land movement at that location). This has been a formidable and sometimes impossible obstacle to effective planning for sea level rise.

In response to this problem, the 2017 NOAA sea level rise report provided data to help determine RSL rise at many specific locations around the entire coast. In order to express these results in somewhat more geographic detail, the Environmental Protection Agency developed sea level rise estimates by 2100 for each state and for each of the six scenarios used in the NOAA report (see app. 2). Looking at the Intermediate scenario, the significance of the state-specific projections becomes apparent. For example, although the Intermediate scenario projection of GLSL sea level rise by 2100 is 3.28 feet, the average rise for most states is higher (e.g., 5.37 feet in Louisiana, 4.33 feet in New Jersey and 4.12 feet in Massachusetts).

Sea Level Rise: The Long View

The year 2100 is a useful point in the future to use to organize projections of sea level rise but there is wide agreement that sea level will continue to rise well after 2100. Speaking to the question of sea level rise after 2100, the third United States *National Climate Assessment* noted in 2014 that "sea level rise will not stop in 2100 because the oceans take a very long time to respond to warmer conditions at the Earth's surface. Ocean waters will therefore continue to warm and sea level will continue to rise for many centuries at rates equal to or higher than that of the current century."[22]

The 2017 NOAA sea level rise report projected global mean sea levels out to the year 2200 with estimates of the physically plausible sea level rise ranging from a low of 1.28 feet to a high of 31.82 feet and an intermediate estimate for global mean sea level of 9.19 feet (see app. 3). Although some sea level rise in future centuries is unavoidable, there is clear evidence that preventing the most extreme warming avoids very damaging sea level rise by 2100 and potentially catastrophic impacts of sea level rising ten feet or more by 2200. Relative sea level rise along most of the American coast is likely to be higher.

Looking beyond 2200 might on first thought seem like an abstract exercise without much practical use for most decisions related to coastal planning. Still, if there were a clear endpoint for future sea level rise, even one very far in the future, that would be helpful to know and could influence some planning, especially decisions concerning relocation of communities or critical infrastructure assets as an alternative to investment in coastal protection structures.

In 2015, Ricarda Winkelmann and coauthors published a paper looking at long-term changes in the Antarctic ice sheet. After noting that the "Antarctic Ice Sheet stores water equivalent to 58 m [190 feet] in global sea-level rise,"[23] there is the comparatively good news in the event that the world were to keep warming within the 2°C cap adopted in the 2015 Paris Climate Agreement, "if the 2°C target . . . were attained, the millennial sea-level rise from Antarctica could likely be restricted to 2 m [6.6 feet]."[24] Failing to live within the Paris Agreement 2°C increase cap, however, results in a more dire prediction: "With unrestrained future CO_2 emissions, the amount of sea-level rise from Antarctica could exceed tens of meters over the next 1000 years and could ultimately lead to the loss of the entire ice sheet."[25]

Achieving International Limits on Warming: An Uphill Battle

As the scientific questions related to sea level rise are gradually resolved, uncertainty about the height of future seas will shift to political questions related to international cooperation to limit the release of greenhouse gases.

Faced with uncertainty over future warming, some plans for managing sea level rise may simply reflect the full extent of possible sea level rise and plan for the worst or near-worst outcome. In other planning efforts, the costs or perceived infeasibility of adapting to the higher-end sea level rise projections may lead to proposals to discount the high end of projections and focus on the lower projections. For example, assuming a future in which warming is limited to about 2°C, consistent with the

Paris Climate Agreement, data from the NOAA 2017 report suggest a likely global sea level rise in the range of 1.6 to 3.3 feet by 2100 and 3.1 to 9.2 feet by 2200. Taking this optimistic view will require some demonstration of confidence in the achievability of the lower-end projections for the warming of the planet.

The extent of future warming will depend on many factors, including the price of energy that is generated with low levels of carbon dioxide emissions compared to energy generated with high levels of carbon dioxide emissions, innovation in the energy storage and transmission sectors, and national policies limiting release of greenhouse gases, such as regulations and energy pricing. The role of the Paris Agreement is to set a global cap on future warming and prompt countries to control greenhouse gas emissions as needed to stay within the cap using the investments or requirements that they prefer.

There is currently little evidence, however, that the world is on a reliable path toward limiting warming to less than 2°C. In addition to rising air temperatures, other indicators, such as the level of carbon dioxide in the atmosphere, are climbing. Levels of carbon dioxide in the atmosphere, measured at 315 ppm (parts per million) in 1958, rose to 405.5 ppm by 2017 and was measured at 415 in May 2019. Many scientists believe that keeping average global temperature below the 2°C warming cap will require keeping carbon dioxide levels under 450 ppm, and some argue that 350 ppm is a better target.

A report from Stanford University points to the scale of the challenge the world faces in limiting warming to 2°C. The authors looked at the International Energy Agency (IAE) estimate of the investment in low carbon energy needed to comply with this warming cap, noting that current investments of about $.75 trillion needed to be tripled.[26] The Stanford researchers concluded that there were several challenges to meeting this investment goal including this one: "Tripling current clean energy spending to $2.3 trillion annually represents a large fraction—perhaps 2/3rds—of the new funds that the world's institutional investors put to work each year across all sectors."[27] Other challenges include the low tolerance for risk among many investors that would make increasing investment in renewable energy projects difficult and the need for major investment in countries where capital is limited.

This daunting picture of the state of low-carbon energy investment brings us back to the question of the likelihood of success of international efforts to limit global warming that are, as of today, expressed in the Paris Climate Agreement. Going forward, might determined efforts by the nations of the world to live within the 2°C cap drive needed changes?

On the positive side of this question, the agreement has the support of most of countries around the world, and preliminary steps by countries to develop plans for limiting the release of greenhouse gases are completed or underway. The purpose of those plans is to create incentives or requirements resulting in dramatically increased investment in low carbon energy or to control the release of carbon dioxide or other greenhouse gases. Each country, however, establishes its own, voluntary "nationally determined contribution" (NDCs) to meeting the agreement goal.

The 2015 Paris Climate Agreement officially took effect in November 2016 after fifty-five countries representing 55 percent of emissions ratified it. Annual Conferences of the Parties, or "COPs," are working out the details of the process intended to result in complete NDCs by 2020. COP 24 in Poland in December 2018 resulted in agreement on a "rulebook" addressing issues such as accounting for emissions, management of carbon market mechanisms, and how to monitor compliance. The hope is that the NDCs will add up to emission reductions sufficient to keep warming below the 2°C increase above preindustrial conditions, or about 1°C (1.8°F) above current conditions, and that nations will then deliver promised contributions using such implementation actions as needed. After that, there are to be "global stocktakes" every five years starting in 2023 to assess progress and make needed adjustments using a "rachet mechanism" in which pledges from each country reflect its "highest possible ambition."[28] So far so good.

Unfortunately, there have been setbacks on the road to implementing the Paris Agreement. Most notably, President Trump announced in June 2017 his intention to withdraw the United States, the second largest source of greenhouse gas emissions after China, from the agreement as soon as a withdrawal is possible in 2020.

Then, in October 2018, the IPCC published a "special report" intended to inform further discussion of the nagging question of whether capping warming at 1.5°C above preindustrial levels would be a more appropriate target than the officially adopted goal of 2°C. After describing a series of "robust differences"[29] of a warming goal of 1.5°C rather than 2° C, including slowing the rate of sea level rise, the authors of the IPCC report note that meeting the lower goal "would require rapid and far-reaching transitions in energy, land, urban and infrastructure (including transport and buildings), and industrial systems (high confidence) . . . and imply deep emissions reductions in all sectors."[30]

How rapid? To stay below an increase of 1.5°C, global net emission reductions from 2010 levels of about 45 percent are needed by 2030, reaching net zero around 2050. To stay below an increase of 2°C, net emissions reductions of about 20 percent are needed by 2030, reaching net zero around 2075. Net zero emission reductions are accomplished using various practices, including carbon capture, and storage and direct removal of CO_2 from the air, which are not yet well developed.

Another setback is that, as of today, there is a gap between the emission reduction plans that countries are developing and the reductions in emissions needed to keep warming under the 2°C cap. In November 2018, the United Nations Environment Program published an annual *Emissions Gap Report*, evaluating progress toward commitments to implement the Paris Climate Agreement, finding three problems. First, the current NDCs are just not aggressive enough: "Pathways reflecting current NDCs imply global warming of about 3°C by 2100, with warming continuing afterwards."[31] Getting back on track will require a whole new level of effort: "This original level of ambition needs to be roughly tripled for the 2°C scenario and increased around five-fold for the 1.5°C scenario."[32] Second, even the current promised reductions are not occurring. Looking at the actions of G20 countries, the United Nations concluded that "the majority are not yet on a path that will lead them to fulfilling their NDCs for 2030."[33] Finally, and perhaps most ominous, the United Nations finds that time is short: "If the emissions gap is not closed by 2030, it is very plausible that the goal of a well-below 2°C temperature increase is also out of reach."[34] It seems obvious, but still necessary to point out, that 2030 is only a decade away.

To states, communities, businesses, and homeowners along the American coast trying to come to grips with the risks they face from storms and rising seas, it is hard to offer confidence that international efforts will keep warming increases under the 2°C warming cap, much less a cap of 1.5°C. Despite the warnings of the scientific community about the consequences for sea level rise of warming beyond 2°C, and a range of other warming impacts, there is yet no clear path to staying below that target. Even if the world rallies to stay within the 2°C cap, there is still the wild card that new research on ice sheets in the Antarctic and Greenland will force an upward adjustment in the range of sea level rise considered likely in the years ahead.

Still, there is reason to hope for significantly increased global commitments to staying within the Paris Climate Agreement cap, continued major reductions in the cost of low carbon energy, substantial increases in investment in low carbon energy production, or a game-changing scientific or engineering breakthrough. But, as the saying goes, hope is not a plan.

There really is no way to sugar coat the news about sea level rise—it is almost all bad. Scientific consensus about the extent of future global sea level rise is steadily trending upward. Relative sea level rise along most of the American coast will be greater than global averages. The commonly reported increases in the height of sea level (e.g., three feet by the year 2100) are not the total increase expected, but simply the first increment of a larger rise that will occur in the decades and centuries after 2100. The extent of the rise in sea levels is linked to the rate and extent of future warming of the planet. International cooperation has resulted in agreement on a cap in warming of 2°C, but commitments by countries around the world to staying within this cap are lagging, and the window for getting back on track is closing within the next decade. The lone positive aspect of the sea level rise picture is that there is still time to prepare for the most significant impacts.

3

Measuring a Shifting Coast

Most people think of the coast as a line drawn on a map or nautical chart—firm and fixed. The idea that the coast has always been there, and is not going anywhere, supports a vast economic investment in coastal homes, communities, and infrastructure. It also perpetuates the conviction that, although a coastal storm may do some damage from time to time, the best response after a storm is to set things right back to the way they were with the help of the many government programs that support that work.

A changing climate pulls the rug out from under that idea. Rising sea level will inexorably shift the coastline inland. More severe coastal storms will push storm surges onto larger land areas causing more extensive damage. These risks will revise the calculation for how to invest in recovery after a damaging storm and focus attention on debates over rebuilding in place or relocating to safer ground. Recognition of the coming inland shift of coastlines will also change how communities and people prepare for such storms in the first place. Still, accepting and acting on the idea that seas will rise and storms will be more damaging can be hard because the specific time and place of storms is impossible to predict, and sea level rise is slow and easy to confuse with changes in the tide.

Many people living along a beach or cliff-front coast are painfully familiar with the concept of coastal erosion and have seen their property wash away during a storm. Coastal erosion, however, is sometimes mistakenly blamed on rising seas when other factors play a larger role, and loss of land that is due to rising seas is sometimes mistaken for simple, old-fashioned coastal erosion. Understanding the subtle relationship between coastal erosion and rising seas is critical to proceeding to think about how to respond to a changing coast.

Finally, as a changing climate drives the coast inland there will need to be a process to measure and map the shifts in the existing coastline. Fortunately, the foundation for such a process already exists. In the past, this process, managed by the National Oceanic and Atmospheric Administration (NOAA), mostly captured the small shifts in the coastline resulting from coastal erosion and accretion. In the future, the business of mapping the shifting coastline will ramp up as relative sea level rise at specific places along the coasts pushes the coastline inland to widely varying degrees.

Tides and Time

A good place to start in thinking about the physical reality of the coast is to understand the daily shifting of the sea and land interface now occurring as a result of tides. Tides result from the changes in gravitational forces as the earth, moon, and sun change position in respect to one another and gravity pulls the water in the oceans so that it rises and falls in relation to fixed coastlines. The predictable effect of these gravitational forces is usually two low tides and two high tides each day in most coastal places.

The vertical range of these daily tides, however, varies considerably from place to place. The shape of continents, coastlines, and the ocean floor modifies tides that would occur in response to astronomical forces alone. The combined effect of all these forces is that some places experience virtually no tide and other places see tide changes of twenty feet.

Not only do some places generally have higher or lower tides than other places, but daily tides at a given place change from one day to the next, and the time of day of a high tide and low tide changes each day. The full range of tidal variation at a given place, accounting for astronomical factors, occurs over an 18.6-year cycle. With a full 19 years of past tidal data, a predictive table of all tides can be determined for any coastal location.

Factors that change cyclically such as distance between and alignment of the moon, earth, and sun also result in tidal changes. For example, "spring tides" occur roughly twice a month in response to phases of the moon. Every twenty-eight days, when the moon is closest to earth, there are "perigean spring tides," and when the two combine to create an unusually high tide, they're unofficially referred to as "king tides."

At most coastal locations, spring tides commonly add only an inch or two to the more common tides and a perigean spring tide might add another inch or so more. Coastal flooding can nevertheless occur as a result of even this small increase in the height of the tides, especially if it is magnified by local conditions such as a stiff wind pushing masses of water toward the shore or heavy rainfall inland filling coastal rivers and estuaries causing flooding because the high tides slow drainage to the sea.

Although this coincidence of astrologically driven very high tide along with storm and rainfall effects may put water in the streets, causing inconvenience and occasionally some damage, the good news is that the high water will recede in a matter of hours as the tide ebbs and storm conditions decline. And, the timing of high tides can be known for certain at any given place. Unfortunately, as sea level gradually rises, the nuisance flooding of today will become more extensive and damaging.

The 2017 NOAA report on sea level rise also looked at the potential for gradually rising sea level to exacerbate coastal flooding, turning minor nuisance flooding into more damaging events. NOAA considers flooding occurring once every five years (i.e., flooding with a 20 percent chance of occurring annually) to warrant a "coastal flood warning."[1] The height of flood waters in these events varies around the coastline but has a median height of 2.6 feet above high tide.

NOAA found that as early as 2030 the "coastal flood warning" level floods will occur five or more times a year rather than once every five years.[2] This early 2030 arrival of much more frequent flooding is expected in the event of sea level rise that is described in the NOAA report as the Intermediate-High case (i.e., sea level rising to almost five feet of increase in global mean sea level by 2100).

The more frequent flooding is expected to begin a few years later in the event of lower rates of sea level rise (i.e., in 2040 in the event of the Intermediate case, and 2060 in the Intermediate-Low case).[3] In early 2018, NOAA updated this work with more localized projections, finding that by 2050, the frequency of high tide flooding would be greatest along the western Gulf of Mexico, followed by the northeast Atlantic, eastern Gulf, southeast Atlantic, and Pacific Coasts.[4]

This flooding may be decades away for some places, but there is evidence that it is already causing economic impacts for some communities. For example, a 2019 study by Miyuki Hino and coauthors found that high tide flooding had reduced visits and economic activity in downtown Annapolis, Maryland.[5]

Coastal Erosion and Coastal Erosion Vulnerability

Even without accounting for rising sea level, the shape of the coastline can change as a result of waves, storms, changes in sediment reaching the coast from rivers, loss of wetlands, land subsidence, and a range of human activities.

Figure 3-1. High tide on a sunny day in 2016 brings flooding to downtown Miami, Florida. Creative Commons photo, Creative Commons Attribution 4.0. https://creativecommons.org/licenses/by-sa/4.0/deed.en.

To track the historical shifts in the coast, the United States Geological Survey (USGS) developed a National Assessment of Shoreline Change, including detailed studies of different segments of the coast. These studies report the average annual rate of shoreline change and the percentage of coastal segments that are eroding over the long term (1800s–1997/2000) and short term (1960/1970s–1997/2000).

USGS determined erosion occurred at more than half of the shore segments over the long and short term along most of the Atlantic and Gulf coasts with the greatest percentage of eroding segments in New England. The highest amount of erosion back from original shoreline was in the mid-Atlantic region, with over eight feet annually over the short term, followed by the Gulf of Mexico, with an annual stepback rate of over five feet (see app. 4). Some specific places along the Atlantic coast, however, are eroding at a rate of almost ten feet per year, including 35 percent of segments in southern Virginia and 16 percent on Cape Cod.[6]

It is important to consider that these data reflect a considerable investment in beach nourishment projects (i.e., adding sand to beaches) by federal, state, and local governments, as well as many coastal protection structures. Had these protection projects not occurred, coastal erosion would likely have been greater.

Coastal storms are generally recognized as the immediate cause of this erosion, but are thought to be acting in the larger context of rising seas: "While sea level rise sets the conditions for landward displacement of the shore, coastal storms supply the energy to do the 'geologic work' by moving the sand off and along the beach."[7] This storm-driven erosion, occurring years prior to sea level rise inundation, can diminish beaches and marshes as well as degrade land-based infrastructure. The results of these USGS studies are generally consistent with the measured changes in sea level and land subsidence around the country, suggesting that, despite the interplay of other factors, rising sea level has gradually pushed back the coast.

To understand the future vulnerability of the coast to changing conditions, USGS combined data on rates of relative sea level rise with information on soils, slopes, and average wave heights into a Coastal Vulnerability Index (CVI). The CVI weighs relative sea level rise and other factors equally and rates each kilometer of the coastline on a risk scale of Low, Moderate, High and Very High (see table 3-1). For example, based on this data, the Gulf of Mexico has the highest number of kilometers in the "very high" vulnerability rating, followed by the Atlantic and Pacific regions.

USGS offered some general conclusions about coastal vulnerability in different regions, pointing to particular areas of concern. In the Atlantic region, Maryland to North Carolina and northern Florida show CVI values of very high vulnerability.[8]

Table 3-1. Coastline in Coastal Vulnerability Index (CVI)

Region	Vulnerability category			
	Low	Moderate	High	Very high
Atlantic	2,326	2,332	2,599	2,215
Gulf of Mexico	647	2,968	1,056	3,387
Pacific	904	1,108	875	1,100
Total	3,877	6,408	7,129	6,702

Note: Numbers in kms.

Source: Data from the "National Assessment of Coastal Vulnerability to Sea-Level Rise," US Geological Survey (CVI websites for each region), results.

In the Gulf of Mexico region, the Louisiana–Texas coast stands out as highly vulnerable.[9] And on the Pacific Coast, concern centers around heavily populated areas like the San Francisco–Monterey Bay area and the area from San Luis Obispo to San Diego.[10] Recent research by Betsey Von Holle and coauthors concluded that the number of segments along the southern portion of the Atlantic coast with very high vulnerability will increase 30 percent by 2030.[11]

Mapping the Shifting Coastline

Preparing for the impacts that more severe storms and higher sea level will have on coastal areas requires basic information like knowing where the coastline is today and how far it will shift inland as sea level rises. NOAA is in charge of mapping changes to the coastline as they occur. Other agencies, including the Federal Emergency Management Agency (FEMA) and USGS map the coastal land forms and elevations that will determine the shape of a new coast as sea level rises. To know the extent of land inundated by two feet or four feet of sea level rise, one needs to know the point from which to measure the two or four feet of water. Studies of sea level impacts commonly measure increases in sea level at a given coastal location from the coast shown on nautical charts produced by NOAA. How is this mapped "coastline" determined?

The answer takes us back to tides. NOAA uses a reference period of about nineteen years, called the National Tidal Datum Epoch (NTDE), as the official time segment over which tide observations are taken and reduced to obtain mean values at a specific location. The NTDE results in data that is used to establish relative elevations for other degrees of the tide at a given place (e.g., Mean High Water is six feet or eight feet above Mean Lower Low Water). More important for a discussion of sea level rise, the NTDE generates a Mean High Water elevation that intersects with the land. NOAA uses the Mean High Water elevation derived from the NTDE to generate the coastline that is mapped on nautical charts and other maps and the Mean Lower Low Water elevation to determine depths.

The present NTDE is based on tidal observations from 1983 through 2001 and, in effect, represents the level of the sea during that period. NOAA considers revision of the NTDE every twenty to twenty-five years and is expected to release updated data for the 2002–2020 period in 2022. If sea level were to be unchanged from the current NTDE, and land did not subside, the intersection of a new Mean High Water elevation with the land would also remain unchanged and the official line of the coast would not move.

If, however, sea level has risen since the 1983–2001 period, the tide data for locations around the country will intersect the land at a point inland from where the intersection occurred in the last NTDE. This new intersection of the land with the new Mean High Water will become the new coastline, located inland from the prior coastline. As explained by NOAA in a 2001 publication:

> At the present rate of 20 cm per century [2 mm per year; but now estimated to be 3.3 mm per year] sea level change could result in a horizontal retreat of the shoreline of 500 to 1500 ft in the next 100 years over some stretches of the U.S. Coast. On East and Gulf Coast beaches, relative sea level rise is accompanied by lateral shoreline retreat orders of magnitude greater than the vertical rise in sea level because of the gentle slope of the coastal plain's surface.[12]

Today's estimate of the annual rise in global mean sea level of about 3.3 mm per year or about .13 inches per year, means that sea level has risen almost 2.21 inches in the seventeen years since the end of the current NTDE. As discussed in the previous chapter, some increases along the American coast may be larger or smaller but this modest degree of change is not likely to cause NOAA to reissue charts with adjusted coastlines for most places. As sea level rise accelerates, however, changes to charts and related products will become more common.

Any change in the location of the coastline is critically important because it is commonly a boundary between private and public ownership of land. In most states, private property ends at the mean high water line. Shifting the ownership of land, or an area that was land, from private to state hands will have substantial implications for assessments of the financial impacts of sea level rise and add controversy to managing these changes. As sea level rises in the decades ahead, NOAA's periodic revision of the NTDE will become a key mechanism for translating measured changes in sea level into inland shifts of the location of the American coast. These periodic coastline adjustments are likely to attract increasing attention and be a catalyst for local conversations about how to manage rising seas and prepare for more extensive changes to come.

So, NOAA will revise the coastline as it changes, but it would be good to know the places that are likely to be inundated by sea level rise and at risk of future storm flooding. The variable geography and elevation of the coast means that the landward extent of inundation by a given amount of sea level rise and storm surge flooding will vary widely from place to place. Sea level rise in the face of the steep and rocky Maine coast will result in limited land area inundated while rising seas in the flatter Florida Keys will result in more extensive inundation. The coastal vulnerability scores developed by the USGS suggest where other factors, such as soils and slopes, open the coast to erosion ahead of inundation.

Fortunately, most of the coast is well mapped and, in many areas, very accurate elevation data resulting from laser mapping from airplanes (LIDAR) is available. Several mapping tools are available on the Internet to illustrate the land area that would be inundated by a given amount of sea level rise. Using these tools, it is possible to estimate the area and population at risk of rising seas (see ch. 4).

Maps simply overlaying a degree of future sea level rise on geography, however, do not tell the story of how more severe storm surges, riding on top of higher sea level, will reach farther inland than current storm surges. The Federal Emergency Management Agency maps areas where there is a one percent chance of flooding in a given year (i.e., the 100-year flood zone) as part of the National Flood Insurance Program (see ch. 9) but does not consider sea level rise. And, NOAA has similar SLOSH maps showing areas expected to flood during storms of differing size categories. Today, however, there is no national map that integrates both sea level rise and storm flooding to show coastal areas at long term risk of both impacts.

It is good news that accounting for the present location of the coast, for tides, and for coastal erosion is pretty well understood. There is even a process in place to map future, incremental shifts in the coast in response to storms and rising seas. Although the country's coastal mapping capacity is impressive, there is room for improvement. For example, different federal agencies mapping the coast need to work together more closely, coordinating investments such as LIDAR mapping. More attention is needed to mapping the combined effects of storms and rising seas. Existing mapping tools can outline land areas to be inundated by a hypothetical amount of sea level rise (e.g., one foot or three feet) but agencies need to cooperate to map areas that they firmly predict will be inundated in the short term (e.g., 20 years) and longer term (e.g., 100 years).

PART I: EPILOGUE

A Warming Planet Drives Coastal Storms and Rising Seas

The point of all the science and data in part 1 is to make the case that the science behind projections of more severe coastal storms and rising seas is sound. The early estimates of past years are now improved and, although research continues, the direction is clear. Separate from well-known changes like tides or coastal erosion, a warming climate is working by somewhat different processes to make coastal storms more intense and to raise sea level, delivering a one-two punch to the coast. Coastal storms will become more damaging in terms of storm surge and rainfall, reaping a growing toll in lives and destruction. Sea level along the American coast is almost sure to rise several feet, may well rise much more by 2100, and keep rising. The bottom line is that sound science is pointing to a real problem for the coast. This is argument number one for a new national program to prepare for more severe storms and rising seas.

PART II

Storms and Rising Seas Disrupt the American Coast

I have seen the hungry ocean gain
Advantage on the kingdom of the shore.

William Shakespeare, Sonnet 64

4

Scale and Cost of the Coming Coastal Inundation

There is great diversity of geography along the Atlantic, Gulf of Mexico, and Pacific coasts of the United States. The beaches of Cape Cod, the wetlands of Louisiana, and the hills of California's Big Sur are iconic coastal images for most Americans. More severe storms and rising seas will be kind to coastal places fortunate to have notable elevation above sea level and will inundate those places with little elevation. Some of these places at risk of inundation are settled communities, both large and small, and others are natural areas, such as marshes and beaches, providing a range of environmental, economic, and other benefits.

New mapping tools, widely available online, provide the means to visualize a given amount of sea level rise or storm surge on a specific coastal area, and these tools have been used to assess the geographic extent of inundation and identify the communities, infrastructure, and ecosystems at risk. From these assessments, researchers have developed several estimates of the economic value of property expected to be inundated, suggesting that property valued at several trillion dollars is at risk. These studies, however, are based on the value of coastal property as it stands today. Over the coming decades, the population of coastal areas is expected to grow steadily, and the new homes and businesses needed by this expanded population will add to the property already at risk of inundation.

Getting a feel for the scale and cost of inundation is important for thinking about how the country should respond to these risks. Is a new national program to prepare for these impacts really needed? What are the differences in impacts from state to state? How do costs of coastal storms compare to those of rising seas? How might coastal population growth change estimates of impacts? Each of these questions is addressed below.

Lands Lost to the Sea

The most obvious way to measure the land area lost as sea level rises is to look at square miles of current land area lost to the sea under different increments of rising seas. Today, with digital mapping databases, projecting the land area to be inundated by a given amount of sea level rise is a pretty straightforward exercise.

In a 2012 paper, Benjamin Strauss and coauthors estimated the land area that is both above mean high tide and below elevations of one, two, or three meters. Strauss found that 12,288 square miles of the contiguous United States will be inundated in the event of a one meter (3.28 foot) rise in sea level, not counting marine or estuarine wetlands. Of this total land area, 3,544 square miles is "dry land" (i.e., land area above mean high water and below one meter of elevation not including freshwater wetlands).[1] In the event of more sea level rise, to the three meter level (9.8 feet), the total land and the dry land lost to the sea increases to 28,385 and 12,871 square miles, respectively (see table 4-1).

How significant is the loss of this amount of land? On one hand, the entire United States is a little over 3.7 million square miles, and the 12,000 square miles projected to be lost to the sea in the event of three feet of sea level rise is a mere .33 percent of that total land area, while even the 28,000 square miles that would be lost in the

Table 4-1. Land area projected to be inundated under different degrees of sea level rise, contiguous US

Sea level rise	Coastal land area affected	
	Total land area inundated, sq. miles	Dry land area inundated, sq. miles
1 m/3.28 ft.	12,288	3,544
2 m/6.5 ft.	20,427	7,876
3 m/9.8 ft.	28,385	12,871

Source: Data from Benjamin H. Strauss et al., "Tidally Adjusted Estimates of Topographic Vulnerability to Sea Level Rise and Flooding for the Contiguous United States," *Environmental Research Letters* 7, no. 1 (March 14, 2012): 9.

event of about ten feet of sea level rise is less than one percent of the entire county. Why is this a problem?

Twelve thousand square miles is about the size of the state of Maryland and larger than eight other states, including Connecticut, New Jersey, and Massachusetts. Even the dry land area of 3,544 square miles is larger than the land area of both Rhode Island (1,044 square miles) and Delaware (1,953 square miles). Twenty-eight thousand square miles is a little bigger than West Virginia and a bit smaller than South Carolina. More important than sheer area lost, however, is the use of this land as home to people and communities as well as ecosystems and natural resources.

Population, Homes, and Communities at Risk from Storms and Rising Seas

Looking at the uses of land projected to be inundated reveals population at risk, the number of homes at risk, and communities, or parts thereof, exposed to storm surge and rising seas.

The Strauss paper includes projections of both population and housing units located in areas along the coast at risk of varying degrees of sea level rise (see table 4-2). This analysis found that some 3.68 million people live on land in the contiguous United States that will be inundated by a bit over three feet of sea level rise and that almost two million housing units are located on this land. In the event of almost ten feet of sea level rise, some twelve million people and six million housing units occupy land that will be under water to varying degrees.[2]

Given that the 2019 population of the United States is about 328 million, 3.68 million is a little over one percent of the entire population, while 12 million people

Table 4-2. Population and housing units in areas projected to be inundated under different degrees of sea level rise

Sea level rise	Population	Housing units
1 m/3.28 ft.	3,682,557	1,946,429
2 m/6.5 ft.	7,710,326	3,999,726
3 m/9.8 ft.	12,112,226	6,102,019

Source: Data from Benjamin H. Strauss et al., "Tidally Adjusted Estimates of Topographic Vulnerability to Sea Level Rise and Flooding for the Contiguous United States," *Environmental Research Letters* 7, no. 1 (March 14, 2012): 10.

is about 3.7 percent. In the case of housing units, there were 135.7 million units nationwide in 2016. Losing close to two million units to sea level rise would be a loss of 1.4 percent.

Another way to think about consequences of rising sea level is to consider how many coastal communities will face inundation. This perspective is important because it focuses on the local governments that will need to plan for and manage more severe storms and rising seas.

In 2017, the Union of Concerned Scientists (UCS) published an analysis of communities likely to be inundated by rising sea levels, focusing on the concept of "chronic inundation" (i.e., when more than 10 percent of a community is expected to flood more often than twice a month as a result of both rising seas and high tides). The UCS found that some 91 communities were already inundated on a chronic basis. In the event of sea level rise of four feet by 2100, some 489 communities would be chronically inundated. In the event of higher sea level rise of 6.6 feet by 2100, 668 communities would be chronically inundated[3] (see table 4-3).

In the same report, UCS researchers used a "Social Vulnerability Index" to look at twenty-nine economic and demographic variables and found that over half of the 167 communities projected to be chronically inundated by 2035 included one or more "socioeconomically vulnerable neighborhoods." They note that these neighborhoods "have traditionally had fewer resources to cope with environmental disasters and change,"[4] and that "this analysis brings attention to the fact that these communities will need more resources and more capacity in order to prepare for the impacts of sea level rise."[5] A subsequent study by the UCS found that coastal communities where demographic data indicated socioeconomic vulnerability were clustered along the Gulf Coast, especially in Louisiana, but that other regions stand out, including the Eastern Shore/Chesapeake coast of Maryland; the mainland side of Pamlico Sound in North Carolina; the New Jersey Shore; Kiawah and Edisto Islands in South Carolina; and the Florida Keys.[6]

Looking more generally at the distribution of sea level rise impacts across the coast, virtually all states and cities will see some impacts. Most of the impacts, however, are clustered in half a dozen states, with by far the greatest impacts in a single state—Florida. Looking at the metrics of population, dry land, and housing units from the 2012 Strauss paper, Florida and Louisiana clearly rank at the top in all three metrics, while rank order of states changes when considering other factors (see table 4-4).

Using the estimate of total population at risk of inundation in the event of one meter (3.28 feet) of sea level rise (i.e., some 3.6 million people), 43 percent are in Florida. Louisiana follows with 24 percent, and all other states are less than 10 percent.[7]

Figure 4-1. High tide puts water in the road on Smith Island, Maryland, in the Chesapeake Bay, 2015. The Chesapeake Bay region has the highest rates of sea level rise on the East Coast due to a combination of sea level rise and land subsidence. Photo by Gary J. Kohn/www.garyjkohn.com.

Table 4-3. Number of communities by estimated degree of inundation at various levels of sea level rise by 2100

Inundation (%)	Present	1.6 ft.	4 ft.	6.6 ft.
10–25	44	112	195	240
25–50	31	61	102	155
50–75	12	58	59	76
>75	4	59	133	197
Total	91	290	489	668

Source: Data from Erika Spanger-Siegfried et al., *When Rising Seas Hit Home: Hard Choices Ahead for Hundreds of US Coastal Communities* (Cambridge, MA: Union of Concerned Scientists, 2017), 17.

Table 4-4. Rank of states facing the most significant impacts of sea level rise, by population, dry land area, and housing units projected to be inundated by 1 m/3.28 ft. rise

State	Population	Dry land area	Housing units
Florida	1	2	1
Louisiana	2	1	2
California	3	6	3
New York	4	12	4
New Jersey	5	11	5
Virginia	6	8	8

Source: Data from Benjamin H. Strauss et al., "Tidally Adjusted Estimates of Topographic Vulnerability to Sea Level Rise and Flooding for the Contiguous United States," Environmental Research Letters 7, no.1 (March 14, 2012): 4.

Future Costs of More Severe Storms and Rising Sea Level

There are two major studies of the economic impacts of a changing climate in the United States that look specifically at losses due to more severe storms and rising seas. The Risky Business Project, sponsored by Bloomberg Philanthropies, published the *American Climate Prospectus* report in 2014. The Climate Change Impacts and Risks Analysis Project, or CIRA, is an ongoing effort of the Environmental Protection Agency (EPA) that supports the *National Climate Assessment*. Both projects offer estimates of the costs of warming in the United States in multiple sectors including coastal communities and property.

Starting with the basic question of the economic value of existing residential and commercial property expected to be lost due to just the rising of sea level, the prospectus offers ranges of cumulative impacts for the years 2050 and 2100. It draws on the idea of "Representative Concentration Pathways" (RCPs) that describe future conditions ranging from effective control of warming (RCP 2.4) to business as usual (RCP 8.5) (see app. 5). For the year 2100, the cost estimate is $430 billion on the low end and $1.144 trillion on the high end[8] (see table 4-5).

The prospectus also finds that economic impacts of rising seas are focused in Florida and Louisiana: "While roughly two-thirds of all current property likely below MSL [mean sea level] by 2050 is in Louisiana and Florida, and three-quarters of all property by the end of the century, Maryland, Texas, Massachusetts,

Table 4-5. Estimated cumulative costs of property damage, in $ billions, due to sea level rise

	Year	
RCP	2050	2100
RCP 2.4 (1.2–2.1 ft.)	$287–$360	$430–$830
RCP 4.5	$294–$366	$759–$926
RCP 8.5 (2–3.3 ft.)	$323–$389	$724–$1,144

Note: See app. 5 for description of RPCs (Representative Concentration Pathways).

Source: Data from Rhodium Group, *American Climate Prospectus: Economic Risks in the United States* (New York: Rhodium Group, 2014), 90.

North Carolina, New York, New Jersey, and California also face meaningful inundation risk."[9]

As communities plan for sea level rise, it is helpful to understand the economic impacts of both sea level rise alone (resulting in permanent inundation), as well as inundation resulting from storm surge that can be expected as a result of major storms (destructive but temporary flooding). The CIRA estimates of impacts of a changing climate on the United States published in May 2017 provide an astonishing picture of cumulative costs of coastal property lost due to both sea level rise and stronger storms: "Without adaptation, cumulative discounted damages to coastal property in the contiguous U.S. are estimated at $3.6 trillion through 2100 under both RCPs [RCPs 4.5 and 8.5]."[10] Authors of the prospectus seem to generally concur with this assessment, pointing out that storm-adjusted costs are roughly "three times as much as higher sea levels alone."[11]

Three trillion is a big number but there are several factors that suggest that it may be too low. Most significantly, the authors acknowledge the recent research related to instability of Antarctic ice sheets, "It should also be noted that none of these analyses . . . incorporate findings from recent Antarctic research suggesting that 6 feet or more of global sea level rise. . . . Consideration of this possibility would roughly double most of the estimates presented in this section."[12] The CIRA report also explains that a range of additional costs of sea level rise and coastal inundation were not considered in this estimate, including impacts on transportation, telecommunication infrastructure, and ecological resources.

Switching from cumulative costs of coastal property loss to annual costs, the

CIRA estimates annual costs under RCP 8.5 to be about $120 billion in 2090, with losses already reaching $75 billion by 2050[13] (see table 4-6). In the case of RCP 4.5, annual losses expected by 2090 are $92 billion with losses reaching $69 billion annually by 2050.[14] Both estimates are for losses due to the combination of sea level rise and more severe coastal storms.

These annual national loss estimates are a bit higher than, but roughly consistent with, the 2016 estimate by the Congressional Budget Office (CBO) of $39 billion in losses per year by 2075 for just damages due to hurricanes made more serious by both sea level rise and population growth. CBO's analysis, however, added two important new perspectives. First, CBO estimated that federal disaster relief spending has historically been roughly 60 percent of damages, which is projected to be $24 billion per year by 2075. So, the annual costs to the country of both hurricane damage and federal relief spending for these storms is estimated to be $63 billion in 2075.[15]

CBO also concluded that the American economy would grow dramatically by 2075, to roughly four times the current size, but that hurricane damages would grow faster, rising from .16 percent of the Gross Domestic Product (GDP) today to .22 percent by 2075, posing an increasing drain on the economy.[16] About 45 percent of this increase is attributed to climate change and 55 percent to coastal development.

The CIRA report authors looked at the geographic distribution of annual losses and found that by far the largest portion are in the Southeast region of the country. State-by-state data are not available, but annual losses in the states from Virginia to Texas are estimated at over $100 billion in 2090 and over $60 billion by 2050.[17] Losses in New England are about $12 billion annually by 2090, of which $10 billion is expected to occur by 2050.[18] Losses on the Pacific Coast make up remainder of all losses nationally.

Preparing for cumulative coastal property losses of over $3 trillion, along with estimated losses of about $100 billion per year by the close of this century is a daunting prospect, but the picture brightens substantially when adjusted to reflect assumptions concerning a range of adaptation actions that property owners and governments can be expected to implement to avoid inundation. The CIRA report authors conclude that "well-timed adaptation measures significantly reduce cumulative discounted costs to an estimated $820 billion under RCP 8.5, and $800 billion under RCP 4.5"[19] And, the cost of the anticipated adaptation measures is included in the roughly $800 billion cost estimates.

What are the costs of these adaptation measures? The CIRA report models

Table 4-6. Estimated annual property loss, in $ billions, due to sea level rise and more severe storms

RCP	Year	
	2050	2090
4.5	$69	$92
8.5	$75	$120

Note: Estimates are provided for several Representative Concentration Pathways (RCPs). See app. 5 for description of RCPs.

Source: Data from United States Environmental Protection Agency, *Multi-Model Framework for Quantitative Sectoral Impacts Analysis: A Technical Report for the Fourth National Climate Assessment* (Washington, DC: Environmental Protection Agency, 2017), 209.

several different adaptation responses to the threat of rising seas and more severe coastal storms, including beach nourishment, property elevation, shoreline armoring, and property abandonment. The CIRA model forecasts an adaptation response for coastal areas at risk based on sea level rise, storm surge height, property value, and costs of protective measures. Based on detailed modeling of seventeen geographic areas, the proportion of losses to costs of adaptation measures varies from place to place but averages roughly 80 percent losses in the form of property abandoned because it was not protected or was damaged by storm surge, and 20 percent costs applied to adaptation measures considered economically feasible.[20]

If this 80/20 proportion of damage costs to adaptation measure costs were applied nationally, adaptation measures costing roughly $2 billion per year (i.e., 20 percent of $800 billion is $160 billion divided by eighty years) might reduce cumulative property losses from over $3 trillion by 2100 to roughly $640 billion. Money well spent. This is, however, an extremely preliminary estimate of the likely cost of adaptation measures and almost certainly too low for several reasons. For example, costs of spending associated with abandoned property, such as relocation, are not considered, nor are costs of planning or costs of mistimed actions.

Still, using the general estimates of over $3 trillion in cumulative damage costs by 2100, the economic value to society of timely and effective response to sea level rise, and the more extensive impacts of storm surges it will bring, nets out to a savings of roughly $2 trillion by 2100. These damage costs do not account for the bill for disaster relief or the intangible costs of lost lives. And, these costs are likely to be higher if sea level rise is faster than now expected (e.g., due to instability of the Antarctic ice sheet, more rapid changes in ocean circulation, or other factors) or new homes and businesses are built in places at risk of inundation in response to significant population growth.

Coastal Population Growth

Population along the coasts is expected to grow significantly, even in low-lying areas, although public awakening to storm surge and sea level rise risks might dampen that growth. As coastal populations grow, investments in housing, infrastructure, and services for these people will grow and the value of all assets at risk will increase. Understanding the scale of projected coastal population increases reinforces the seriousness of the potential impacts and is part of the argument for a national program to prepare for these risks.

Many of today's estimates of impacts of rising seas—whether they be for population, housing units, or more general economic losses—are commonly based on just the people and property existing along the coast today, rather than the population and assets that will actually be inundated in the future. For example, the authors of the American Climate Prospectus identify the difficulty in predicting future development along the coasts: "Rather than attempt to predict how the built environment will evolve in the decades ahead, we assess the impact of future changes in sea level and storm activity relative to the American coastline as it exists today."[21] The CIRA report authors make a similar statement, although they accounted for appreciation of existing property.[22]

How is the population of coastal areas likely to change in the decades ahead? The United States Census Bureau reports a total United States population of 328 million in 2019 and projects steady growth out to the year 2060 to about 416 million, an increase of 30 percent. The rate of annual increase is a little under one percent, or about 2.5 million people annually.[23] The Census Bureau does not publish projections past 2060, but the United Nations issued a report in 2015 describing population growth in countries around the world, projecting a rise in United States population to 389 million in 2050 (a bit below the Census Bureau estimate for that year of 398 million) and 450 million in 2100.[24]

How many of these future citizens will be living on the coast or be at risk of rising sea level? In 2013, National Oceanic and Atmospheric Administration (NOAA) reviewed new census data from 2010 and concluded that "if current population trends continue, the already crowded U.S. coast will see population grow from 123 million people [in 2010] to nearly 134 million people by 2020."[25] This projection, however, assumes a growth rate that is slightly lower than the annual rate for the nation as a whole for the 2010 to 2020 period. Is the coastal population really likely to grow slower than the national population? Well, no.

The complicating factor here is that NOAA is reporting data from "coastal counties," including counties along the Great Lakes (i.e., the "inland coasts") that are not at risk of sea level rise. Subtracting out the populations of counties along the Great Lakes—about 18 million—the adjusted saltwater coastal county population in 2010 was about 105 million. In addition, the annual growth rates for the Great Lakes states are generally much lower than for saltwater states (e.g., under .5 percent in Great Lakes states, while the saltwater state growth rate averages a little over 2 percent). At a 2 percent growth rate, the 105 million saltwater residents in 2010 grows to over 130 million by 2020.

Coastal counties, however, are large, and most people in these counties are not living near the water or in places at risk of storm surges or sea level rise. What light can be shed on the question of how many people in coastal counties are in low lying areas and how this population might grow?

In 2015, Barbara Neumann and coauthors reported on total population growth estimates for countries around the globe as well as projected population growth specifically in the narrow coastal area that is below ten meters (or 32.8 feet) of elevation, termed the Low Elevation Coastal Zone (LECZ), and the 100-year floodplain. In the case of the United States LECZ, Neumann estimates a population of 23.4 million in 2000, growing to 34 million in 2030 and 44 million 2060. This is less than the annual growth rate for coastal counties as a whole but above the national rate. In the 100-year floodplain, Neumann reports 3.5 million in 2000, growing to 5.3 million in 2030, and a bit over seven million in 2060.[26]

Can coastal population growth in the coming decades continue as in recent years even as public understanding of storm and sea level rise risks grows? Surely some people will see the risks and decide not to move to the coast, and some will move back from the coast. The CBO analysis of hurricane impacts assumed that both population and per capita income growth would slow in the most vulnerable coastal counties. Other variables driving population changes include the firmness of government policies discouraging locating in risky areas and whether major, damaging storms happen to hit highly populated areas.

It seems clear, however, that population right along the coast will increase, perhaps almost doubling the 2000 population by 2060, and that the new homes, infrastructure, and services these people bring will be at risk of more severe storms and rising seas. The costs of damages and government relief efforts will go up.

Significant land area and communities that are home to millions of Americans will be inundated by rising seas. This inundation will be gradual, appearing first as more frequent flooding at high tide in the most vulnerable areas. Some areas are likely to be damaged by more severe storms before permanent inundation occurs. The projected costs of loss of existing coastal property to damages of more severe storms and rising seas by the end of the century is several trillion dollars. If rapid population growth right along the coast continues, these costs will be much higher. Brightening this picture somewhat is analysis suggesting that prompt and thoughtful preparation can dramatically reduce costs, and that investments in imagining future communities that are well adapted to the new coast would be money well spent.

5

Coastal Storm and Sea Level
Rise Risks to Critical Infrastructure

Sea level rise and more severe coastal storms pose a risk to critical infrastructure, such as transportation and water treatment facilities, which are necessary for the day-to-day operation of society. Military facilities, essential to national security, are also at risk. A national program to prepare for more severe storms and rising seas needs to address these critical infrastructure assets, and this requires a basic understanding of the risks these facilities face.

Unfortunately, there are few national studies of these sectors focused on storms and rising seas. Some national studies look at diverse climate change impacts on an entire infrastructure sector without focusing directly on coastal issues, and some studies look at sea level rise and storm risks for just specific facilities or clusters of facilities. Even these glimpses, however, are enough to suggest significant damages and the potential for wider costs to society, including disruption of daily life, when this infrastructure is out of operation or operating below par.

The range of critical infrastructure along the coast is diverse in terms of purpose, value, and risk of damage. In planning to protect critical coastal infrastructure, the country will need to consider how to build an understanding of storm and sea level risks into siting decisions for new coastal infrastructure. In the case of existing facilities, hard decisions will need to be made on whether to protect an asset in place or relocate to a safer site. Of course, these decisions need to be made in the context of storm and sea level rise plans for critical infrastructure in other sectors as well as plans for communities and ecosystems.

Federal agencies have the capacity to develop sector-specific plans for infrastructure in their areas of responsibility and can work with other governments and the private sector to make timely investments to protect or relocate the facilities most at risk. Federal agencies have developed broad plans for climate change adaptation generally but these plans commonly focus on agency missions rather than infrastructure sectors. Although the agencies acknowledge the risks of storms and sea level rise to infrastructure, they rarely go the next step to define preparedness actions.

This chapter focuses on transportation, water treatment, and military infrastructure, but the country will need to plan for impacts of severe storms and rising seas in other sectors, including emergency service facilities, hospitals and other health care facilities, and general government buildings. Most of this infrastructure is publicly held. Risk of more severe storms and rising seas to energy infrastructure and tourism and fishing assets along the coast, much of which is privately held, is discussed in chapter 7.

Transportation Infrastructure

The 2014 *National Climate Assessment* states that "sea level rise, coupled with storm surge, will continue to increase the risk of major coastal impacts on transportation infrastructure, including both temporary and permanent flooding of airports, ports and harbors, roads, rail lines, tunnels, and bridges."[1] Protecting transportation networks is especially challenging, the authors note, because "inundation of even small segments of the intermodal system can render much larger portions impassable, disrupting connectivity and access to the wider transportation network."[2] A tip of the iceberg problem.

A 2008 report by the National Research Council looking at all climate change-related threats to the transportation system found that sea level rise was a top concern: "Potentially, the greatest impact of climate change for North America's transportation systems will be flooding of coastal roads, railways, transit systems, and runways because of global rising sea levels, coupled with storm surges and exacerbated in some locations by land subsidence."[3]

Several reports look at specific components of the transportation network. For example, a 2014 report by the Federal Highway Administration repeated 2008 estimates of 60,000 miles of roads at risk of coastal flooding and sounded a warning of future impacts of coastal storms and rising seas on roads and bridges.

The US transportation system is vulnerable to coastal extreme event storms today and this vulnerability will increase with climate change. Hurricane Sandy caused over $10 billion in damage to coastal roads, rails, tunnels, and other transportation facilities in New York and New Jersey. . . . Hurricanes Ivan (2004), Katrina (2005), Ike (2008), and other storms have also caused billions in damage to coastal roads and bridges throughout the Gulf Coast....This vulnerability will increase as sea levels rise.[4]

More recently, authors of a 2018 study of coastal roads in states along the Eastern Seaboard found that tidal nuisance flooding "threatens 7,508 miles (12,083 km) of roadways including over 400 miles (644 km) of interstate roadways. . . . With sea level rise, nuisance-flood frequency is projected to grow at all locations assessed."[5] Today, over 100 million hours of vehicle delay occur due to tidal nuisance flooding, but these delays are projected to multiply dramatically to 1.2 billion hours in 2060 and 3.4 billion by 2100 under an Intermediate low sea-level-rise scenario.[6] The states of Florida and New Jersey have the largest percentage of roads at risk (4.7 percent and 4.6 percent, respectively) and, of all the road miles at risk, 60 percent are located in Florida, Georgia, South Carolina, and North Carolina.[7]

The Eastern Seaboard, more specifically the Northeast Corridor, also faces flood risks to rail infrastructure. In late 2018, Bloomberg News reported on an internal Amtrak study that found that coastal storms and rising seas threaten to erode track, signals, power poles, and power substations and that parts of the corridor are at risk of "continual inundation."[8] A ten-mile stretch of track around Wilmington, Delaware is of special concern, but track along the Connecticut coast and at other locations faces flood risks and, because some of these sections lack alternative routes, this flooding could result in disruption along the larger corridor.

There is general recognition that ports and harbors "will need to be reconfigured to accommodate higher seas,"[9] but there has been little assessment of specific ports, timeframes, or costs. The 2014 *National Climate Assessment* notes that "most ocean-going ports are in low-lying coastal areas, including three of the most important for imports and exports: Los Angeles/Long Beach (which handles 31% of the U.S. port container movements) and the Port of South Louisiana and the Port of Galveston/Houston (which combined handle 25% of the tonnage handled by U.S. ports)."[10] Some additional complications for ports to address are bridge clearances for large vessels and elevation of access roads or rail lines, sometimes miles away, which may be inundated even if port facilities are protected.

Figure 5-1. Hurricane Sandy damaged transportation infrastructure and homes in Mantoloking, New Jersey, 2012. Photo by Greg Thompson, US Fish and Wildlife Service.

In the case of airports, the 2014 assessment reports that "thirteen of the nation's 47 largest airports have at least one runway with an elevation within 12 feet of current sea levels, . . ."[11] which is "within the reach of moderate to high storm surge."[12] Airports in the New York City area (JFK, LaGuardia, and Newark), Florida (Fort Lauderdale, Tampa, and Miami), and San Francisco area (Oakland and San Francisco) are on this list. Although most of these at-risk airports are at over five feet of elevation, Louis Armstrong International, serving New Orleans, is at minus 1.7 feet.

Federal agencies have identified a special concern for the multiple modes of transportation that come together on the coast of the Gulf of Mexico. A 2008 detailed assessment of all forms of transportation in the Gulf Coast region between Galveston, Texas, and Mobile, Alabama, offered a sobering assessment.

Relative sea level rise will make much of the existing infrastructure more prone to frequent or permanent inundation—27 percent of the major roads, 9 percent of the rail lines, and 72 percent of the ports are built on land at or below 122 cm (4 feet) in elevation. Increased storm intensity may lead to increased service disruption and infrastructure damage: More than half of the area's major highways . . . almost half of the rail miles, 29 airports, and virtually all of the ports are below 7 m (23 feet) in elevation and subject to flooding and possible damage due to hurricane storm surge.[13]

Despite the concern for sea level rise impacts on transportation systems, planning to respond to these risks at the national level is limited. The 2014 Department of Transportation *Climate Change Adaptation Plan*[14] mentions sea level rise among other climate risks but does not lay out a strategy to respond to these risks specifically. The plan commits to including climate change considerations in investment planning and asset management and this will include some consideration of rising sea level. To that end, the 2014 Federal Highway Administration report provided technical guidance and methods for assessing the vulnerability of coastal transportation facilities with a focus "on quantifying exposure to sea level rise, storm surge, and wave action."[15] The plan also recognizes the importance of understanding the implications of changing climate for entire transportation systems, rather than just individual facilities, but does not specifically mention sea level rise in this context. Despite some evidence that officials perceive the problem, any effort to get beyond assessments and devise actual remedies to storm and sea level rise risks, at individual facilities, or for the transportation sector more broadly, is still to come.

Water Infrastructure

Drinking water treatment plants and sewage treatment plants provide essential services, and the interruption of service due to flooding of either type of water facility can cause immediate hardship for communities, undercut the local economy, and present serious public health and water pollution issues.

Sewage treatment facilities are especially vulnerable to storm surge and sea level rise because they are commonly located at a low elevation and next to a waterbody receiving a discharge of treated water. Flooding of a treatment plant interrupts treatment and results in discharge of untreated sewage, sometime for extended periods. In the case of drinking water facilities serving coastal communities, storm surge and rising seas pose a risk of inundation of treatment plants as well as a risk of saltwater intrusion to freshwater supplies pumped from coastal aquifers. Although the greatest risks to these facilities lie in the future, damages are already occurring. Hurricane Florence, for example, caused releases of sewage from the Wilmington, North Carolina, wastewater treatment plant, and its storm surge flooded the Onslow County wastewater treatment plant causing $7 million in damages.

Unfortunately, the national picture of water treatment system risk to storm surge and sea level rise is still hazy. In early 2018, researchers at the University of California at Berkeley, led by Michelle Hummel, published the first national study of sea level rise risk to sewage treatment plants.

> Across the United States, 60 wastewater treatment plants, serving over 4 million people, are exposed to flooding with 1 ft of SLR [sea level rise]. The largest increases in exposure occur from 3 to 4 ft of SLR, when an additional 83 plants serving 5.9 million people become exposed, and 4 to 5 ft of SLR, when an additional 91 plants serving 9.9 million people become exposed. By 6 ft of SLR, a total of 394 plants is exposed, and over 31 million people could be impacted by loss of wastewater services.[16]

In the case of drinking water systems, a study focusing on just the Atlantic and Gulf Coasts and on systems drawing water from surface water influenced by the tide, found twenty public water systems serving over one million people at risk of saltwater intrusion to freshwater supplies in the event of sea level rise, with five of these systems serving over 100,000 people "highly vulnerable."[17] The study points out that, although "this is relatively good news, especially considering the numbers of people that live in coastal communities, . . ."[18] more work is needed to assess saltwater intrusion risks to water systems drawing from groundwater, which are common in Florida, and to refine geographic data on the precise location of intakes and groundwater wells. For example, rising sea level poses especially complex challenges for the drinking water supply for the Miami–Dade County area. As sea level rises, saltwater moves underground toward the wells that draw water from the Biscayne Aquifer. Water from a treatment plant built in 2013 that treats salty water through "reverse osmosis" costs "two and a half times as much to process as water from the Biscayne Aquifer."[19]

Another consideration is that many coastal metropolitan areas are served by large water systems, and the interruption of treatment due to flooding can have ripple effects for inland communities that would otherwise face no storm surge or sea level rise risk. For example, the Deer Island Wastewater Treatment Plant on the shores of Boston Harbor treats waste from several million people in the greater Boston area.

Recognizing the risks, many of these large water systems, including systems serving New York, San Francisco, and Seattle, have developed plans specifically focused on flood and sea level rise inundation challenges. Large municipal water systems have organized a Water Utility Climate Alliance to share information and practices for dealing with climate change risks, including sea level rise. Several of these systems, such as the San Francisco Public Utilities Commission, have focused on including sea level rise risk assessments in capital planning.

Smaller systems, however, often lack resources to address sea level rise and storm surge risks. The Environmental Protection Agency (EPA) has developed an online tool to help water systems evaluate and address climate risks, including sea level rise (i.e., the Climate Resilience Evaluation and Awareness Tool). As part of this work, the agency has provided online mapping of storm surge and hurricane risk data that local drinking water and sewage treatment systems can use to understand and address risks.

In addition to water treatment plants, waste storage facilities, including hog waste lagoons and coal ash ponds, are at risk of flooding due to coastal storms and rising seas. Hurricane Florence, like Hurricane Floyd in 1999 and Matthew in 2016, demonstrated that heavy rainfall occurring over days in areas like eastern North and South Carolina can cause coastal rivers to rise and flood these facilities. These floods release toxic contaminants from coal ash ponds and biological and nutrient pollutants from hog waste lagoons. Initial reports after Hurricane Florence indicate that wastewater spilled from at least one coal ash pond, and thirty-two hog waste lagoons had overtopped, including a massive spill of hog waste of some seven million gallons. Flooding from Hurricane Florence also killed some 3.4 million chickens and turkeys and 5,500 hogs.[20]

Finally, storm surges and rising seas pose a threat to waste disposal sites, including Superfund sites, brownfield sites, solid waste facilities, and numerous other solid and chemical storage sites such as gas stations with underground tanks. Residential septic systems are at risk of both inundation and rising groundwater levels due to saltwater intrusion driven by rising seas. There is no national assessment of storm surge and sea level rise risk to all such sites, but the Environmental Protection Agency (EPA) released a screening analysis of climate change risk to 1,639 Superfund sites nationwide and identified 49 sites at risk of rising seas and 24 sites as needing closer evaluation.[21] These facilities are commonly in areas with disadvantaged populations and pose environmental justice concerns.

The 2014 *Climate Change Adaptation Plan* for the EPA includes chapters focusing on topics such as water, air, and wastes, as well as different regions of the country. The water chapter of the plan identifies storm surge and sea level rise as among six top climate change threats to water quality. The plan identifies some actions related to sea level rise and storm surge, including encouraging water systems to use the Climate Resilience Evaluation and Awareness Tool and development of "initial screening criteria to identify water and wastewater facilities on the Atlantic and Gulf Coasts that may be at risk of inundation in the event of a storm surge comparable to Hurricane Sandy."[22]

Some nonprofit organizations are pressing for a bolder agenda to protect water infrastructure from more severe storm surge and rising seas. For example, in a 2014 report, the Center for American Progress (CAP) urges EPA and states to "consider elevating, relocating, and building any new wastewater treatment plants farther from flood- or storm-surge-prone waters"[23] and to "prioritize improvements in wastewater systems that are most vulnerable to the impacts of a changing climate, including sea-level rise, drought, and extreme rainfall events"[24] when making low-interest loans to water systems.

In October 2018, Congress passed water infrastructure legislation taking small steps toward improving resilience of water infrastructure, but just for drinking water systems and focused on security and storm risks rather than rising seas. Water systems nationwide are directed to assess risks from "malevolent acts and natural hazards"[25] and other risks to system resilience and to develop emergency response plans based on these assessments. Natural hazards are defined to include floods and "hydrologic changes" without any direct mention of sea level rise. Another section of the bill provides supplemental funding for Drinking Water State Revolving Loan Funds of $100 million to assist small drinking water systems in recovery after a declared disaster.

These limited steps may prompt some needed preparation but run the risk of misdirecting investments due to lack of consideration of risks of rising seas. And, taken together, they fall short of a national plan for assessing risks to all coastal water facilities from more severe storms and rising seas and defining protection or relocation actions for specific facilities in the context of broader state and local plans.

Military Coastal Infrastructure

The coastal storm and sea level rise risks faced by Department of Defense (DoD) facilities located in the United States are well studied, perhaps with greater intensity than other infrastructure sectors. Despite progress in assessing risks, plans to protect these facilities are still emerging.

A 2013 report by the Department of Defense (DoD) Strategic Environmental Research and Development Program summarized sea level rise risk to domestic military facilities noting that "about 10 percent of DoD coastal installations and facilities are located at or near sea level and are already vulnerable to flooding and inundation.

Rising sea levels and more intense heavy downpours will make these conditions worse. The National Intelligence Council (NIC) estimated that more than 30 military installations in the continental United States were already facing elevated levels of risk from sea level rise (NIC 2008)."[26]

A 2011 report by the National Academy of Sciences (NAS) focused on Navy installations and quoted from the 2010 DoD Quadrennial Defense Review vulnerability assessment that

> identified 128 DOD installations that could be affected by a sea-level rise of equal to or greater than 1 meter. Fifty-six of these installations (or 43 percent of the total) were Navy installations. This number represents more than 50 percent of the 103 Navy installations that reported. Roughly $100 billion is the estimated dollar value of U.S. Navy installations that are at risk due to this one facet of climate change.[27]

DoD is not the only source of assessments of sea level rise impacts on military facilities. The Union of Concerned Scientists (UCS) looked at military bases and evaluated likely impacts of rising seas in 2015, finding that "the military is at risk of losing land where vital infrastructure, training and testing grounds, and housing for thousands of its personnel currently exist."[28] The UCS evaluated eighteen military installations, looking at risks related to the frequency and extent of tidal flooding as well as permanent loss of land due to sea level rise and concluded that "by 2050, in just 35 years, three installations in Florida and Virginia are projected to lose more than 15 percent of their land area to the sea according to the intermediate scenario [3.7 feet of sea level rise]. By 2100, nearly half of the sites studied lose 25 percent or more of their land area in the intermediate scenario and 50 percent or more in the highest scenario [6.3 feet of sea level rise]."[29] Some of the hardest hit installations lose between 75 percent and 95 percent of land area by the end of the century in the highest scenario, including the Naval Air Station Key West in Florida, Joint Base Langley-Eustis and Naval Air Station Oceana in Virginia, and the Marine Corps Recruit Depot on Parris Island in South Carolina.

Although less studied than sea level rise, coastal storms are a risk to military facilities. In October 2018, for example, Hurricane Michael caused "widespread and catastrophic damage"[30] to Tyndall Air Force Base near Panama City, Florida. As a result of the storm, "every single building and residence at Tyndall Air Force Base...is destroyed or so badly damaged that it is unlivable."[31] Estimates of rebuilding costs range from $3.6 to $5 billion.[32] Just a month earlier, Hurricane Florence damaged hundreds of homes on Marine Corps Base Camp Lejeune on the coast of North Carolina.

Although bases in different regions are at risk, the UCS notes special concern for bases in the mid-Atlantic region due to "the low elevation and land subsidence in the mid-Atlantic states (New Jersey through Virginia) and the faster rate of sea level rise on the East Coast."[33] The authors note that "many of the military installations in this area are of crucial importance to the armed forces: NS Norfolk (Virginia), the largest naval base in the world and home to the US Fleet Forces Command; the US Naval Academy (Maryland), one of the highest-ranked public colleges in the country and a National Historic Landmark; and Joint Base Anacostia-Bolling (Washington, DC), home to the Defense Intelligence Agency."[34]

This is strong evidence of serious potential impacts and raises the question of why DoD has not developed a comprehensive assessment of coastal storm and sea level risks or response plans for the facilities most at risk. The NAS called for such efforts in 2011, recommending that military leadership "ensure that a coordinated analysis is undertaken to address naval-installation vulnerability to rising sea levels, higher storm surges, and other consequences of climate change."[35] The NAS also opined that "a broader and more detailed assessment will provide a foundation, but there is a clear need for a more detailed global analysis and an action plan to address the vulnerabilities of those coastal installations identified as being at 'very high risk' and at 'high risk.'"[36]

In 2014, DoD published a "Climate Change Roadmap" that served as the department's climate change adaptation plan. Like the Department of Transportation, DoD concluded that sea level rise and associated storm surge was one of the top four risks related to a changing climate that it is facing, along with rising global temperatures, changing precipitation patterns, and increasing frequency or intensity of extreme weather events.[37] In a 2014 report, the Government Accountability Office (GAO) evaluated DoD preparations for climate change noting optimistically that "the department's intention is to conduct screening-level assessments in two phases: coastal and noncoastal. DOD has, according to officials, prioritized its initial assessment efforts on 704 coastal installations and sites."[38]

Perhaps with this goal in mind, in 2016, DoD released *Regional Sea Level Scenarios for Coastal Risk Management* and compiled an accompanying database on regionalized sea level scenarios for 1,774 DoD sites worldwide that are coastal or tidally influenced. The purpose of the report is "to enhance and increase the efficacy of screening-level vulnerability and impact assessment for DoD coastal sites worldwide containing permanent or enduring assets."[39] This document offers a lucid and detailed description of sea level rise, storm surge, and methods of assessment, and bodes well for the critical task of assessing risks at specific facilities and drawing facility-specific assessments together for a picture of risk to United States military facilities. Such a picture of risks would provide a solid foundation for measures to protect or relocate facilities.

Just when the stage seemed to be set for a comprehensive assessment of sea level rise risk to military facilities, in December 2017 the risk picture darkened when President Trump removed climate change from the list of threats to national security in an update to the National Security Strategy. Then, in January of 2018, DoD released a report summarizing results of a survey of climate-related impacts on installations around the United States. Unfortunately, the survey questions "focused on observed effects from past severe weather events"[40] and did not even address rising sea level. Looking backward on past impacts to facilities, the report identified inland flooding, wind, or drought impacts at over 700 facilities and reported flooding from storm surges at just 225 facilities. The *Washington Post* reported, however, that the term "climate change" appeared twenty-three times in the 2016 draft of the report but only once in the final version. More important, references to sea level rise in the original survey and maps of its impacts in the draft report were removed from the final report.[41]

It may be premature to worry that the first major DoD report on climate risks to facilities released by the Trump administration barely mentions impacts of future sea levels and storm surges. But a failure to follow through on the solid technical foundation laid in the 2016 coastal risk management report would be a major step backward. Contention over reporting of risks at military facilities continued in 2019 when congressional Democrats found major flaws in a new DoD report on climate impacts, including lack of attention to immediate climate impacts and demanded revisions. The revised report was criticized for lack of cost information and "opaque" methodology.[42]

An optimist can find some hope that the national security implications of climate change and rising seas are still in the sights of the defense establishment. For example, a Military Expert Panel made up of retired flag rank officers working with the Center for Climate and Security focused directly on sea level rises, concludes that "sea level rise, coupled with projections of increasingly frequent and intense storms, present significant risks to critical coastal military installations at home and abroad. Further, this report concludes that these climate risks are already being experienced and are likely to increase significantly during all the timescales explored in the recent literature—between 2035 to 2100."[43]

Figure 5-2. Naval Station Norfolk, Virginia, the world's largest naval station, currently floods about nine times a year during high tides. Photo by D. E. Erickson, US Navy.

The Military Expert Panel highlighted that rising seas have implications for national military strategy: "Sea level rise at just one site can have a significant impact on strategy. Hampton Roads, Virginia, dubbed 'the greatest concentration of military might in the world' by former secretary of defense Leon Panetta, is by itself an invaluable operational and strategic hub for both the United States and its allies."[44]

The panel offers a clear-eyed assessment of military preparedness for rising seas to date, observing that "despite these strategic concerns, a comprehensive assessment of sea level rise and broader climatic impacts on U.S. military and national security strategy has not yet been conducted."[45] The panel boldly asserts that, "conducting assessments of strategic impacts will, however, not be sufficient,"[46] and recommends that "information on the implications of sea level rise risks to military installations, and how it affects military and national security strategy, will need to find its way to senior leadership in order to drive high-level adjustments in strategic thinking about climate impacts."[47]

This message seems to be getting through to the Congress. The fiscal year 2019 National Defense Authorization Act, enacted in August 2018, includes several new directions to prepare military facilities for climate change including sea level rise.

New military facilities are to be built two feet above base flood elevation or three feet in the case of critical facilities. Congress also provided new authority to improve access roads outside a base at risk of flooding or inundation by rising seas. These are small steps in the right direction, but still well short of a national plan to protect or relocate facilities and operations at risk.

Senior military leaders are promising to do better. At his confirmation hearing in April 2019 to be the next chief of naval operations, Admiral Bill Moran noted "We are largely a waterfront service, so climate change when there's rising waters are going to be a problem for us if we don't address them. So we are in the planning stages to look at how to reinforce those areas."[48] At the same hearing, General David Berger, nominated to be the next Marine Corps commandant, said "The two biggest challenges are the rising water levels and the severe storms that roll up the coast and through our bases and stations."[49] In a step toward backing up these promises, the Navy is seeking $21 billion over twenty years to upgrade four shipyards, including Norfolk Naval Shipyard, making work on dry docks a priority.[50]

Stepping back to look at the big picture of storm and sea level rise risks to critical coastal infrastructure, several concerns stand out. In the case of transportation, water, and military facilities, significant risks are documented. But, there is little evidence of response actions or policies to reduce risks to existing facilities or avoid risks as new facilities are developed. Information on risks to other infrastructure types (e.g., telecommunications, public safety, energy grids) is sketchy. Perhaps most important, work to prepare one type of critical coastal infrastructure for more severe storms and sea level rise needs to be done in coordination with other types of infrastructure and the communities and businesses that are served. This coordinated planning for secure coastal infrastructure has not yet emerged but will be essential to successful preparation for more severe storms and rising seas.

6

Coastal Ecosystems Facing Inundation: Wetlands and Beaches

Damage or loss of property and critical infrastructure facilities due to coastal storms and rising seas is unavoidable but can be substantially reduced with prompt and thoughtful adaptation actions, including building coastal protection structures and relocating assets. Coastal ecosystems, including wetlands, beaches, dunes, marshes, and mangrove forests, will also be damaged by storms and inundated as sea level rises. Sustaining these diffuse natural systems as storms intensify and seas rise is much more difficult than protecting human-made infrastructure because of the large land areas involved and the natural and human-made constraints on landward migration of the ecosystems.

Like critical coastal infrastructure, coastal ecosystems provide valuable benefits to society. Wetlands, beaches, and mangrove forests provide important buffering to moderate impacts of storms and flooding. Marshes provide nurseries for fisheries and habitat for wildlife. Beaches also provide recreational benefits and support tourism that is a critical part of many coastal economies.

Many ecosystems around the coast are the subject of local studies of climate change impacts generally and sea level rise or storm impacts more specifically. Some of these local assessments provide hints of the national scale of potential losses but it is hard to generalize because methods differ (e.g., use of different sea level rise projections).

Sadly, coastal ecosystems do not have a federal agency dedicated to their health and resilience in the same way that transportation facilities are the responsibility of the Department of Transportation or military facilities are the responsibility of the Department of Defense. Without a clear mandate for the collection and assessment of information about the condition of coastal ecosystems and the risks they face from more severe storms and rising seas, federal agencies have not developed the plans and programs needed to sustain these resources and support successful landward migration in the decades to come.

Wetlands and Marshes

Coastal wetlands include salt marshes, bottomland hardwood swamps, freshwater marshes, and mangrove forests. The list of valuable functions or "ecosystem services" that coastal wetlands perform is impressive. In addition to providing a buffer to reduce storm surge and help minimize coastal flooding, coastal wetlands absorb ocean energy to minimize coastal erosion, support fisheries and commercial and recreational fishing, provide critical habitat for birds and other species, and even sequester carbon."[1]

A major assessment of coastal wetlands by the United States Fish and Wildlife Service and the National Oceanic and Atmospheric Administration (NOAA) in 2013 looked at the status of wetlands in coastal watersheds and found about 41 million acres of wetlands in these watersheds, including freshwater wetlands scattered across the watersheds and saltwater wetlands. About 8 million of these 41 million acres are in Great Lakes watersheds. These 41 million acres are about 38 percent of the total wetlands in the lower forty-eight states and over 80 percent are located in the Southeast.[2]

The area of these coastal wetlands, however, is shrinking fast. Between 2004 and 2009, "the average annual rate of change was an estimated loss of 80,160 acres, a 25% increase in the rate of wetland loss from the previous reporting period."[3] Not surprisingly, "seventy-one percent of the estimated wetland losses were in the coastal watersheds of the Gulf of Mexico."[4]

Although knowing the distribution of all wetlands throughout a coastal watershed is interesting, in the context of preparing for storm surge and sea level rise, knowing the amount and condition of saltwater wetlands is more important. The 2013 coastal wetlands report found 6.4 million acres of saltwater wetlands, which are about 16 percent of all coastal wetlands.[5] Some 3.4 million acres of these saltwater wetlands are in watersheds along the Gulf of Mexico with 2.4 million acres in Atlantic Coast watersheds, and just over 700,000 acres in Pacific watersheds.[6]

This total saltwater wetland acreage is a decline of about 1.5 percent over the 2004–2009 period, and the rate of decline was 35 percent greater than the rate over the

1998–2004 period. Most of these losses occurred along the Gulf of Mexico as wetlands were converted to open saltwater.[7] As to the cause of these losses, the report states that "saltwater wetland losses in the Gulf of Mexico have been attributed to the effects of severe coastal storms such as Hurricanes Katrina and Rita in 2005 and Hurricane Ike in 2008, which inundated wetlands with storm surge, abnormally high tides, increased rainfall, runoff, increased sediment and debris deposition and erosion."[8]

Along the Atlantic Coast, the story is more complicated. Some freshwater wetlands were inundated by sea water to become saltwater wetlands, and some saltwater wetlands were lost. The report suggests the cause: "The majority of the losses in estuarine marsh areas were attributed to erosion and/or inundation related to increases in sea level."[9]

On the Pacific Coast, recent research on sea level rise impact on wetlands by Karen Thorne and coauthors concluded that "tidal wetlands are highly vulnerable to end-of-century submergence, with resulting extensive loss of habitat. Using higher-range SLR [sea level rise] scenarios, all high and middle marsh habitats were lost, with 83 percent of current tidal wetlands transitioning to unvegetated habitats by 2110. Tidal wetland loss was also likely under more conservative SLR scenarios. . . . Horizontal migration of most wetlands was constrained by coastal development or steep topography."[10]

As conditions on the Pacific Coast illustrate, a key question that arises in evaluating sea level rise risks to wetlands and marshes is whether sea level will rise gradually enough so that the wetland can naturally move back as waters rise. A 2010 study by Mathew Kerwin and colleagues at the United States Geological Survey (USGS) found a tipping point at a sea level rise rate of about a meter per 100 years.

If global temperature warming follows conservative IPCC projections and ice sheets contribute little water to the oceans, our model experiments indicate that many marshes will accrete vertically and maintain their position within the intertidal zone. However, if temperature warming follows more rapid scenarios and/or ice sheets contribute significant water such that sea level rises by more than a meter in the next 100 years, our models indicate that most marshes will permanently submerge despite their tendency to accrete more quickly.[11]

The authors caution that each marsh is different and some may be lost to slower rates of sea level rise. Modeling the interplay of erosion, accretion, soil saturation, salinity, organic matter, and sediment loading at different degrees of sea level rise at different coastal sites is a complex job. Fortunately, there is a model that looks

Figure 6-1. Kids look for wildlife at Sweetwater Marsh, a unit of the San Diego Bay National Wildlife Refuge. The marsh has limited space to migrate landward due to adjacent development. Photo by Lisa Cox, US Fish and Wildlife Service.

at all these factors and describes the migration of wetland boundaries as sea level rises in both numerical and map outputs. The Sea Level Affecting Marshes Model, or SLAMM, developed by Jonathan Clough and colleagues, offers a rapid and low-cost way to visualize future inland migration of marshes and wetlands. The SLAMM is available for the entire coastlines of nine states with partial coverage of an additional twelve states and two United States territories.

A related decision-support tool is the Coastal Resilience website operated by a public–private partnership led by The Nature Conservancy (TNC), a national non-profit organization. It presents sea level rise information by project area and can provide an assessment of "Future Habitat," such as marshes, for the area, often drawing on the SLAMM. For example, the Florida Keys site provides a tool that reports that one meter of sea level rise by 2060 would result in the loss of over 63,000 acres of regularly flooded marsh, or 4 percent.

Despite the evolution of helpful tools, to date there is no national analysis of the potential for landward migration of saltwater wetlands and marshes as sea level rises, no estimate of the net loss or gain in saltwater wetlands that might result, and little understanding of the ecological significance of potential losses. Some losses of

Table 6-1. Percent of wetlands globally converted to open water due to sea level rise

Year	Conversion if 1.6 ft. by 2100	Conversion if 3.2 ft. by 2100
2050	22%	32%
2080	32%	44%

Source: Data from Loraine McFadden, Tom Spencer, and Robert J. Nicholl, "Broad-scale Modelling of Coastal Wetlands: What is Required?" *Hydrobiologia*, (2007): 14.

freshwater wetlands will also occur as sea level rises and the extent and significance of these losses is also unknown.

Loraine McFadden and coauthors attempted to quantify global wetland conversion to open water as a result of sea level rise, estimating losses of between 20 percent and 40 percent, although some estimates of losses in the event of 3.2 feet of sea level rise are over 60 percent with better accounting for constraints on landward migration[12] (see table 6-1).

The application of methods similar to those used in the global analysis to the Gulf of Mexico results in similar loss estimates. Assuming a rate of sea level rise resulting in a 3.2 foot rise by 2100, the percentage of coastal wetlands expected to become open water by 2060 is 37 percent in Texas, 32 percent in Florida, and 26 percent in Alabama and Mississippi.[13]

These wetland loss estimates are not based on the more detailed methods in the SLAMM, and there is reason to hope that the continued diligent application of the SLAMM to many more sites will focus the wetland-loss picture nationally. In addition, the Department of the Interior Climate Science Center in the Southeast region has a study underway of marsh migration potential for the entire Gulf of Mexico with a goal of producing "customized landscape conservation-design products focused on identifying landward migration routes for coastal wetlands."[14]

In another signal of growing recognition of sea level rise impacts on wetlands, and wetlands management projects, the National Academy of Sciences in 2018 called for a "mid-course assessment"[15] in the multibillion dollar effort to restore the Everglades. The academy noted that existing project plans are based on conditions that existed between 1965 and 2005 but that a changing climate is altering the ecosystem and there is "compelling recent evidence that sea-level rise is accelerating."[16] Noting that current plans "do not adequately account for these changes," the academy recommended "systemwide modeling of interactions among both authorized and planned projects under scenarios of future possible climate and sea level-rise conditions."[17]

A somewhat more optimistic view comes in a 2016 article by Mathew Kerwin and coauthors arguing that a site-specific assessment is important and that "marsh vulnerability tends to be overstated because assessment methods often fail to consider biophysical feedback processes known to accelerate soil building with sea level rise, and the potential for marshes to migrate inland."[18] Kerwin concluded that "these examples suggest that marsh survival is possible even at high rates of SLR [sea level rise], as long as the marsh is allowed to transgress inland and compensate marsh erosion at the ocean boundary."[19]

Finally, it is important to remember that mangrove forests are especially valuable in reducing the impacts of severe storms, and that protecting mangroves is recognized as an alternative to engineered coastal protection structures.[20] There are some 620,000 acres of mangrove forests in the United States, virtually all in South Florida. The good news is that mangrove forests are more resilient to sea level rise than most other types of wetlands. The bad news is that, as sea level continues to rise, they too will convert to open water.

Mangrove forests have a small footprint in the United States but they have significant benefits in terms of both storm protection and sequestration of carbon. Wetlands, salt marshes, and mangrove forests sequester carbon in soils, and wetland plants add carbon to soils each year at rates significantly greater than tropical forests. A 2018 report by the National Academy of Sciences concluded that "negative emission technologies," such as forest management, agricultural land management, and direct air capture are critical tools to meeting the goals of the Paris Climate Agreement. Although the total carbon sequestered by coastal wetland management is small compared to forest and agriculture practices, the costs are low, and the academy recommended that "coastal blue carbon" practices be supported, noting the "potential to more than double the current [sequestration] rate through several CO_2 removal approaches that restore and create coastal wetlands."[21] Planning for the successful migration of coastal wetlands inland as sea level rises could also enhance carbon sequestration—a win-win proposition.

Wetlands are fortunate to be protected under the Clean Water Act and several related laws. Section 404 of the act requires permits that limit development in wetlands, and the Environmental Protection Agency (EPA) adopted a goal in its 2014 *Strategic Plan* to "achieve a net increase of wetlands nationwide, with additional focus on coastal wetlands."[22] Unfortunately, this goal is deleted in the new EPA strategic plan developed by the Trump administration. And, in yet another blow to wetlands protection, the

Trump administration proposed in late 2018 to dramatically reduce the range of waters protected by the Clean Water Act, including reducing the wetlands covered by the act. The percentage of wetlands losing protection under the proposal may be as high as 50 percent nationwide,[23] although the percentage of coastal wetlands losing protection may be lower. The proposal is sure to be challenged in court.

EPA also makes grants to states to operate wetland protection programs and some states have adopted wetland protection plans. Some of these wetland plans, including those in Delaware, Virginia, and California, address sea level rise. The state of Delaware's plan, for example, sets a goal of planning for the impact of sea level rise on wetlands, including to "identify preservation areas for potential wetland migration."[24] In general, however, most coastal states are not looking at the long-range future of coastal wetlands, and there is no national program or plan to assess the timing of inundation of specific wetlands or the feasibility of facilitating land-ward migration, or the measures that will be needed to support successful migration.

Other federal agencies, including NOAA and the Fish and Wildlife Service, have very limited funds for acquisition of wetlands. In general, however, federal agencies responsible for wetlands protection have not yet considered how emerging tools for evaluating potential landward migration of saltwater wetlands can support policies and investments to protect existing wetlands at risk of inundation and the dry lands that will become wetlands as sea level rises.

Beaches and Dunes

Of the roughly 12,383 miles of marine coastline in the United States,[25] about 5,400 miles is in the form of a beach, such as a mainland beach, barrier beach, or linear beach along a bluff.[26] Although there are multiple measurements of the linear miles of coastline and tidal shoreline, there is no national database of the area of beach systems including supporting dunes and marshes. And, there is no national assessment of the geological and other constraints on beaches indicating which beaches will be able to naturally migrate landward as sea level rises and which are likely to disappear because they are hemmed in by a cliff face, other geological features, or human development.

In California, a state where there is strong interest in both beaches and a changing climate, there is special concern for the impact of sea level rise on beaches and information suggesting that most will not survive higher sea levels. A USGS study in 2017 noted that "significant impacts to the shoreline will occur due to accelerated sea-level rise, with 31% to 67% of beaches in Southern California lost by 2100 under the 0.93 to 2.0 m [3–6.5 feet] SLR [sea level rise] projections."[27] Asked to estimate how much sand may be needed to nourish the beaches above the substantial beach nourishment already occurring, USGS study coauthor Patrick Bernard commented this: "My sense is that it's an order of magnitude larger—you might need 10 times the amount of sand than what's been placed before to maintain beaches. . . . It's going to take a much larger effort."[28]

Building on the USGS assessment of beaches, the California State Coastal Conservancy and The Nature Conservancy in California developed a more comprehensive assessment of sea level rise impacts on coastal habitats in the state. Some 25 percent of public conservation lands are at risk of inundation by five feet of sea level rise, as are 58 percent of rocky intertidal habitats and regularly flooded estuarine marshes.[29] About half of the conserved coastal lands are resilient to sea level rise, and the authors identified some eighty square miles of "potential future habitat"[30] that could help mitigate habitat losses. But each region of the coast has highly vulnerable conservation lands, with the San Francisco Bay delta by far the most vulnerable.

Unlike wetlands, beaches are not protected under the Clean Water Act, and no federal agency speaks to supporting landward migration of beaches as part of its climate adaptation planning. It is true, however, that NOAA supports state coastal management programs, and some state programs are focused on beach condition and impacts on beaches. Although EPA works with states to monitor water pollution at beaches and notify the public of pollution problems posing a swimming risk, it does not evaluate sea level rise risks or develop beach migration plans. The Army Corps of Engineers has a significant role in beach nourishment, but this work is focused on restoring sand to beaches that communities consider important economically rather than preserving beaches based on their ecosystem values or helping them migrate landward.

For federal agencies or states, taking a more proactive role in beach protection will be an uphill fight. Tools for evaluating the potential for landward migration of beaches are not as developed as for wetlands. In addition, geological barriers to beach migration and the intensity of development on beaches, and on uplands behind beaches, means that they commonly have more limited options for landward migration than do wetlands.

A final challenge to successful landward migration of both beaches and wetlands is hardening of the coastline in the form of seawalls, bulkheads, and related structures. These hardened shorelines are intended to protect property and limit erosion but also "support reduced diversity and abundances of marine fauna when compared to natural shorelines,"[31] and reduce nutrient filtration, carbon storage, and recreational value. Rachel Gittman estimated that 14 percent of the shoreline is hardened today, mostly along the Atlantic and Gulf Coasts, and that these structures are associated with housing density and storm frequency.[32] She estimated that nearly 30 percent of coastline will be hardened by 2100 if the current rate of hardening continues. Accelerating sea level rise and more severe storms, however, are likely to prompt an acceleration of these projects, further limiting landward migration of wetlands and beaches. Although these structures require a federal permit, and some states have adopted limitations (see ch. 14), permit decisions do not now occur in the context of a larger plan for responding to storms and rising seas, opening a path to much increased coastal armoring, damage to ecosystems, and limits on landward migration of ecosystems.

Coastal National Parks and Protected Areas

Although national parks are not ecological resources per se, they are generally recognized as places of outstanding ecological and cultural importance.

A 2015 study by the National Park Service (NPS) looked at the potential impacts of a 3.3 foot rise in sea level on coastal national parks concluding that "results from this study show that well over one-third of the FMSS [Facilities Management Software System] listed assets within 40 coastal NPS units are at risk to long-term SLR [sea level rise]. These assets have combined value of over $40 billion and many of the high exposure assets provide essential day-to-day functions, such as visitor access."[33]

The sixteen parks identified as having high exposure to sea level rise include iconic places such as Everglades National Park and Assateague Island National Seashore. Eleven parks face the loss of all identified assets within the park. And, report authors offered this caution: "The SLR exposure analysis is likely a fairly conservative estimate of the number of assets at risk over the next 100 years, as storm impacts (especially within the units along the east coast) may be a more imminent threat to NPS property."[34] A follow-up study by the Park Service in 2018 used the most recent sea level rise and storm surge data to evaluate 118 Park Service units and found "considerable risks to infrastructure, archeological sites, lighthouses, forts, and other historic structures,"[35] with significant impacts on the National Capitol Region around Washington, DC.

A study in 2016 of the sixteen units of the National Estuarine Research Reserve System, managed by NOAA, described a "Marsh Resilience to Sea Level Rise" (MARS) index with up to ten metrics and concluded that "this assessment revealed moderate resilience overall, although nearly all marshes had some indication of risk. Pacific marshes were generally more resilient to SLR [sea level rise] than Atlantic ones, with the least resilient marshes found in southern New England."[36]

Although the estimates of vulnerability of these protected coastal places to sea level rise provide a foundation for response actions, it is not clear what action federal agencies will take. The National Park Service is conducting studies of some thirty additional park units but has not announced development of response plans. The authors of the NOAA study of estuarine research reserves noted the following: "One certainty that applies to all tidal marshes is the need for continued long-term monitoring and study, both to understand how these important ecosystems respond to SLR and other stressors associated with climate change and to evaluate the management actions implemented to protect them."[37]

Beaches and wetlands have the potential to be afterthoughts in work to prepare for more severe storms and rising seas. There is an unfortunate tendency to assume that these resources will simply fend for themselves as waters rise and will transition to upland areas where favorable geography makes this possible. In addition, there is sometimes an implicit assumption that, as sea level rises, the needs of communities and critical infrastructure will have priority over ecosystems in competition for optimal land area. Although coastal parks have an advocate in the form of the National Park Service, most coastal ecosystems do not have a federal government agency charged with helping them survive storms and rising seas. Taken together, these circumstances do not point to a healthy future for coastal ecosystems. Sadly, degradation of these ecosystems poses risks for key sectors of the coastal economy, including tourism and fishing.

7

Private Sector Losses as Seas Rise: Tourism, Fishing, and Energy

Coastal communities, beaches, and publicly owned infrastructure come up first in any conversation about damages expected from more severe storms and rising seas. But it is important to remember that private sector enterprises and assets play an essential part in sustaining coastal communities and are also at risk.

Key economic sectors with substantial private assets at risk are tourism and commercial and recreational fishing. Tourism is an essential component of the economy of many coastal communities and several states. Commercial and recreational fishing industries have a smaller, but still significant, economic footprint and make important contributions to the identity of many coastal communities.

In addition, a number of industries rely on major facilities that are right on the coast (e.g., shipbuilding and chemical plants). These industries employ thousands of people in coastal communities and support local economies. Other industries, such as telecommunications, are at risk because just part of their assets are in risky coastal areas. For example, a 2019 analysis found that "sea level incursions could have a devastating impact on Internet communication infrastructure even in the relatively short term" and that "1,186 miles of long-haul fiber conduit and 2,429 miles of metro fiber conduit will be underwater in the next 15 years."[1] But, the industrial facilities along the coast with the most to lose to more severe storms and rising seas are in the energy sector. Impacts of storms and rising seas on oil refineries are especially significant as they pose risks for corporate owners, for the communities that are home to these facilities, for the environment, and for the larger national economy.

Who can be counted on to help these private enterprises understand the risks they face from more severe storms and rising seas and sort out options to manage and reduce impacts?   The story is different in each case. For tourism, there is growing evidence of the potential economic harm but little progress in coming to grips with these threats. In the case of commercial and recreational fishing, the National Oceanic and Atmospheric Administration (NOAA) is working with state and local partners to protect fisheries, including "Essential Fish Habitat," from various threats including rising seas and more severe storms. In the case of energy facilities, the Department of Energy has evaluated risks and industry associations are encouraging preparedness. Some communities that are home to these facilities, as well as investors in companies that own the facilities, are also showing interest in understanding and addressing these risks.

Although there are some positive developments in terms of preparing privately held assets for more severe coastal storms and rising seas, it is important to recognize that these efforts fall well short of a comprehensive assessment of risk that informs a coordinated plan to prepare a sector to meet flood and inundation risks.

Coastal Tourism and Rising Seas

The economic value of coastal tourism is impressive. The *National Climate Assessment* reported in 2014 that, "coastal recreation and tourism comprises the largest and fastest-growing sector of the U.S. service industry, accounting for 85% of the $700 billion annual tourism-related revenues, making this sector particularly vulnerable to increased impacts from climate change."[2]

Much of this tourism is focused on beaches, which are directly threatened by rising seas. In 2013, James Houston focused on the value of beaches and estimated that beaches alone contribute $225 billion annually to the United States economy. Houston also reported that 200 million Americans made 2.2 billion beach visits in 2010.[3]

Florida is the state with the most to lose from loss of tourism due to more severe storms and rising seas. Julie Harrington of the Center for Economic Forecasting and Analysis at Florida State University, and coauthors, reported that "in 2015, a record 106.3 million tourists visited Florida, with an economic impact of ~$90 billion. Over the last five years, tourism has averaged about 6% growth annually. . . . Tourism accounts for more than one million direct jobs and an additional 1.5 million indirect jobs."[4] Looking more directly at beach-oriented tourism, Harrington concluded that "Florida beaches had approximately 810 million beach day visits in 2012, the most of any state or country in the world. Also in 2012, the beach-oriented visitors to Florida totaled 38.4 million or 42% of the total visitor population to Florida."[5]

In a 2007 assessment of climate change impacts on the state of Florida, Elizabeth Stanton and Frank Ackerman reported that "tourism, one of Florida's largest economic sectors, will be the hardest hit as much of the state's wealth of natural beauty—sandy beaches, the Everglades, the Keys—disappears under the waves. . . . Costs of inaction are projected to total $9 billion by 2025, $40 billion by mid-century, and $167 billion at the end of the century."[6]

Several studies have focused on the impacts of more severe storms and rising seas on the tourism economy of the Outer Banks in North Carolina. David Edgell Sr. and Carolyn McCormick reported that "North Carolina's Outer Banks accounts for expenditures of over one billion dollars and 20,000 jobs. The visitors to the Outer Banks are most interested in the natural, cultural and historic resources; primarily the 130 mile stretch of beaches of the Outer Banks."[7]

A study by researchers at several North Carolina colleges concluded that "by 2080, 14 of the 17 recreational swimming beaches in southern NC are projected to have eroded all the way to the road, making beach recreation no longer possible,"[8] and defined potential future economic losses: "The lost recreation value of climate change-induced sea-level rise to local beach goers is projected to be $93 million a year by 2030 and $223 million a year by 2080 for the southern North Carolina beaches.... Reduced opportunities for beach trips and fishing trips are projected to result in lost recreational benefits totaling $3.9 billion for the southern North Carolina beaches over the next 75 years."[9]

Beach loss and the resulting economic impact on tourism is also a concern in California. Commenting on the US Geological Survey (USGS) study finding significant long-term losses of California beaches, study coauthor Sean Vitousek noted that, "beaches are perhaps the most iconic feature of California, and the potential for losing this identity is real."[10] In 2016, the state of California offered a general estimate of beach tourism value: "It is estimated that of California residents only, there are about 15 million users and 150 plus million visits to beaches annually. This results in billions of dollars annually that flow into city, county, state, and federal economies."[11]

A few years earlier, in 2012, the state of California commissioned a study of the economic impacts resulting from loss of just a few of the major beaches along the coast and calculated costs in terms of lost tourism spending and local and state revenue out to the year 2100 due to narrower beaches attracting fewer visitors, including the following:

- Venice Beach, Los Angeles County, $439.6 million;
- Zuma Beach and Broad Beach, Malibu, Los Angeles County, $498.7 million;
- Carpinteria City and State Beach, Santa Barbara County, $164.7 million; and
- Torrey Pines City and State Beach, San Diego County, $99 million.[12]

Farther west, rising seas are raising concern for tourism in Hawaii. A 2017 series of articles by *Hawaii News Now* reported that, "a summer with record-breaking king tides . . . have swallowed up stretches of Waikiki Beach and pushed water into parking lots and roads."[13] A 2008 study of Waikiki concluded that, "our analyses estimate that nearly $2.0 billion in overall visitor expenditures could be lost annually due to a complete erosion of Waikiki Beach."[14] Mufi Hannemann, president and CEO of the Hawaii Lodging and Tourism Association, is quoted saying, "You don't want to wait for a catastrophe to happen before you start to react and say, 'Man, we should have done this.'"[15]

These estimates of losses to the tourism economy from rising seas are still fragmentary. Projections of just the beach-related losses are limited to beach-oriented states, differ in method and timing, and generally do not address losses that might be associated with more severe coastal storms. More comprehensive assessments of tourism losses due to severe storms and rising seas might look at losses of resort properties, potential loss of foreign tourist spending, lost state and local revenues, impacts on transportation and accommodation, impacts on historical and cultural sites, and losses of natural features on the coasts other than beaches.

The literature on tourism and sea level rise conveys a sense of both urgency and frustration with limited options. In a comprehensive evaluation of sea level rise impacts on tourism in the Caribbean, Daniel Scott and coauthors thoughtfully articulated their view of the urgency of the situation.

> While this study focused on the Caribbean region, it is clear that SLR [sea level rise] will similarly represent the preeminent long-term threat to the sustainability of coastal tourism . . . and indeed coastal tourism destinations worldwide. It is, therefore, imperative that the coastal tourism research community together with coastal management professionals initiate the process of adapting to SLR, not because there is an impending catastrophe, but because there will be important opportunities to avoid adverse impacts in the near-term, and because the complex planning required for coastal management and protection (environmental assessments, financing, land acquisition and construction) often takes decades to complete.[16]

Closer to home, in the conclusion to their study of tourism on the North Carolina Outer Banks, David Edgell and Carolyn McCormack observe that, "we no longer have the luxury of debate. The impact of climate change on the Outer Banks is now and it is necessary to work toward immediate climate change adaptation strategies."[17] In Florida, Julie Harrington noted that "to ensure that Florida remains one of the top tourist attractions and destinations in the U.S. and worldwide, the state should continue making improvements in building capacity towards adaptation, specifically in the tourism industry."[18]

Unfortunately, calls for attention to the damage that more severe storms and rising sea will cause to the American tourism economy have gone largely unheeded. One possible reason for the lack of a more focused response is the decentralized structure of the tourism sector. Tourism, as it exists in countless places along the American coastline, is made up of hundreds of types of businesses, many of which are small, and tens of thousands of business owners. This diversity makes delivering information about more severe storms and rising seas very challenging, and building consensus in support of response actions difficult.

Another consideration is the concern that any suggestion of future beach loss will deter tourists and damage local economies. Still another reason for the reluctance to engage the tourism and sea level rise problem is frustration with costly, short-term solutions. Beach nourishment and protection structures are commonly identified actions suggested as responses to impacts of rising seas on tourism. Yet, in some places, such as Southern California, there seems to be some recognition that just adding sand to beaches as seas rise is an expensive and ultimately futile strategy.

The alternative, however, involving stepping tourism back from the shore, can be hard to contemplate. As Danial Scott explains, "Coastal retreat is a largely untested strategy for coastal tourism destinations. The planning required for coordinated retreat would be highly complex, severely challenging local governments and planning authorities and would very likely result in major legal disputes, substantial impacts on property values, a reduced taxation base and a decline in tourism activity during this transitional phase."[19]

Helping thousands of coastal communities, and businesses large and small, all with economic survival on the line, face up to the reality of more severe storms and rising seas, and the need to make peace with the reality of a new coast is an intimidating challenge. Who will take up this challenge, and what role governments will play, needs to be sorted out. The economic stakes, and the difficulty factor, multiply as time passes.

Rising Seas Threaten Commercial and Recreational Fishing

Compared to the coastal tourism sector, the commercial and recreational fishing industry has a smaller footprint, in terms of money and jobs, and is fortunate not to face an existential threat from more severe storms and rising seas.

NOAA keeps detailed accounts of fish landings and reported that "commercial landings . . . were 9.6 billion pounds . . . valued at $5.3 billion in 2016."[20] The value of fish landed, however, is just a small part of the value of the entire industry. NOAA reported that in 2014 the entire commercial harvest sector "supported 1.4 million full- and part-time jobs and generated $153 billion in sales, $42 billion in income, and $64 billion in value-added impacts nationwide."[21]

A first thought of coastal storms and rising seas threatening commercial fishing might be for damage to port infrastructure and on-shore processing facilities. Indeed, a 2016 study by Lisa Colburn and coauthors identified some 175 coastal communities on the Atlantic and Gulf of Mexico coasts deemed "highly engaged and/or reliant commercial fishing communities." Colburn evaluated sea level rise risk to these communities with a "sea level rise risk index," and found that seventeen of these communities were at high risk, located mostly in the mid-Atlantic region and Florida. Seafood commerce businesses in these communities are also at risk of rising seas. For example, with three feet of sea level rise, Colburn found some 125 businesses with revenues of over $82 million at risk and, with six feet of sea level rise, over 400 businesses with revenue of $395 million were deemed at risk.[22]

Another significant aspect of sea level rise on the fishing industry is the potential loss of coastal marshes and estuaries functioning as near-shore nurseries to support healthy fisheries. The gradual degradation or loss of these nurseries as sea level rises is one factor among several related to climate change—including warming water temperatures and ocean acidification—that put healthy ocean fish populations at risk.

Although it is hard to sort out the relative impacts of a changing climate, fishing practices, and other factors on the viability of fisheries or the fishing industry, protection of nursery and spawning areas has long been a priority, and there is evidence that sea level rise poses a risk to these critical areas. The guiding legislation relating to fisheries is the Magnuson-Stevens Act that, among many other things, calls for Regional Fisheries Management Councils to develop Fisheries Management Plans and to identify "Essential Fish Habitat" within those plans.

Essential Fish Habitat (EFH) is defined as "those waters and substrate necessary for fish spawning, breeding, feeding, or growth maturity."[23] Fisheries Management Plans are to prevent adverse effects on EFH, using fishing related restrictions and restrictions on nonfishing activities that may result in "conversion of aquatic habitat that may eliminate, diminish, or disrupt the functions of the EFH."[24] Federal agencies are to advise NOAA of federal actions that may adversely affect essential habitat, and NOAA is directed to work with other federal agencies and states to "further the conservation and enhancement of EFH."[25]

As indicated in chapter 6, sea level rise and more damaging coastal storms will result in losses of coastal marshes and wetlands. A review of climate and fisheries management issues by EcoAdapt summarized the scientific literature, "Rising sea level is likely to lead to changes in important habitat for fish and shellfish in various life stages. Although in certain areas, such as the Gulf of Mexico, coastal inundation could initially increase coastal habitat and nursery areas . . . sea level rise is also likely to lead to flooding, saltwater intrusion, and degradation of key breeding nursery ecosystems, such as coral reefs, mangroves, estuaries, and tidal marshes."[26]

NOAA developed an "EFH Mapper" online that lets a user look at mapped essential habitat by region and species, but NOAA has not provided a national summary of the area of all the habitat defined in the mapper. Without that data, it is hard to estimate the national implications of rising seas for this habitat. In general, NOAA reports that "fisheries and the councils have described EFH for multiple life stages of nearly 1,000 federally managed species and designated more than 100 HAPCs [Habitat Areas of Particular Concern]."[27]

Although the data on areas protected as important habitat for fish is a bit sketchy, NOAA officials see protection of this essential habitat as an important tool for protecting fisheries and the fishing economy. NOAA senior manager Elaine Sobeck observed in 2016, "Today, overfishing has been largely addressed. Commercial and recreational fishermen should be proud of the sacrifices they have made in order to get fish stocks to a point where they aren't subject to overfishing. This gives us some breathing room to focus on habitat."[28] Sobeck added, "The challenge for EFH is how to deal with the fact that the ocean environment is changing."[29]

In 2016, NOAA sponsored an "Essential Fish Habitat Summit" in Annapolis, Maryland, and the report from that meeting noted that, "managers agree that EFH needs to be updated to account for climate change effects, especially since EFH consultations try to anticipate future impacts and habitat use."[30] This direction is consistent with elements of 2015 NOAA *Fisheries Climate Science Strategy,* and supporting Regional Action Plans, that call for climate vulnerability assessment for all living marine resources and ecosystem indicators in order to track climate-driven changes.

As it is a challenge to estimate the relative importance of fish habitat in the larger context of other climate risks and other fishery management challenges, it is also hard to estimate the net change in the area and quality of this habitat as a result of rising seas. As in the case of beaches or wetlands, habitat may smoothly transition landward as seas rise in some places, be blocked by geographic features in other places, and be blocked by development in still other places. Will this sifting of gains and losses net out to an increase or a decrease? Will habitat areas that are gained function as well for fish as the areas lost?

For the past several years, Congress considered legislation to reauthorize the Magnuson-Stevens Act, and in June 2018, the House of Representatives passed a reauthorization bill, largely along party lines, that is "expected to provide more flexibility in stock rebuilding schedules as well as ease other regulatory burdens."[31] Hearings on the legislation passed mostly without reference to essential habitat, sea level rise, or climate change, with the notable exception of testimony by Dr. John Quinn, chair of the New England Fishery Management Council, who spoke on behalf of the Council Coordination Committee made up of the leadership of the eight Regional Fishery Management Councils.

> Addressing climate change will require establishing the support to enable fishery managers to develop creative solutions to new challenges. Fishery managers also will need a strong scientific foundation to support climate-ready fisheries management. Managing climate-ready fisheries is a long-term endeavor that will require investing in the information needed to support informed decision-making, along with a commensurate shift in resources and attention.[32]

Finally, it is also important to remember, amid all NOAA's fishing statistics, the less tangible consideration that the fishing industry defines the character of many coastal communities. In 2016, Sorna Khakzad and David Griffith looked at contributions that the fishing industry made to building a sense of community concluding that, "fishing material culture, including fish houses, boats, docks, etc., are significant for fishermen and their communities in the sense that they represent their authentic activities, and they feel these items and places are repositories of history and memory, representing their individual and community's identity and sense of place. These buildings and sites are landmarks that form their traditional environment."[33]

A decline in the fishing industry as a result of warmer, more acidic oceans, coupled with gradual damages of rising seas, is a double hit to fishing communities and could be a triple hit if a major storm happens to strike. Steps to help the industry weather the challenges of a changing climate have the double benefit of supporting a valuable industry while also helping sustain the values of coastal communities as they adapt to rising seas.

Coastal Energy Infrastructure

One of the five "key messages" in the energy chapter of the 2014 *National Climate Assessment* is that, "in the longer term, sea level rise, extreme storm surge events, and high tides will affect coastal facilities and infrastructure on which many energy systems, markets, and consumers depend."[34] The report goes on to note that, "in particular, sea level rise and coastal storms pose a danger to the dense network of Outer Continental Shelf marine and coastal facilities in the central Gulf Coast region. Many of California's power plants are at risk from rising sea levels, which result in more extensive coastal storm flooding, especially in the low-lying San Francisco Bay area. Power plants and energy infrastructure in coastal areas throughout the United States face similar risks."[35]

In July 2015, a Department of Energy report offered a more detailed risk assessment, looking at different elements of energy infrastructure and the combined impacts of storm surge and sea level rise: "As recent hurricane events have demonstrated, this study found that an extensive amount of U.S. energy infrastructure is currently exposed to damage from hurricane storm surge. Furthermore, between 1992 and 2060, the number of energy facilities exposed to storm surge from a weak (Category 1) hurricane could increase by 15 to 67 percent under a high sea-level rise scenario from the recent *National Climate Assessment*."[36]

This Department of Energy report looked specifically at oil refineries, power plants, and natural gas terminals and reported the comparatively good news that, even in the event of thirty-two inches of sea level rise, less than 10 percent of all major energy facilities nationally for the three types of facilities are at risk[37] (see table 7-1).

The report points out the high risk that energy facilities already face today from a Category 3 hurricane. Adding thirty-two inches of sea level rise has a minor impact on the existing, underlying storm surge risk, increasing the number of oil refineries at risk from thirty-four to thirty-nine, the number of power plants from 549 to 597, and the number of natural gas terminals from twenty-four to twenty-five.

These Department of Energy findings, however, suggest a lower level of impact than reported in a 2012 study by the nonprofit organization Climate Central, which assumed a slightly higher degree of sea level rise and found more facilities at risk.

This analysis identifies 287 facilities less than 4 feet above the high tide line, spread throughout the 22 coastal states of the lower 48. More than half of these are in Louisiana, mainly natural gas facilities. Florida, California, New York, Texas, New Jersey each have 10 to 30 exposed sites, mainly electricity

Table 7-1. Energy facilities at risk of sea level rise and Category 3 hurricanes

Energy sector	Sea level rise in inches			
	0	10	23	32
Oil refineries				
Number	34	34	36	39
Percent of total	4.9	4.9	5.2	5.6
Power plants				
Number	549	562	589	597
Percent of total	4.4	4.5	4.7	4.8
Gas Terminals				
Number	24	24	25	25
Percent of total	8.6	8.6	9.0	9.0

Source: Data from James Bradbury et al., *Climate Change and Energy Infrastructure Exposure to Storm Surge and Sea-Level Rise* (Oak Ridge, TN: United States Department of Energy, 2015), 15–16.

in the first three states and/or oil and gas in the last two. All told, this brief catalogues 130 natural gas, 9 electric, and 56 oil and gas facilities built on land below the 4 foot line. Below the 5 foot line, the total jumps to 328 facilities with similar geographic and type distribution.[38]

Looking at the more local implications of sea level rise impacts on electric power generation, a 2015 study by R. Bierkand and coauthors found that the share of power produced by facilities at risk from a 100-year storm, assuming sea level rise by 2100, varied strongly by state: "For Delaware it is 80% of the mean generated power load. For New York this number is 63% and for Florida 43%."[39] The consequences of lost electric power are diverse and, in addition to personal inconveniences, include lost business production and degradation of critical services, such as health care.

In Florida, the Turkey Point Nuclear Generating Station just south of Miami, Florida, is the poster child of energy facilities at risk of storms and rising seas. Hurricane Andrew passed directly over the facility in 1992, and although no radiation was released to the environment, "the hurricane knocked out all offsite power for the plant for more than five days, caused the total loss of the plant's communication systems, blocked the access road to the site with debris, and damaged the fire protection and security systems as well as the warehouse facilities."[40]

Florida Power and Light proposed to add two new nuclear power units to the site in 2009, arguing that "the units would be elevated approximately 26 feet above sea level and are specifically designed to withstand natural disasters such as hurricanes, tornados, earthquakes, flooding, and tidal surges."[41] Opponents of the project, such as state Senator Jose Javier Rodriguez, are concerned, noting that, "they're only taking into account one foot of sea-level rise in the future."[42] The Nuclear Regulatory Commission had scheduled a hearing on the project in late 2017 but Hurricane Irma delayed the meeting. The commission announced approval of the licenses in April 2018. Applying this same thinking nationally, in early 2019 the commission voted 3 to 2 for final national regulations that removed some requirements related to flooding. Commissioner Jeff Baran argued this will allow plants to prepare "only for the old, outdated hazards...calculated decades ago when the science of seismology and hydrology was far less advanced than it is today. This decision is nonsensical."[43]

Petroleum production and refining is the element of the energy sector where assets are most likely to be privately held. The Department of Energy reviewed the exposure of oil refineries to coastal storms and rising seas and found that thirty-four refineries, constituting 4.9 percent of United States refining capacity, are currently exposed to storm surge inundation from a Category 3 hurricane. With sea level rise of twenty-three and thirty-two inches, the number of facilities at risk of a Category 3 storm increases to thirty-six and thirty-nine, respectively.[44]

Refineries along the Gulf of Mexico are most at risk. Ten of the sixteen oil refineries that are exposed to storm surge from Category 1 hurricanes are located in the Gulf Coast region, and most of the refineries that are not currently exposed to a Category 1 storm surge but will become exposed as a result of sea level rise are located in Louisiana and east Texas (Galveston Bay and Port Arthur).[45]

In a 2015 evaluation of storm and sea level rise risks to refineries, the Union of Concerned Scientists noted that coastal storm damage to refineries can have consequences for the American economy: "In 2005, for example, Hurricanes Katrina and Rita devastated the Gulf coast, shutting down 23% of the U.S. refining capacity, causing a significant drop in gasoline production and resulting in a 50% jump in the weekly average spot price of conventional gasoline."[46]

Some communities are looking at ways to do more to prepare for storm surges and rising seas that address the risks related to major industrial facilities, including refineries, located in a geographic cluster. One example is work sponsored by Rice University and other partners to develop plans to protect the Houston and Galveston area, home

Figure 7-1. Turkey Point Nuclear Generating Station in Homestead, Florida, the sixth largest power plant in the United States, is vulnerable to both storm surge and rising sea level. Photo by Acroterion, Creative Commons License 3.0. https://creativecommons.org/licenses/by-sa/3.0/deed.en.

to dozens of refineries, storage terminals, and other facilities. Philip Bedient, an engineering professor at Rice University leading the work noted this: "We're kind of sitting ducks until we start to take this more seriously and do something about it. . . . It's already bad. . . . If you're going to have sea-level rise, it's going to be worse."[47]

Hurricane Harvey seems to have been something of a wake-up call for major oil companies with facilities along the Texas Gulf coast. In August 2018 the state of Texas sought some $12 billion to strengthen coastal protection structures along the Texas coast, including most of the thirty refineries in the state. The wisdom of this investment might be questioned, but the proposal suggests that the legendary influence of oil companies on public policy in Texas is undiminished. The Associated Press reported that "the proposals approved for funding originally called for building more protections along larger swaths of the Texas coast, but they were scaled back and now deliberately focus on refineries."[48]

While sea level rise poses a threat to refineries on shore, more intense storms are a threat to off-shore production facilities. In August 2005, Hurricanes Katrina and Rita resulted in "the destruction of 115 platforms, damage to 52 others, damage of 535 pipeline segments, and near total shut-down of the Gulf's offshore oil and gas production."[49] Chris Oynes, then director of the United States Minerals Management Service concluded that, "the overall damage caused by Hurricanes Katrina and Rita has shown them to be the greatest natural disasters to oil and gas development in the history of the Gulf of Mexico."[50] In 2008, Hurricanes Gustav and Ike destroyed another sixty platforms.[51]

Local community activism and powerful national industry associations are distinctive elements of the coastal energy management landscape. Another key feature of this landscape is the emerging role of corporate shareholders in climate change preparedness. As discussed in greater detail in chapter 17, investor resolutions pressing for better preparedness for storms and rising seas are slowly gaining traction. And, global frameworks for consistent reporting and disclosure of corporate climate change preparedness are emerging (e.g., the new reporting mechanism developed by the Financial Stability Board).

Stepping back to look at coastal energy facilities more generally, the Department of Energy deserves some credit for making modest headway in defining risk from climate change generally and coastal storms and rising seas more specifically. The Department of Energy 2014 *Climate Change Adaptation Plan* provides a general overview of the department's response to a changing climate, although it does not focus on sea level rise. [52] It mentions a 2013 department report looking at the vulnerabilities of the energy sector to climate change and extreme weather. This report describes some storm surge and sea level rise impacts and lists "opportunities" for future work: "Better characterization at the regional and local levels of climate change trends relevant to the energy sector, including water availability, wind resources, solar insulation and cloud cover, and likelihood and magnitude of droughts, floods, storms, sea level rise and storm surge."[53] A pretty full plate.

Will the energy sector be able to take the next step and translate coastal energy infrastructure risk assessments into action plans? There is reason to hope so. The stakes for local communities, companies, and the national economy are high. A combined effort by the Department of Energy, engaged local communities, shareholders, industry associations, and industry leaders has the potential to advance the sector toward preparedness for coastal storms and rising seas. Fortunately, energy companies sit on the other end of the spectrum from the tourism industry in terms of ownership structure. While the tourism industry is highly decentralized, ownership or management in the energy sector is highly centralized. Should the handful of people who manage corporations owning these assets have an epiphany on the need to prepare for more severe coastal storms and rising seas, this highly centralized structure would support prompt and effective implementation of response actions.

The coastal tourism and fishing sectors will have a harder time than the energy sector preparing for more severe storms and rising seas. Their fortunes are tied more directly to the uncertain future health of coastal ecosystems. And, management in both sectors is generally less centralized than in the energy sector and less equipped to engage in long-range planning. Still, tourism and fishing are critical to coastal economies and to the identity of coastal communities. Energy, tourism, and fishing all need to be well represented in efforts to prepare for more severe storms and rising seas, but creative efforts will be needed to effectively engage tourism and fishing sectors.

PART II: EPILOGUE

Storms and Rising Seas Disrupt the Coast

More severe storms and rising seas will bring economic, environmental, and social disruption to the coast on an unprecedented scale. Millions of people and hundreds of coastal communities face risks of more extensive storm flooding followed by permanent inundation with potential losses of coastal property running into trillions of dollars. Critical public infrastructure facilities are at risk along with important ecosystems and private sector assets supporting tourism, fishing, and coastal energy. Impacts in each of these sectors will rebound on other sectors, multiplying the disruption. There is evidence that better preparing for these impacts could significantly reduce costs and losses. Yet, in each of these areas, most governments are either not yet engaged or are still assessing risks rather than implementing response actions. Nothing about this story is likely to attract the interest of the evening news. Still, it is an important story, and the scale of impacts, and the importance of assets at risk, is the second basic argument for a national program to prepare for more severe storms and rising seas.

Part III

A Nation Unprepared for Coastal Storms and Rising Seas

Floods are "acts of God,"
but flood losses are largely acts of man.

Gilbert F. White

8

The Politics of Coastal Storms and Rising Seas

Coastal communities, critical infrastructure, ecosystems, and significant private assets are all at risk from more severe storms and rising seas. Understanding what can be done to respond to these risks requires knowing what tools and resources are already at hand and how they might be updated to better meet the challenges of a transition to a new coast.

The existing tools and resources are substantial. The United States has a long-standing commitment to provide federal flood insurance to property owners in flood-risk areas, including coastal communities. The federal government commits billions of dollars to help states, local governments, and private citizens recover from disasters, including coastal storms. Federal agencies work with states to implement several programs focused on protecting coastal communities and managing coastal natural resources. And, until very recently, the federal government led a major effort to help states and communities adapt to a changing climate. Each of these major national program efforts is described in this part to support the conclusion that, despite these commendable efforts, the country is not prepared for more severe storms and rising seas.

Before reviewing the national-level tools and resources available to respond to these challenges, it is important to sketch the political background in which these programs operate and to answer some key questions. What do we know about public opinion on the topics of coastal storms and rising seas? What political issues have come up in the context of state actions to respond to coastal storms and rising seas? What do the new directions of the Trump administration mean for the existing national coastal programs and for flood insurance, disaster assistance, and climate adaptation planning? How has the Congress addressed coastal storms and rising seas?

On the whole, the answers to these questions paint a sorry picture of inattention and misplaced priorities. The Trump administration is abandoning coastal protection and climate change programs, including climate change adaptation planning, associated with the Obama administration. As of this writing, Congress has given little attention to coastal issues beyond approving massive supplemental appropriations for recent major hurricanes and considering modest changes to the flood insurance program without adapting it to the challenges of storms and rising seas. Several states that face some of the most significant impacts of coastal storms and rising seas turned their backs on the problem. There is, however, a ray of hope when turning to the limited public opinion research on the topic.

Public Opinion on Climate Change, Storms, and Sea Level Rise

Opinion polling indicates that the public agrees with the idea that the country needs to address climate change and prepare for more severe coastal storms and sea level rise. This public support validates the idea of updating existing programs to fit the evolving understanding of the risks to the coast and adding to these programs as needed.

Numerous national polls point to public belief that the climate is changing. For example, a 2018 poll on climate change found that 73 percent of Americans "think global warming is happening,"[1] a slight increase from the 70 percent of that view in a 2016 edition of the poll.[2] The percentage of people that understand that global warming is mostly human caused rose from 55 percent in the 2016 poll to 62 percent in 2018.

There is, as of yet, little national polling on the specific subjects of coastal storms and sea level rise, but in 2013 researchers at Stanford University conducted a poll of public perspectives on these topics. The study found that the public

- "believes that global warming will cause sea levels to rise (73 percent);
- believes that this will be a serious problem for the U.S. (76 percent); and
- overwhelmingly supports preparing now for the impacts of global warming (82 percent), rather than waiting (16 percent)."[3]

When asked "How much do you think the federal government should do to reduce the effects of rising sea level in the future?" 85 percent of respondents favored action of some sort:

- A great deal, 18 percent;
- Quite a bit, 27 percent;
- Some, 30 percent;
- A little, 10 percent;
- Nothing, 14 percent.[4]

Public opinion on sea level rise can be complicated by opinions on the related topic of preparing for coastal storms and hurricanes. Some people may think generally about coastal flooding and make little distinction between temporary storm surge flooding and permanent sea level inundation. In the Stanford poll, however, 72 percent of respondents indicated that global warming will cause storms to be more damaging, about the same percentage that are concerned about sea level rise.

Although the polling numbers tell a positive story about public understanding of the risks posed by coastal storms and rising seas, it is important to remember that for many people there remains serious doubt about global warming and rising seas and a general mistrust of the media providing this information. Tom Horton, an author with a background in writing about Chesapeake Bay, talked with people living with rising seas around the Bay and reported in 2018 that "we knew going in that phrases like global warming and sea level rise would never pass the lips of most of our interviewees. . . . Eventually she said what was on her mind: It would be too easy for us to 'put me between two PhDs and make me look like a fool.' Besides, 'I just hate how the media puts fear in our hearts, talking about the land sinking and humans changing the climate.'"[5]

States Struggle with the Politics of Sea Level Rise

Although sea level rise has not emerged as a political issue at the national level, it has become controversial in some states and localities. Unfortunately, even a small dose of political controversy, when added to the mix of difficult new issues related to rising seas, can freeze up the workings of government.

State government experiences addressing sea level rise have been both good and bad. Looking at these experiences collectively, there is creativity and success to report in some coastal states (see ch. 14). Unfortunately, some of the states where sea level rise is expected to occur soonest and be most damaging have struggled to move past political controversy. The poster children of contentious sea level rise planning are North Carolina and Florida.

The saga of sea level rise debates in North Carolina garnered wide public attention, including some ridicule. The state initially took the commendable step of forming a scientific panel to look at sea level rise impacts. The panel based its report on the best science then available and noted that sea level rise in North Carolina was expected to be 39 inches by 2100.

> A one meter (39 inch rise) is considered likely in that it only requires that the linear relationship between temperature and sea level that was noted in the 20th century remains valid for the 21st century. . . . This level of rise is consistently encapsulated within all of the projections reviewed, and is not located at the upper or lower extremes of the projections. Given the range of possible rise scenarios and their associated levels of plausibility, the Science Panel recommends that a rise of 1 meter (39 inches) be adopted as the amount of anticipated rise by 2100, for policy development and planning purposes.[6]

This modest recommendation, however, triggered a backlash from coastal counties and development interests concerned that it would undermine property values. These interests created a new organization—termed "NC-20"—just to take on this issue, and pressed the legislature to reject the report and pass legislation changing course. The bill enacted by the North Carolina Legislature (H819), called for eliminating any use of the 39-inch estimate until a new study could be developed to reexamine the issue.

Supporters of the law questioned the use of models to make projections and argued for relying only on historical trends. In addition, the bill provides that any estimate of future sea level in the revised report could be for no longer than thirty years. In opposing the bill, state Representative Deborah Ross said that "by putting our heads in the sand literally, we are not helping property owners. We are hurting them. We are not giving them information they might need to protect their property. Ignorance is not bliss. It's dangerous."[7]

The national press reported on H819 as evidence of denial of climate change, and late night television host Stephen Colbert mocked the legislature for passing a law defying science: "I think this is a brilliant solution . . . If your science gives you a result that you don't like, pass a law saying the result is illegal. Problem solved."[8]

Warring parties established a truce of sorts when the revised report required by the law was issued in 2015. Among other things, the report found that, for the thirty-year projection mandated by the law, sea level rise was estimated to rise just eight inches. North Carolina Public Radio reported that sea level rise critics were expressing satisfaction: "'We believe that the report before you today is a much better and thorough report that encompasses not only a scientific approach but just plain common sense that is applicable in today's development world,' Heather Jarman, a lobbyist for real estate and development in Wilmington, told the Coastal Resources Commission last week in Dare County."[9]

The final result, however, did not strike all observers as common sense. Orrin H. Pilkey, a noted expert on sea level rise and professor emeritus of geology at Duke University's Nicholas School of the Environment, writing in the *Charlotte Observer* in 2017, commented that "unfortunately, the coastal management program of North Carolina is going in the wrong direction, almost as though an important sea level rise isn't just around the corner. Our management policies are harnessing our future generations to a disaster. The CRC [Coastal Resources Commission] must first and foremost take a longer view of the rising sea and trash the ill-considered 30-year, 8-inch projected rise that currently governs their actions."[10]

A poll taken in North Carolina shortly after Hurricane Florence in 2018 indicates some shift in public attitudes with 52 percent of people thinking climate change "very likely" to have negative impacts on coastal communities, compared to 45 percent just 16 months earlier. Surprisingly, 62 percent thought climate change should be considered in local planning and 76 percent thought restricting real estate development in flood prone areas a good idea.[11] In late 2018, Governor Roy Cooper recognized this shifting landscape, issuing an executive order that, among other things, called for a statewide climate resiliency plan without mentioning sea level rise.

In Florida, a state with much at stake in responding to sea level rise, the Florida Center for Investigative Reporting, which is supported by major newspapers across the state, reported in 2015 that state employees of the Department of Environmental Protection were directed "not to use the term 'climate change' or 'global warming' in any official communications, emails, or reports."[12] The center quoted University of Miami professor Harold Wanless saying that "the state government needs to acknowledge climate change as settled science and as a threat to people and property in Florida. . . . You have to start real planning, and I've seen absolutely none of that from the current governor. . . . It's beyond ludicrous to deny using the term climate change. It's criminal at this point."[13]

As recently as 2017, as Florida faced several major hurricanes, criticism of the absence of state leadership on climate change has continued. Kathy Baughman McLeod, a conservation expert who served on the Florida Energy and Climate Commission, which was effectively dismantled after Governor Rick Scott took office in 2011, stated the following: "The science has been brought on a silver platter to Governor Scott, and he's chosen not to do anything. . . . If there is climate action, it's all coming from local and regional collaboration. There is no state leadership on climate

Figure 8-1. The *Charlotte Observer* published this cartoon on the topic of sea level rise and coastal property values in 2018. Cartoon by Kevin Siers, reprinted by permission.

change in Florida, period."[14] Florida seemed to have voted for more of the same in 2018, electing as governor Ron DeSantis, who stated during the campaign, "I am not a global warming person."[15] In a surprising and potentially positive development, however, Governor DeSantis issued a directive in January 2019 creating a new Office of Resilience and Coastal Protection charged with helping prepare Florida's coastal communities and habitats for sea level rise.

Although the political fires over climate change and sea level rise have not died down at the state level in Florida, there is notable progress at the local and regional level across the state, including in South Florida. In addition, some states, including California, New York, Maryland, and New Hampshire have developed and are implementing strong sea level rise adaptation programs without major controversy. As University of New Hampshire Professor Cameron Wake commented, "Unfortunately, there's a long-standing cultural divide around climate change. On a political level, this has made it difficult for coastal states to act on—or even acknowledge—the growing risk of coastal flooding from climate change. New Hampshire, however, is an exception. The state has passed legislation and made rule changes designed to better prepare the state for the damage from storm surge and rising seas."[16]

At the local level, communities ranging from Miami to Norfolk to Los Angeles have comprehensive sea level rise assessment and planning processes underway (see ch. 14 for more information). As in the case of state-level planning, these programs apply a variety of planning models and policies to reduce the impacts of storms and rising sea level. And, like state-level planning, there are impressive cases of creative approaches. But, the majority of coastal communities have no plans in place, and many of the communities at the greatest risk of sea level rise do not have plans. Most communities, however, have adopted Hazard Mitigation Plans that address response to conventional storms (see ch. 10).

Interestingly, most local sea level rise planning does not seem to have generated controversy and drama comparable to that at state levels in Florida or North Carolina. This civility may be because many local plans are still under development, and the pain of regulations or higher taxes needed to implement the plans has not yet hit residents. Another possible explanation is that the practical and tangible challenges of rising seas presented at the local level tend to focus discussion on finding achievable solutions.

Trump Administration Handicap

The "long-standing cultural divide over climate change" described by Professor Wake makes governments at all levels hesitant to act on climate change, and by extension, sea level rise. Although most of the national fighting over climate change is focused on proposed actions to reduce emissions of greenhouse gases and science issues other than sea level rise, these clouds of controversy can make getting the political approval and financial resources needed to just get started on planning for more severe coastal storms and sea level rise an uphill battle.

The Trump administration marched into this already unstable situation casting doubt on the science, announcing plans to back out of the Paris Climate Agreement, revoking initiatives to adapt to a changing climate, and cutting funding for coastal programs. This opposition to anything to do with climate change is especially jarring as it followed years of steady progress of efforts to reduce greenhouse gases and prepare for climate change impacts during President's Obama's two terms.

President Trump and senior administration officials, such as former Environmental Protection Agency administrator Scott Pruitt, have expressed strong doubts about climate change science. For example, in March 2017, Pruitt said, "I think that measuring with precision human activity on the climate is something very challenging to do, and there's tremendous disagreement about the degree of impact. . . . So, no, I would not agree that it's a primary contributor to the global warming that we see."[17] President Trump offered a similar, if pithier, conclusion when asked about federal agency climate findings in the November 2018 *National Climate Assessment:* "I don't believe it."[18] The editorial board of the *Washington Post* looked at actions rather than words and concluded in June 2018 the following: "Yet even though evidence continues to mount about the direct and dangerous disruptions humans are imposing on the climate system, the federal government's highest leadership refuses to acknowledge the problem, let alone do anything about it."[19]

The debate over whether climate change even exists extends into a passionate fight over federal government programs and policies to limit warming by reducing release of greenhouse gases. This debate has focused on the president's decisions to withdraw from the Paris Climate Agreement, to withdraw the regulation to reduce release of carbon dioxide from power generation facilities, to roll back regulations requiring greater fuel efficiency in passenger cars, and to drop efforts to control emissions of methane from oil and gas production sites.

Despite numerous public statements on climate change, neither President Trump nor other senior administration officials have spoken formally about risks of more severe coastal storms or rising seas. Still, he has informally expressed skepticism about rising seas. The mayor of Tangier Island, Virginia, reported his 2017 conservation with the president: "He said not to worry about sea-level rise. . . . He said, 'Your island has been there for hundreds of years, and I believe your island will be there for hundreds more.'"[20]

Unfortunately, the president has not been content to simply oppose efforts to reduce greenhouse gases but has struck down initiatives from the Obama administration to adapt to the impacts of a changing climate. On the same day as the announcement of the withdrawal of the Clean Power Plan, President Trump also issued Executive Order 13783, addressing a range of other Obama administration actions related to climate change, including revoking Obama's Executive Order 13653: *Preparing the United States for the Impacts of Climate Change.* The Obama order laid a foundation for a federal government effort to plan for the impacts of a changing climate, stating that "the Federal Government must build on recent progress and pursue new strategies to improve the Nation's preparedness and resilience."[21]

Further evidence of Trump administration backtracking on climate change adaptation comes in the form of a relentless deconstruction of policies developed during the Obama administration. It has revoked guidance on addressing climate change in environmental assessments in April of 2017 (see ch. 12), revoked the Federal Flood

Risk Management Standard in August of 2017 (see ch. 12), and revoked the National Ocean Policy executive order in June of 2018 (see ch. 11).

On another front, federal programs related to climate change, and more specifically, coastal protection, were targeted for dramatic funding reductions in the administration budget proposals. For example, for the fiscal year 2018, the administration budget proposed to eliminate funding for Coastal Zone Management Program grants to states, the national Sea Grant College Program grants to universities, and the National Estuaries Program. Fortunately, Congress has not adopted the proposed cuts.

Looking back over the wreckage of Obama administration climate initiatives derailed in the past several years, it is hard to be optimistic about any federal efforts to adapt to climate change generally or sea level rise more specifically. Still, there is a bit of good news. For example, the federal agency climate adaptation plans called for in the Obama executive order on climate change adaptation still exist. Although these plans do not focus specifically on sea level rise, they are a starting point for agency adaptation planning, and some agencies may continue in the direction these plans set.

In addition, some federal agency efforts to integrate sea level rise science in agency operations are proceeding. The Army Corps of Engineers issued a sea level rise policy in 2011, and a more detailed technical letter, "Procedures to Evaluate Sea Level Change Impacts, Responses and Adaptation"[22] in 2014; these documents continue to guide planning.

Other federal agencies are carrying on with useful efforts. The National Oceanic and Atmospheric Administration (NOAA) continues to conduct research on coastal storms and sea level rise, to provide its "Sea Level Rise Viewer" to the public, and to offer small grants for coastal resilience projects. The National Aeronautics and Space Administration continues to provide critical data related to changes in the climate. The Environmental Protection Agency (EPA) continues to provide technical assistance to water utilities and community organizations around critical coastal estuaries to help them prepare for climate risks, including coastal storms and rising seas.

Finally, federal agencies working through the Global Change Research Program received approval to publish a new *Climate Science Special Report* in 2017 and, in November 2018, they released an ambitious *National Climate Assessment*, the fourth such comprehensive report, as required by the Global Change Research Act.

United States Congress Avoids Storms and Rising Seas

Meanwhile, as the executive branch retreats from climate change, the United States Congress has invested considerable energy in disaster relief legislation following Hurricanes Harvey, Irma, and Maria, and in reauthorizing the National Flood Insurance

Program. Unfortunately, neither of these efforts is focused on preparing coastal states and communities for the overlapping challenges of more severe storms and rising seas or laying a foundation for a transition to a new coast.

Congress was able to enact three separate appropriations bills of $15, $20, and $84 billion in September and October of 2017, and February of 2018, providing appropriations for hurricane recovery and an additional appropriation of $19 billion for hurricane response and other disasters in June 2019. These staggering sums are on top of $50 billion appropriated for recovery from Hurricane Sandy in 2012 and a total of $121 billion appropriated following Hurricanes Katrina, Rita, and Wilma in 2005 and Hurricanes Gustav and Ike in 2008. It is remarkable that after such effulgence, Congress has not summoned the curiosity and initiative to take a hard look at the causes of such widespread, deadly, and costly damage and set to work crafting a long-term solution that would, of necessity, also address rising seas.

Congress also worked diligently on the small piece of the coastal inundation puzzle that is the National Flood Insurance Program (see Chapter 9). Both the House and Senate have considered flood insurance reauthorization bills with modest improvements to the program but, in May 2019 Congress passed the eleventh short term extension to continue the existing program until September 30. Giving credit where credit is due, during the first two years of the Trump administration, Congress enacted some modest, positive measures in this area. For example, the 2018 water resources legislation provided new authority for drinking water systems to develop plans to prepare for "natural hazards" (see ch. 5). The Disaster Recovery Reform Act provided useful increases in predisaster plan funding (see ch. 10). And, as mentioned in chapter 5, the 2019 National Defense Authorization Act included some helpful authority to prepare military bases for more severe storms and rising seas. Useful as these efforts are, they fall well short of a coordinated look at the problem with an eye toward enacting legislation to set a smarter course.

In the House of Representatives, committees with jurisdiction over coastal issues have held limited hearings on disasters and coastal storms and almost none on rising seas. A notable exception to this rule was testimony on sea level rise provided by oceanographer and sea level rise expert John Englander, in 2015, to the Committee on Natural Resources, in the context of a review of energy issues in the Coastal Zone Management Program. Although somewhat off the committee's script, Englander provided detailed testimony on the risks of rising seas, noting that "the sea is rising and the shoreline is shifting. We have time to adapt, but no time to waste."[23]

The House has given more attention to disaster policy. In addition to enacting the Disaster Recovery Reform Act, several committees considered H.R. 4177, the Prepare

Act, which proposed establishing an interagency council of federal agencies to coordinate priorities related to extreme weather and called on federal agencies to develop plans to managing extreme weather. But the bill was not enacted.

The Climate Solutions Caucus, a group of about seventy-two members evenly divided between House Democrats and Republicans, traces its origins to the Citizen's Climate Lobby that focuses on a carbon fee and dividend approach to managing United States greenhouse gas emissions. In the 115th Congress, the caucus was cochaired by congressmen from South Florida, Democrat Ted Deutch and Republican Carlos Curbelo, both of whom have expressed concern for rising sea level. In May 2017 the caucus, although not a formal committee of Congress, held a hearing on coastal issues that included testimony on ocean acidification, coastal flooding, and sea level rise, but it has yet to fully explore the topic or articulate a response to rising seas.

In 2015, two congressmen from New Jersey, Democrat Frank Pallone Jr. and Republican Frank LoBiondo, launched the Congressional Coastal Communities Caucus they describe as "a bipartisan group designed to highlight the unique concerns of those that live, work, and do business along America's coasts."[24] Rep. Pallone stated at the time that "from mitigating the effects of future storms, to ensuring a vibrant tourism sector, and protecting our fragile marine ecosystems, New Jersey's coastal residents have a unique set of priorities and concerns."[25] Topics related to coastal storms and sea level rise seem applicable to the mission of the caucus, but to date they have not been addressed in hearings or legislative proposals.

In the Senate, the last national-level attention to sea level rise came at a hearing of the Committee on Energy and Natural Resources in 2012 when the committee heard testimony from the federal government and other witnesses on sea level rise science and impacts. Senator Bill Nelson (D-Florida) held two field hearings of the Commerce Committee, focusing on the impacts of rising seas in southern Florida, one in Miami Beach in 2014, and one in West Palm Beach in 2017, where one witness stated that "we know a lot more and we're a lot more scared."[26] These hearings have not yet resulted in further hearings or proposed legislation.

Senators created a Senate Oceans Caucus in 2011 with a mission to "raise awareness of coastal and marine issues and find common ground on legislation that affects our oceans, Great Lakes, coasts, and the communities and businesses that rely on these resources."[27] The caucus has focused on domestic and international fishing issues and, in early 2017, announced new priorities including a focus on the issue of marine debris.[28] A similar Oceans Caucus exists in the House of Representatives, also focusing on marine debris.

In early 2017, the Joint Ocean Commission Initiative, a bipartisan group chaired by Norman Mineta and Christine Todd Whitman, proposed an *Ocean Action Agenda*,[29] outlining ocean priorities for the Trump administration and Congress. The agenda is very broad, with recommendations ranging from port infrastructure to leadership in the Arctic. Sea level rise and coastal storms are not a focus of the report, and none of the almost thirty specific recommendations speaks directly to these challenges. Several general recommendations, however, have a coastal inundation element, including investing in coastal infrastructure designed with "resilience" in mind, reforming the National Flood Insurance Program, and funding for protection of natural features from various hazards, including sea level rise.

Although the agendas of the coastal-oriented committees or caucuses in Congress do not demonstrate a sustained interest in either increases in coastal storm severity or rising sea level, several individual members have spoken to sea level rise issues. For example, Senator Sheldon Whitehouse (D-Rhode Island) gave a speech on the Senate floor in 2016, one of many he has made on climate change issues, reviewing in detail the risks of rising seas, calling for expanded support for coastal preparedness planning, and concluding that "the rising tide calls for increased investment in coastal resiliency around the country. . . . We have a moral obligation to pluck our heads from the sand and get to work. The oceans warn; it is time we woke up and listened."[30]

Senator Whitehouse also championed one of the few recent legislative accomplishments related to ocean policy—enactment of a bill to create a National Resilience Coastal Fund, allowing the National Fish and Wildlife Foundation and the NOAA administrator to make grants to states to "support programs and activities intended to better understand and utilize ocean and coastal resources and coastal infrastructure."[31] (See ch. 11 for more information.)

In another notable example of individual initiative, Representative Salud Carbajal (D-California) introduced a bill in 2017 calling for a coastal climate change adaptation preparedness and response program (HR 3533). The bill would "provide assistance to coastal states to voluntarily develop coastal climate change adaptation plans"[32] and provide new grant support for both planning and implementation of projects through the Coastal Zone Management Program managed by NOAA. The bill, however, did not advance in the House. Several other positive bills, addressing living shorelines and corporate climate risk disclosure, also did not advance.

On the whole, Congress has to date shown a willingness to respond to calls for disaster assistance and to make only minor amendments to the National Flood Insurance Program, but little interest in investigating new ideas with the potential to better prepare for continued expensive disasters from coastal storms and the enormous costs of coastal inundation from rising seas.

The 2018 midterm election results were a mixed bag when it comes to responding to more damaging storms and rising seas. Control of the House of Representatives shifted to Democrats who formed a new "Select Committee on the Climate Crisis" that, along with several other House committees, offers a venue for hearings on climate change generally, as well as on adaptation to impacts like more severe storms and rising seas. The much discussed "Green New Deal," however, makes only a passing reference to more severe storms and rising seas. And, over a dozen congressional Republicans that had expressed interest in addressing climate issues were defeated, including Representative Curbelo of Florida.

Infrastructure financing is a top priority for Democrats and funding for projects to adapt to more severe storms and rising seas is sometimes mentioned as part of an infrastructure initiative. For example, presidential candidate Senator Amy Klobuchar's infrastructure proposal calls for helping "states and cities plan for the impacts of climate change by building stronger, more resilient transportation networks and public infrastructure to withstand rising sea levels, a changing climate, and extreme weather"[33] among an array of other priorities. Presidential Candidate Beto O'Rouke released a climate proposal in April 2019 calling for, among other things, a dramatic increase in pre-disaster mitigation funding, but not focusing on sea level rise. Former vice president and now presidential candidate Joe Biden released a climate proposal in June 2019 that includes building resilience to storms, floods, and sea level rise and commits to defining a climate adaptation agenda.

But Democrats in Congress have not unveiled a climate legislative agenda, and the chance of enacting climate or infrastructure legislation is limited by Republican control of the Senate and a likely presidential veto. Although 2019 offers new opportunities, it seems clear that it will take a major effort to shift Congress' benevolent attention to the topic of preparing for more severe storms and transitioning to the new coast that will come with rising seas.

The political battlefield on which efforts to respond to coastal storms and rising seas will need to succeed is surprisingly promising. The political controversy over storms and sea level rise in key states like Florida and North Carolina has not disappeared but is abating while some states are making solid progress with both studies of the problem and actual response actions. The outlook for attention to the topic in Congress has dramatically improved with the 2018 elections, although legislation may be years away. Most promising of all is the solid majority of the public supporting preparation for climate change, including the coastal storms and rising seas. The darkest cloud on this scene is the aggressive animosity of the Trump administration to any effort to assess or adapt to the impacts of a changing climate. Today, the challenge is to mobilize strong public support behind additional affirmative government responses to more severe storms and rising seas.

9

National Flood Insurance Program: Coastal Misdirection

The United States has a long history of damaging, sometimes catastrophic, flooding. Today, the principal federal government program to manage flooding and help homeowners recover from a flood is the National Flood Insurance Program (NFIP). The most serious flood events are recognized as disasters eligible for federal assistance (see ch. 10).

Unfortunately, the NFIP is struggling to keep up with challenges posed by conventional flooding and is a mismatch with the problem of more severe coastal storms and rising seas. It provides an insurance benefit for, and a tacit endorsement of, decisions to live in places at risk of flooding. This policy could be sustained, financially and morally, when flood losses were predictable, modest, and mostly rare. But, with the advent of huge losses from severe coastal storms and gradual but permanent inundation of some coastal properties due to rising seas, reasonable premiums can't offset losses. Locked into an unsustainable insurance program with large existing losses, and larger losses to come, the country as a whole is subsidizing coastal policyholders, effectively diverting scarce federal resources from more constructive preparation for storms and rising seas.

Not only has the NFIP become financially unsustainable and counterproductive, it is morally unsustainable. The current NFIP makes it national policy to send people the wrong signal about living in risky coastal areas. It provides a financial incentive for people to stay in places that, with more severe storms and rising seas, are no longer just risky but are becoming increasingly dangerous and on a path to being uninhabitable. The government should instead be pointing people away from these areas and toward safer ground. Correcting this misdirection will be difficult but is necessary.

This chapter describes the origins of the flood insurance program, how it accounts for storm surge and rising seas today, and the long and slippery slope the program followed toward insolvency. Some of the changes being debated to improve the operation of the program are also described. Proposals for more fundamental changes needed to the flood insurance and disaster assistance programs are described in chapter 18.

Origins of the National Flood Insurance Program

For much of the nation's history, the solution to flooding was thought to be building flood control structures such as dams and levees. Flooding along the lower Mississippi River was a regular problem in the nineteenth century. Despite federal investments in flood control, the "Great Flood of 1927" breached the levee system along the Mississippi River causing catastrophic flooding. As James Wright reported in his history of flood control policy, "At the flood's highest point, the river spread 50 to 100 miles wide in a 'chocolate sea.' The official death toll was 246 but may have reached 500."[1]

At this time, flooding was understood to occur primarily in the context of river systems, rather than along the coasts. Although the 1927 flood clearly deserves its name, it occurred nearly three decades after the Galveston hurricane of 1900 that resulted in an estimated 6,000 to 12,000 deaths and is still the single most deadly coastal storm in United States history. The damage and death in the Galveston storm resulted largely from a storm surge estimated at fifteen feet. Although the Galveston hurricane was not recognized as a part of a larger national flooding problem, the response was essentially the same as that applied to river flooding—structural protection. The city of Galveston built a seventeen-foot seawall[2] and pumped sand to raise the elevation of much of the community.

Federal government investments in the decades following the Great Flood were still guided by the view "that we could build our way out of almost any problem. . . . "[3] But, the national understanding of flood control solutions started to shift in the middle of the twentieth century. In 1942, Gilbert White wrote his seminal book titled *Human Adjustment to Floods* in which he argued for "adjusting human occupancy to the floodplain environment so as to utilize most effectively the natural resources of the floodplain, and at the same time, applying feasible and practicable measures for minimizing the detrimental impacts of floods."[4]

Despite a growing awareness of the need to add management of floodplains to the toolbox of flood protection, changes came slowly: "In the three decades since the Flood Control Act of 1936, the nation had relied almost entirely on engineering solutions to solve its flood problems, yet overall flood losses were not reduced."[5]

James Wright concluded that the NFIP, established in 1968, represented a "quantum shift in policy."[6] "Congress established the NFIP as a 'quid-pro-quo' program. Through it, relief from the impacts of flood damages in the form of federally backed flood insurance became available to participating communities contingent on flood loss reduction measures embodied in state and local floodplain management regulations."[7]

Operation of the National Flood Insurance Program

The NFIP, managed by the Federal Emergency Management Agency (FEMA), has been repeatedly tweaked and prodded over the years but the core program has remained essentially unchanged. Communities at risk of floods from inland and coastal flooding are encouraged to join the program. Once a community adopts a local ordinance meeting minimum requirements related to flood mitigation, owners of residential and commercial property can purchase flood insurance backed by the federal government, sometimes at subsidized rates.

An extensive program of mapping areas at risk of flooding forms the foundation of the program and focuses on areas that will be inundated by the flood event having a 1 percent chance of being equaled or exceeded in any given year, termed the 100-year floodplain. Today, property owners with a federally insured mortgage are required to have flood insurance if located in a 100-year floodplain mapped by FEMA.

This floodplain mapping work is a big job and predictably controversial. FEMA has struggled to adjust maps to reflect the most recent data. A 2017 report by the Department of Homeland Security found that only 42 percent of maps were up-to-date.[8] In, addition, FEMA is required to consult with local stakeholders in developing maps, and some local governments have lobbied to shrink the size of floodplains to help homeowners avoid flood insurance requirements or to maintain the development potential of land.[9] Perhaps most important, the maps have been criticized on the grounds that they understate coastal flood risk because they do not recognize potential future flooding as a result of more severe storms or rising seas.

Communities electing to participate in the NFIP have to meet some basic minimum requirements, including those that apply specifically to coastal areas. These requirements are spelled out in detailed regulations but boil down to some core elements.[10] Communities must adopt the most recent FEMA Flood Insurance Rate Map

(FIRM) for the 100-year floodplain. Communities must also adopt a permit program that requires proposed development be protected to at least the base flood elevation (i.e., the elevation to which floodwater is anticipated to rise during a flood having a 1 percent chance of being equaled or exceeded in a given year). The local permit must also avoid an increased flood hazard risk for other properties and require that new, substantially improved, or substantially damaged properties be protected to the base flood elevation.

Storm Surge and Sea Level Rise in the Current NFIP

As the NFIP developed, three key elements related to coastal storms and sea level rise emerged. FEMA identified areas at risk of storm surge, termed "V zones," and developed policies to protect these areas. Second, communities were encouraged to adopt flood protection measures beyond the minimum requirements, including recognition of sea level rise impacts, in exchange for insurance premium reductions under the Community Rating System. And, Congress enacted the Coastal Barrier Resources Act, prohibiting federal flood insurance for new construction on barrier islands and related lands as a means of discouraging development of these sensitive areas. Understanding these three program elements is necessary to considering how the NFIP should be changed.

High Risk V Zones: Designated on local flood maps, V zones are "high hazard areas along coastlines that are subject to flooding from storm surge and wave impacts during coastal storms and hurricanes."[11] These V zones, however, do not include land that will be permanently inundated as sea level rises and, because they refer only to historical storm data, do not account for an increase in coastal storm intensity as a result of climate change.

V-zone standards apply to new construction and substantial improvements, but not to existing buildings. And, while there is no prohibition against locating a new or substantially improved building in the V zone, it must at least be landward of the mean high tide line and not over water. In terms of construction standards in V zones, buildings must be elevated on pilings so that the lowest horizontal structural member is above the base flood elevation. In addition, areas underneath an elevated building need to have open areas allowing flood water to "flow under the building without placing additional loads on the foundation."[12]

In addition to tougher standards for location and design of new or substantially improved construction, policyholders in V zones pay higher premiums. Calculation of NFIP premiums is a fine art requiring intimate knowledge of extensive tables and other rules. A rough estimate, however, is that policies for buildings in V zones that predate the development of the flood map for the community (i.e., "pre-FIRM" properties) are about 50 percent higher than for policies for buildings in the mapped 100-year floodplain but that are not in the V zone. For newer buildings built after the flood maps were in place for the community (i.e., "post-FIRM" properties) premiums are about double. FEMA reports that the average cost of flood insurance in 2017 is about $1,000.[13]

Even at these higher rates, the price of flood insurance in V zones is not likely to often be a constraint on property owners wanting to build or substantially improve structures in these zones. And, NFIP premiums are still a bargain compared to private insurance. A 2011 report by the Property Casualty Insurers Association of America concluded "that the federal government is providing overall flood insurance at one-half the true-risk cost; specifically, in higher-risk areas, it is providing flood insurance at one-third the true-risk cost."[14]

The total amount of insurance available to residential and nonresidential policyholders does not change for V zones (i.e., in general, limits of $250,000 for residential buildings and $100,000 for residential contents; $500,000 for nonresidential buildings and $500,000 for nonresidential contents).

Community Rating System: The Community Rating System (CRS), implemented in 1990, offers individual communities already participating in the NFIP program a chance to lower flood insurance premiums paid by local policyholders by voluntarily adopting a range of additional flood mitigation measures. Today, some 1,444 communities participate in the CRS. Although these communities represent only a small percentage of the over 22,000 communities that have joined the NFIP, more than 69 percent of the 5.1 million residential and commercial NFIP policies are written in CRS communities.[15]

Communities in the program are divided into nine "classes" based on their adoption of flood control measures from a menu of options. Each measure earns a specific number of points, and 500 points moves a community from one class to the next, toward Class 1. Each class step means that policyholders in that community get a flood insurance premium discount of five percent up to a total 45 percent discount. Points can be earned by actions such as improving public information, enforcing higher regulatory standards, and developing a comprehensive floodplain management plan.

In 2017, FEMA published a new manual for CRS program coordinators describing in detail how points would be awarded to communities in each of these areas. More important, the new manual speaks directly to steps a community can take that recognize or adapt to rising sea level. Some examples of sea level rise credits include the following: advising prospective buyers of a property of the potential for flooding due to climate changes and/or sea level rise; basing a community regulatory map on future-conditions hydrology, including sea level; and addressing the impact of sea level rise in a Watershed Master Plan.

In addition, the manual refers to the 2017 NOAA report outlining projections of future sea level rise by 2100 (see ch. 2) and adopts the Intermediate-high case (i.e., 4.92 feet of sea level rise by 2100) for planning purposes, stating that because "there is uncertainty inherent in estimating future sea levels, the CRS has adopted a base minimum projection for sea level rise for the purposes of CRS credit and meeting CRS prerequisites."[16]

Including sea level rise in the new CRS *Coordinator's Manual* is an important step, and adopting the Intermediate-high projection from the 2017 NOAA sea level rise report provides helpful guidance to communities as they consider which of the six scenarios in the NOAA report to focus on for planning purposes. Unfortunately, the number of points available for most sea level rise planning is relatively small and, taken together, would not be enough to advance a community from one CRS class to the next. More important, by discounting premiums in return for modest storm and sea level rise preparation actions, the government is effectively endorsing a property owner's decision to remain in a risky location, albeit now marginally safer than without the additional measures. This compounds the original misdirection by the NFIP.

Coastal Barrier Resources Act: The Coastal Barrier Resources Act (CBRA), created in 1982, discourages development on sensitive coastal lands by prohibiting flood insurance under the NFIP for new structures on these lands and limiting federal disaster assistance in the event structures are damaged. Although it withdraws federal financial incentives, it does not directly regulate how property owners develop their land within or near CBRA units. Today, the John H. Chafee Coastal Barrier Resources System, named after noted environmentalist and Rhode Island Senator John Chafee, includes some 862 units, with about 3.5 million acres of land and associated aquatic habitat.

The United States Fish and Wildlife Service manages the CBRA program and in 2002 issued a report describing its accomplishments. It looked at the range of savings associated with avoided development on CBRA lands, including infrastructure investments and disaster payments saved, and found $684 million saved from 1983 to 1996, and projected an astonishing cumulative total of almost $1.3 billion in savings from 1983 to 2010.[17]

Even this huge savings projection, however, turned out to be a significant underestimate. Research published in March 2019 by Andrew Coburn and John Whitehead found that CBRA reduced federal expenditures associated with damage from coastal storms by $9.49 billion (in 2016 dollars) between 1989 and 2013, noting that the Fish and Wildlife Service has "significantly underestimated the extent and degree to which developed U.S. shorelines would be impacted by coastal storms after 2002, as well as the federal costs associated with those impacts."[18] Looking to the future, the authors project that disaster recovery costs avoided could exceed $100 billion over the next fifty years without accounting for future sea level rise.

Congress never intended CBRA to be a solution to sea level rise, but it does offer a model that might be applied to the now emerging understanding of the risks of sea level rise-driven inundation of a wider range of coastal lands.

Major Hurricanes Destabilize the National Flood Insurance Program

The NFIP was steaming along pretty well in the 1990s. Some 19,000 communities participated at that point, and 4.2 million policies were in place,[19] all while operating on a sustainable basis financially.[20] In 2005, Hurricanes Katrina and Rita hit the Gulf Coast causing widespread damages and throwing the program into a financial tailspin. As of December 2006, payments to meet these massive losses left the NFIP $16.75 million in debt.[21] Sadly, this was just the beginning of losses that exceeded the income from premiums and surcharges.

From 2003 to 2012, annual losses or paid claims through the NFIP averaged $4 billion a year while the average amount of premiums paid was $2.6 billion.[22] Then, in late October 2012, Hurricane Sandy crashed into New Jersey, New York, and southern New England, pushing a storm surge measured at over nine feet at Battery Park at the tip of Manhattan and causing widespread damages. Heading into the 2017 hurricane season, the program was $24.6 billion in debt with just $5.8 billion in borrowing authority.[23]

In September 2017, the Congressional Budget Office (CBO) issued a damning report reviewing the dire financial straits the NFIP was in and found three major problems. First, CBO reviewed all five million policies, the NFIP rate structure, and estimated potential future losses, including hurricane losses not accounting for sea level rise, and concluded that, "overall, considering all expenditures and premium income, the program had an expected one-year shortfall of $1.4 billion."[24] About half of this annual shortfall was payments exceeding premiums, and the other half was costs related to operations, such as mapping, and interest payments on the debt.

Second, CBO found that policies in coastal counties were generating the losses while policies in inland counties were actually generating a small surplus: "The agency estimates that the shortfall for the NFIP program as a whole stems largely from premiums falling short of expected costs in coastal counties, rather than in inland counties."[25]

Finally, not only were coastal policyholders mostly generating the losses for the program, CBO pointed out that 85 percent of policyholders in high risk V zones do not pay rates that reflect their actual flood risks: "The result is that most policyholders whose property is at risk of wave damage from storm surges do not pay premiums that cover their expected costs. Instead, the additional expected costs from wave damage are spread broadly among the NFIP's policyholders, resulting in a cross-subsidy from inland counties (on average) to coastal counties."[26]

Recognizing these financial challenges, the federal Government Accountability Office (GAO) identified the NFIP as one of thirty-four governmentwide programs deemed to be a "high risk" in 2006, and the program has remained on this list through 2017. GAO reported that the program "likely will not generate sufficient revenues to repay the billions of dollars borrowed from the Department of the Treasury to cover claims from the 2005 and 2012 hurricanes or potential claims related to future catastrophic losses."[27]

On top of all that, starting in late August of 2017, Hurricane Harvey hit the Texas coast, and Hurricanes Irma and Maria hit Florida, Puerto Rico, and the Virgin Islands. Estimates of NFIP losses from these storms were not immediately clear but were expected to far exceed borrowing authority. In October 2017, Congress passed a disaster relief bill that, for the first time, simply forgave $16 billion of NFIP debt to offset the losses expected. At the end of 2017, the program had $20.5 billion in debt and $9.9 billion in borrowing authority.[28] For FY 2018, FEMA projected revenues of $5.6 billion and expenses of a whopping $15.5 billion.[29]

Congress Reforms the Flood Insurance Program

Congress attempted to deal with the financial shortcomings of the NFIP when it passed the Biggert-Waters Flood Insurance Reform Act in 2012. The act addressed the problem of funding shortfalls with the obvious solution of raising rates, or more specifically, phasing out subsidized rates and shifting to risk-based rates with annual increases capped at 25 percent each year. It is important to remember that "risk-based rates" means risks from storms that are understood by looking backward at the historical record of storms without considering any increased risk associated with more intense coastal storms or sea level rise.

About 20 percent of all policies (some one million policies, mostly older structures predating the NFIP) benefited from significant subsidies and the Biggert-Waters Act focused on policies for second homes, properties with repetitive losses, and nonresidential structures. Subsidized rates were to be phased out as communities adopted updated flood maps. Policyholders in primary residences, however, could keep subsidized rates until the property was sold or a severe loss occurred.

In a demonstration of the difficulty of changing broad-based public subsidy programs, the public backlash against Biggert-Waters was swift. Policyholders facing dramatic rate increases pressed Congress for relief, and Congress retreated, passing the Homeowner Flood Insurance Affordability Act in 2014. The new law goes so far as to provide for refunds of increased premiums that some policyholders were charged under Biggert-Waters, but it did not give up altogether on risk-based rates. Some rate increase provisions were retained in a modified form (e.g., slower rates of premium increase).

Conservatives in Congress argued to hold the line. Rep. Jeb Hensarling (R-TX), who chairs the House Financial Services Committee, said the bill "would postpone actuarially sound rates for perhaps a generation. . . . We'd kill off a key element of risk-based pricing permanently, which is necessary if we are to ever transition to market competition."[30]

Republicans with constituents hit by rate increases and most Democrats, however, supported the revisions. Rep. Shelley Moore Capito (R-WV) summed up the desperation many in Congress were hearing from constituents: "In some cases, their only choice was to either spend their life savings on their flood insurance bills or walk away from their house, ruining their credit."[31] The precarious financial condition of many homeowners suggests the difficulty in changing these policies but also the risk of much wider financial harm that will come as homes are increasingly lost to more severe storms and rising seas.

The legislation enacted in 2014 did not deal with the longer-term reauthorization of the NFIP and, in November 2017, the House of Representatives passed the 21st Century Flood Reform Act. The bill proposed to continue the retreat from Biggert-Waters started in 2014 with another round of adjustments. So the idea of restoring financial soundness by ramping up rates to more fully cover risks was not dead but was also not seen as sufficient to balance the books.

The House bill, however, took another shot at strengthening financial soundness with new authority for expanded entry of private insurers in the flood insurance market. Representative Sean Duffy (R-WI), original sponsor of the bill, noted that "we set up a private market. Now, you don't have to take the private market, but you have an option to get a private plan that might have a better rate than the government offers you. You have a choice. . . . And by the way, when we get the private market in, we offload our risk to the private sector. . . . We have private companies in play. That's a great thing."[32]

Legislation introduced in the Senate by Senators Cassidy and Gillibrand also sought to encourage a greater role of the private sector. Senate bill 1313, however, took a more cautious approach. The official bill summary states the following: "Such market entrance by private insurers should be gradual and preserve the availability and affordability of flood insurance coverage for all consumers while continuing the investment in floodplain mapping and management."[33] Of course, private sector insurers already participate in the NFIP in a substantial way in that they have contracts to manage NFIP policies and claims. Although not enacted, the House and Senate bills would have taken the big additional step of encouraging private companies to write flood insurance policies and pay claims from their own funds, rather than government funds.

But private sector flood insurance is growing even without incentives from Congress. Private flood insurance is offered both in place of an NFIP policy and to supplement the NFIP coverage now capped at $250,000. Private market flood coverage grew by 51 percent in 2017 and "private cover now represents nearly 15% of all flood premiums nationwide,"[34] valued at over $623 million. In March 2019, FEMA announced its plan to act without Congress to revise rate structures to better reflect risk. Starting in 2020, rates developed under "Risk Rating 2.0" will reflect detailed risk data. Rates for riskiest properties might double while rates for lower risk properties might be cut by 50 percent.

Given the recent high debt of the NFIP, it is not surprising that there is enthusiasm in Congress for shifting some of the risk exposure from the NFIP to private insurers. Those who see the huge debt and the annual operating shortfall of the current program as unsustainable politically, hope that private investment will pick up the shortfall left by subsidized rates and reduce or avoid periodic federal bailouts. The argument here is that the private funds can help keep the NFIP going and sustain both the remaining rate subsidy and the flood management practices that participating communities commit to in order to participate in the program. FEMA is also gradually shifting some risk to the private sector through reinsurance agreements.

Not everyone thinks the benefits of bringing the private sector more fully back into the flood insurance market would work out well in the long run. The Center for Economic Justice has expressed concerns that private insurers would "cherry pick" the less risky policies leaving the government to insure the higher risk properties. With fewer policies paying into the NFIP fund, program costs, and thus rates for the remaining policyholders, would go up. Other questions raised by expanded private sector flood insurance are availability of coverage year to year

and potential erosion of community commitment to flood mitigation requirements of the federal program.

Most observers recognize that the shift away from subsidized rates and toward
financial soundness and private sector coverage will raise rates and make insurance
unaffordable for some people. There is less agreement about what to do about this
problem.

The Trump administration proposed establishing a means-tested affordability
program for low-income policyholders. People with incomes less than 80 percent
of area median income, some 26 percent of policyholders in flood zones, would be
shielded from substantive rate increases that they would otherwise experience under
existing law.[35] Senate bill 1313 would provide vouchers of various amounts to offset
flood insurance costs if they pushed housing costs more generally above 40 percent
of household income.

The reform bill passed by the House of Representatives gives states the option of
operating affordability programs under narrower eligibility criteria. More important,
the bill passes decisions about how much subsidy help to provide to the state and
makes the state cover any costs with a surcharge on all other flood insurance policies
in the state. This approach protects the commitment to financial soundness but will
drive costs for those not covered by an affordability program higher.

Proposals to Adapt Flood Insurance to More Severe Storms
and Rising Seas

These attempts to improve operation of the NFIP do not directly address the challenges of more serve storm surges and rising seas and are not sufficient to meet these
challenges. There are, however, a range of policy proposals that speak directly to the
risks that rising seas pose to the flood insurance program or have multiple benefits,
including benefits for adapting to more severe storms and rising seas. These ideas
include providing information about sea level rise on flood maps, prohibiting flood
insurance for new construction, and expanding authority to buy properties that generate repetitive losses.

Sea Level Rise Risk Maps: In 2015, an advisory committee to FEMA recommended
that it "provide future conditions flood risk products, tools, and information,"[36]
including information about future population, land uses, and climate change. The
committee boldly called on FEMA to "incorporate Local Relative Sea Level Rise scenarios into the existing FEMA coastal flood insurance study process,"[37] and to "prepare map layers displaying the location and extent of areas subject to long-term
erosion and make the information publicly available."[38]

By map layers, however, the committee did not mean the official maps that define floodplains where insurance is required. The committee cautioned that "maps displaying the location and extent of areas subject to long-term coastal erosion and future sea level rise scenarios should be advisory (nonregulatory) for federal purposes. Individuals and jurisdictions can use the information for decision-making and regulatory purposes if they deem appropriate."[39]

FEMA has deferred any decision on these recommendations until after implementation of pilot studies. But, if adopted, this recommendation for active provision of information about sea level rise scenarios and projected changes in population and land use could greatly improve local flood mitigation and help communities understand long-term risks of rising seas.

Prohibition on Insuring New Construction: A second reform to the NFIP that would have multiple benefits, including preparation for rising seas, is a prohibition on insuring new construction in floodplains. The Trump administration letter to Congress on improvements to the NFIP in October 2017 proposes this approach, effectively taking the basic concept of the Coastal Barrier Resources Act of prohibiting federal flood insurance for new construction on barrier islands and sensitive lands and applying it much more widely.[40] Part of the reasoning is that, because new construction in floodplains must already comply with flood resistant building standards, the private market will have an interest in insuring these properties.

Of course, this prohibition would apply to inland as well as coastal property, and there is no mention of sea level rise in the Trump administration position paper. Still, it seems clear that, at least for most places at risk of storm surges and rising seas, insuring new construction would not be a good investment. Banning federal flood insurance for new construction in these areas at risk of sea level rise would likely reduce development pressure on these risky areas and make the eventual task of relocating from these areas less complex.

Avoiding Repetitive Losses: A third key idea for reform to the NFIP is to reduce payments for existing properties that are damaged, then rebuilt with flood insurance funds, and then damaged again, often repeatedly. These repetitive loss properties make up about 1 percent of the NFIP insured properties but result in an amazing 38 percent of all losses.[41]

At one level, reducing these recurring losses would be smart because it would improve NFIP operational efficiency and increase financial soundness. At another level, it would prepare the program for more severe storms and rising seas. A common strategy for reducing repetitive losses is to offer to buy a storm damaged property and then tear it down. Owners facing a long and expensive rebuilding process are sometimes more willing to sell than owners not facing rebuilding. Owners taking a buyout offer might buy another property at risk of flooding but might also relocate to a safer area. Because more severe storms and sea level rise will result in a steady increase in repetitive losses, strengthening the NFIP's capacity and resources for buyouts promotes relocation to safer areas and reduces the density of properties in risky areas that will eventually be inundated by rising seas.

Unfortunately, outside of a major disaster where Congress provides funds directly for buyouts, funding for buyouts of repetitive loss properties is scarce. Today, the grant program best able to fund repetitive losses is the Hazard Mitigation Grant Program authorized under the Stafford Act (see ch. 10). More limited funding is now available under the Flood Mitigation Grant Program managed under the NFIP, and this program has made buyout of repetitive loss properties a priority in the past.

The problem is that the number of repetitive loss properties is huge (tens of thousands of properties, depending on how they are counted), and the current grant funds available for buyouts make only a small dent in addressing high-loss properties that exist today, and the growing number that are likely to need attention in the near future. One solution is to simply increase the funding for buyouts. Senate bill 1313 would have directed $400 million from insurance policy surcharges to the Flood Mitigation Grant Program, more than doubling the current funding. The House-passed NFIP reform bill would have authorized $225 million annually for flood mitigation assistance grants focused on repetitive loss properties.

In addition, there are several new ideas for supporting an expanded effort to buyout repetitive loss properties. For example, in 2015, Becky Hayat and Robert Moore at the Natural Resources Defense Council proposed that, rather than focusing on paying to rebuild properties with significant damage, the NFIP "should be geared toward making relocation the easiest and most attractive option for property owners to pursue."[42] More specifically, they suggest the NFIP adopt and promote a new, reduced rate structure applying to policyholders who agree to sell their home to FEMA if it suffers damages greater than 50 percent of the market value. Savings from avoiding costly repetitive losses would help offset losses due to lower premiums.

FEMA also is trying to push states toward more active use of a strategy for addressing repetitive losses as a condition of eligibility for a 90 or 100 percent federal share in property acquisition projects funded with Hazard Mitigation Grants. And, FEMA awards points to communities that develop a repetitive loss analysis for the Community Rating System.

In the context of a hundred years of past practice dedicated to structural solutions to flooding problems, the NFIP was a ray of sunshine. It provided affordable, reliable flood insurance for individuals at little cost to the government while prompting communities to reduce flood risk by implementing local mitigation measures. Sadly, the tables have turned for the NFIP. Major hurricanes have increased coastal property losses, making the program financially unsustainable. The emergence of the risk of permanent inundation of insured properties due to rising seas further weakens the financial structure of the program while increasing the moral hazard associated with insuring these homes, all while facilitating new development in these risky areas. The challenge going forward is finding a way to disengage from insuring property at risk of permanent inundation by rising seas in a gradual transition that recognizes the often precarious financial situation of many homeowners and the consequences for coastal communities.

10

Coastal Disaster Planning: Preparing for the Wrong Hazards

The United States is fortunate to have a strong and broadly capable network of state and federal disaster response programs, built around the Stafford Disaster Relief and Recovery Act (1988). The Stafford Act authorizes programs that complement the National Flood Insurance Program (NFIP) by providing federal technical, financial, logistical, and other assistance in the event of major disasters requiring a response beyond the resources of a state. These programs apply to disasters other than flooding, including earthquakes and wildfires. Unfortunately, the costs of disasters are increasing steadily, largely driven by the very high cost of the impacts of major hurricanes on coastal communities.

Although this broader disaster assistance program is primarily designed to support recovery after a disaster, it includes a modest investment in disaster preparedness planning. Criticisms are building, however, that spending for disaster recovery will grow to unmanageable levels without a much more substantial investment in disaster preparation and planning. This limited investment in disaster preparation, at a time of rapidly increasing disaster recovery costs that are likely to grow as coastal storms become more severe and seas rise, puts the country on a path toward dramatically escalating disaster recovery costs, as well as escalating losses of life and property.

A related concern is that a large part of today's disaster planning investment is focused on postdisaster planning—looking back at a disaster that just happened and planning to avoid it in the future. Some critics argue that disaster planning should look forward, rather than back, addressing a wider range of hazards, including hazards resulting from a changing climate. In effect, they argue that there is not enough investment in disaster preparation, and much of the planning that is occurring is preparing for the wrong hazards.

Core Elements of Disaster Relief

Named after venerable Senator Robert T. Stafford of Vermont, the Stafford Act (1988) is built around a Disaster Relief Fund that stands ready to support disaster costs as a need arises. Congress maintains the fund with annual appropriations that can be carried over from year to year, and the Federal Emergency Management Agency (FEMA) draws on the fund in response to disaster declarations by the president. FEMA also manages Presidential Policy Directive 8, addressing national preparedness and covering everything from natural disasters to acts of terrorism and pandemics.

In fiscal year 2016, Congress appropriated $661 million in base funding to the Disaster Relief Fund but can, and often does, provide additional funding as needed through supplemental appropriations. The Government Accountability Office (GAO) reported that, between 2005 and 2014, FEMA spent $104.5 billion from the fund,[1] or about $10 billion a year. If the huge, $40 billion cost of Hurricane Katrina is excluded, however, average annual spending from the fund was about $5 billion.[2]

The act is intended to provide federal aid when the resources of local and state governments can't meet a need. Generally, a state governor requests federal support, and the president approves the request, making a "disaster declaration" after considering factors such as the scale of the disaster, the expected cost of relief, other recent disaster events in the state, and a state's preparedness efforts. In the case of assistance to governments, as opposed to individuals, FEMA considers the estimated costs of assistance relative to the population of the state and generally would support a request where damages are expected to exceed a "per capita damage indicator," which is adjusted annually and was set at $146 per capita in 2018. A state might also seek more limited federal assistance by asking for declaration of an emergency rather than a major disaster.

Once a federal disaster declaration is issued, the federal government coordinates the response effort. Federal aid can include distribution of food and medicine, housing assistance, as well as funding to local, state, or tribal governments and some nonprofit organizations to provide emergency services, conduct debris removal operations, and repair or replace damaged public infrastructure such as roads, bridges, and water facilities. If a local government has suffered a major revenue loss, aid might include a loan to support local government operations. Assistance also goes to affected households and can take the form of housing assistance, crisis counseling, case management services, legal services, and disaster unemployment assistance. The federal government pays between 85 and 100 percent of these costs.

This conventional model of disaster response is changing in reaction to more severe coastal storms, with Congress providing major supplemental appropriations after a storm. Three separate supplemental appropriations acts provided $120 billion, mostly in response to the three major 2017 coastal storms. Congress directed much of this funding to diverse federal agencies, such as the Army Corps of Engineers and the Department of Housing and Urban Development (HUD). The HUD Community Development Block Grant Disaster Recovery program, for example, can support restoration of infrastructure, housing, and economic revitalization in response to a disaster.

Addressing Coastal Storms and Rising Seas in Hazard Mitigation Planning

There is strong evidence that investment in disaster preparation saves money in the long run. A 2005 report by the Multi-hazard Mitigation Council of the National Institute of Building Sciences found that "on average, a dollar spent by FEMA on hazard mitigation (actions to reduce disaster losses) saves the nation about $4 in future benefits."[3]

The Hazard Mitigation Grant Program (HMGP) is the largest grant program supporting disaster planning and provides grants to state and local governments in which disasters have been declared with the goal of reducing the impact of future disasters. FEMA guidance for the HMGP indicates that "the key purpose of HMGP is to ensure that the opportunity to take critical mitigation measures to reduce the risk of loss of life and property from future disasters is not lost during the reconstruction process following a disaster."[4]

Funding for these grants is from the Disaster Relief Fund and is usually 15 percent of the total amount of federal assistance provided to a state, territory, or tribe following a major disaster declaration. This funding adds up over the years. FEMA reports spending over $13 billion on the HMGP program from 1989 to 2016.[5] The funding supports projects such as property acquisition, structure elevation, flood proofing structures, and purchase of generators as well as floodplain restoration and green infrastructure practices for stormwater management.

FEMA requires state, tribal, and local governments to develop and adopt hazard mitigation plans as a condition of receiving disaster assistance, including funding for specific mitigation projects that can reduce disaster vulnerability. These plans are intended to address all risks in a state or locality, not just flooding. All states and over 22,000 local jurisdictions have hazard mitigation plans. Jurisdictions must update their hazard mitigation plans and resubmit them for FEMA approval every five years to maintain eligibility for disaster assistance.

FEMA's mitigation planning regulation and 2015 supporting guidance for developing hazard mitigation plans specifically requires assessment of the probability of various hazards in the future, such as those caused by climate changes, including sea level rise: "FEMA recognizes challenges posed by climate change, including more intense storms, frequent heavy precipitation, heat waves, drought, extreme flooding, and higher sea levels. . . . FEMA encourages Recipients and subrecipients to consider climate change adaptation and resiliency in their planning and scoping efforts."[6]

Hazard mitigation plans are supposed to include mitigation strategies that are tied to the assessments, but while the requirement for assessment of climate risks is clear, the guidance suggests more flexibility in terms of mitigation strategies for future events generally, and for a large and complex challenge like sea level rise more specifically. In a 2017 paper assessing hazard mitigation plans, Dr. Melissa Stults observed that hazard planners were paying more attention to predisaster planning, but that a continuing focus on "use of previous occurrences of hazards as a foundation for estimating the probability of future hazard events"[7] was a problem, concluding that "in practice, this means that communities are using past events to predict the future."[8] Because of their scope addressing all hazards, and the challenges of focusing on long-term risks like rising seas, these plans are rarely the best forum for considering the combined risks of storms and rising seas, especially in places where these risks are significant. They should, however, reflect assessments and decisions made in other planning processes.

Reviewing the broad picture of federal planning for disaster mitigation, the GAO pointed to a fundamental problem of the lack of a strategy: "There is no comprehensive, strategic approach to identifying, prioritizing and implementing investments for disaster resilience, which increases the risk that the federal government and non-federal partners will experience lower returns on investments or lost opportunities to strengthen key critical infrastructure and lifelines."[9]

The Association of State Floodplain Managers, reporting from its Fifth Gilbert H. White National Policy Forum in 2015, concurred with this assessment, pointing out the unsustainable financial costs of the current approach and adding a call for action: "As the disaster damage exposure to the federal taxpayer continues to grow, the nation can no longer afford to continue following a policy of buying its way back from disaster with emergency funding. Instead, steps must be taken earlier, during planning and implementation, to limit the impact of future disasters."[10]

Shifting Planning Focus to Future Disasters

It seems clear that current disaster planning could be strengthened by rethinking the focus on avoiding the last disaster (i.e., postdisaster planning) and shifting attention to investing in predisaster planning that is more likely to recognize emerging threats like more severe storms and rising seas.

The existing Pre-Disaster Mitigation Grant Program (PDM) is a much smaller effort than the Hazard Mitigation Grant Program and is intended to fund the planning and preparation actions that will generally protect communities from future disasters rather than disasters already experienced. These Pre-Disaster Mitigation grants are not related to major disaster declarations and can fund both planning and projects that mitigate diverse threats, including flooding of power and water systems, medical facilities, and transportation systems.

Funding is also available for property acquisition, and the program guidance also mentions protection of natural infrastructure in coastal areas including projects to "increase community resilience by reducing risk through green and natural infrastructure or natural defenses, including building coastal wetlands to absorb destructive forces from wave action and to protect the shoreline from erosion; . . . and relocating/acquiring large tracks of land and properties to create coastal wetlands, marshes, a buffer zone, or a natural recreation area."[11]

The Pre-Disaster Mitigation Grant Program is clearly on the right track, but there are two problems. First, is limited funding. The Hazard Mitigation Grant Program grants are funded directly out of the Disaster Relief Fund; the Pre-Disaster Mitigation grants are funded by annual appropriations that have been inconsistent from year to year, as low as $25 million in several recent years and generally not more than $150 million.[12] Funding in 2017 for the entire country, not just the coast, was just $90 million.

Second, FEMA policy specifically directs the limited predisaster planning funds away from flood related events. FEMA sets priorities annually for the competitive predisaster funding, "with priority given to applications from applicants that have little or no disaster funding available through the Hazard Mitigation Grant Program (HMGP), and to project applications for nonflood hazard mitigation activities, such as wildfire, drought, and seismic and wind mitigation, which cannot be funded through the Flood Mitigation Assistance program."[13]

Yes, there is a third grant program. In addition to the big Hazard Mitigation Grant Program and smaller Pre-Disaster Mitigation program, FEMA also administers the Flood Mitigation Assistance Grant Program. This program focuses specifically on flood preparation, rather than disaster preparation generally. It is authorized under the NFIP, rather than the Stafford Act, and it provided $160 million in grants to communities all across the country in 2017 and 2018. This program is focused directly on the ambitious and important goal of "reducing or eliminating claims under the National Flood Insurance Program (NFIP)."[14] Reduction of repetitive losses has been a priority, but in 2017, funds for repetitive loss-properties were capped at $70 million because of built up demand for more general flood mitigation on a communitywide basis.

In 2017, FEMA proposed to make Flood Mitigation Assistance Program grants for both "advance assistance," to develop flood mitigation strategies, and "community flood mitigation projects," that integrate cost-effective natural floodplain restoration solutions and improvements to NFIP-insured properties.[15] Examples of communitywide projects include planning and projects for protection of infrastructure, restoration of wetlands and floodplains, and protection of water and sewer facilities. These funds are less tied to preventing the last flood but not yet fully focused on the challenges of a changing climate generally or more severe storms and rising seas more specifically.

Despite the good intentions of the Flood Mitigation Assistance Program, the GAO looked at the overall picture of federal investments in hazard mitigation and concluded that a shift in priorities is needed.

> Most federal funding for hazard mitigation is available after a disaster. For example, from fiscal years 2011–2014, FEMA obligated more than $3.2 billion for HMGP post-disaster hazard mitigation while the Pre-Disaster Mitigation Grant Program obligated approximately $222 million. There are benefits to investing in resilience post-disaster. Individuals and communities affected by a disaster may be more likely to invest their own resources while recovering. However, there are also challenges. Specifically, the emphasis on the post-disaster environment can create a reactionary and fragmented approach where disasters determine when and for what purpose the federal government invests in disaster resilience.[16]

Congress seems to have taken this criticism to heart. It provided more funds—$235 million—for predisaster mitigation grants nationwide in 2018. And, as part of supplemental disaster assistance appropriations focused on the 2017 hurricanes, Congress provided the Department of Housing and Community Development with almost $16 billion for disaster mitigation. This is a major investment in thinking ahead, but half the funding is targeted to Puerto Rico with most of the rest going to Louisiana and Texas.

Then, in October 2018, Congress enacted the Disaster Recovery Reform Act that included the key idea to expand mitigation investments by reserving not less than 6 percent of total funding for seven disaster programs from the Disaster Relief Fund for mitigation grants. Annual spending varies, but the average is close to $10 billion, and 6 percent would be $600 million. Congress, however, took its foot off the gas a bit when it changed the mandatory reservation of funds in earlier drafts of the bill to be discretionary, meaning that the $600 million could be reduced from year to year. Still, there is reason to hope that the ground is shifting in favor of investment in avoiding future disasters, including more severe storms and rising seas.

Other Proposed Improvements to Disaster Assistance

In July of 2018, FEMA issued an "after action" report on the 2017 hurricane season with eighteen findings and thirty recommendations. The strength of this report is a close examination of operational issues during the 2017 relief efforts and definition of corrective actions ranging from workforce training, to smarter logistics, to better shelter and housing services.

In a larger sense, FEMA's take away message from 2017 hurricanes is summarized in Administrator Brock Long's statement opening the report, calling for reducing operational complexity, building a culture of preparedness, and readying the nation for catastrophic disasters. Key ideas here are improving coordination with critical infrastructure sectors and state and local governments, and that "governments need to be better prepared with their own supplies . . . and to be ready for the financial implications of a disaster."[17] Sadly, missing from the FEMA report is any suggestion that the 2017 hurricane season had anything to do with a changing climate or that improving future performance will require rethinking disaster relief programs with more severe storms and rising seas in mind.

In September 2018, the GAO was a bit more critical in its evaluation of FEMA's response to disasters in 2017, including hurricanes and wildfires. Among other issues, GAO found the workforce was overwhelmed and "at the height of FEMA workforce deployments in October 2017, 54 percent of staff were serving in a capacity in which they did not hold the title of "Qualified."[18] GAO pointed to other issues including housing and debris removal while recognizing the logistical challenges associated with disaster response on islands rather than the mainland.

Beyond FEMA's self-assessment and the GAO's audit, there are several other criticisms of the nation's disaster relief programs, including concern for the growing share of federal spending relative to spending by states and the need for local governments to adopt the most current building codes.

Disaster relief programs have been effective in helping communities recover but demand for this federal assistance is growing steadily. Fiscal conservatives are concerned about steadily increasing costs and are interested in limiting this spending. The Cato Institute, a conservative think tank, reported with some exasperation, that "the number of disaster declarations has soared in recent decades. The annual average number was 51 in the 1970s, 29 in the 1980s, 74 in the 1990s, 127 in the 2000s, and 139 so far in the 2010s."[19]

One way to look at the problem of the growing number of disasters and costs of disaster response is that a changing climate is causing more flooding, drought, and severe storm events, and thus greater costs. Another view, offered by fiscal conservatives, is that states are taking advantage of generous federal policies to shift costs from local and state taxpayers to the federal government, driving up federal costs but also costs generally on the theory that states spending their own funds would be more frugal.

A related concern is that the federal share of assistance is too high compared to the share a state provides, making it too attractive to states. The Heritage Foundation, another conservative think tank, has recommended lowering the federal share to 25 percent of costs, arguing that "the result has been that states now request federal help whenever they can, since it will bring federal dollars. This creates a vicious cycle as states respond to increased federalization of disasters by preparing less and setting less funding aside for disasters. As a result, states are less prepared for disasters, they request more government help, and thus the cycle is perpetuated."[20]

Striking the right balance between federal and state shares of disaster response is both important and difficult because of the varying capabilities of states. Rather than simply slashing the federal share, the National Infrastructure Advisory Council proposed in 2016 a "disaster deductible"[21] for each state. Under this model, every state or region would start with a fixed deductible (e.g., a state or region would cover 40 percent of disaster costs compared to federal payment of 60 percent). FEMA would outline a range of mitigation actions and states that adopt some or all those actions could improve their federal share of costs. Much would depend, however, on whether higher levels of federal aid were traded for meaningful upgrades to state preparedness or whether this bar is set very low.

The reports and studies of disasters and needed improvements would fill a small library with volumes on Hurricanes Katrina and Sandy being prominent. For example, the House and the Senate both issued reports in 2006 following Hurricane Katrina focusing on operational and organizational changes, including a Senate recommendation to abolish FEMA in favor of a more broadly powerful entity. The *Hurricane Sandy Rebuilding Strategy,* issued in 2013, focused more on improving response to disasters in the New York/New Jersey region than on new national policies. Sea level rise is largely absent from these assessments.

Finally, local building codes are an issue in disaster recovery because there is often a lag in the adoption of the most current building codes by local communities, including building standards related to flooding. This topic gets some attention in the Disaster Recovery Act, passed by Congress in 2018, which provides several incentives to encourage communities to stay current on building codes and new authority for FEMA to require use of the most current code in grant-funded projects.

Building codes adopted in coastal communities, however, can be something of a two-edged sword. It is true that up-to-date building codes improve the resilience of buildings to conventional flooding and storm surges, reducing damage costs to property owners, claims to the NFIP, and disaster recovery costs. But, building codes commonly focus on construction standards, rather than location, and can convey too strong a sense of invulnerability (i.e., "I built to the latest code so I must be safe."). In the case of inundation from rising seas, stronger building construction, or even elevation of buildings, is of little use when roads and utilities are permanently under water. Codes that call for safer construction and are also tied to policies that point development to safer places are a better response to rising seas.

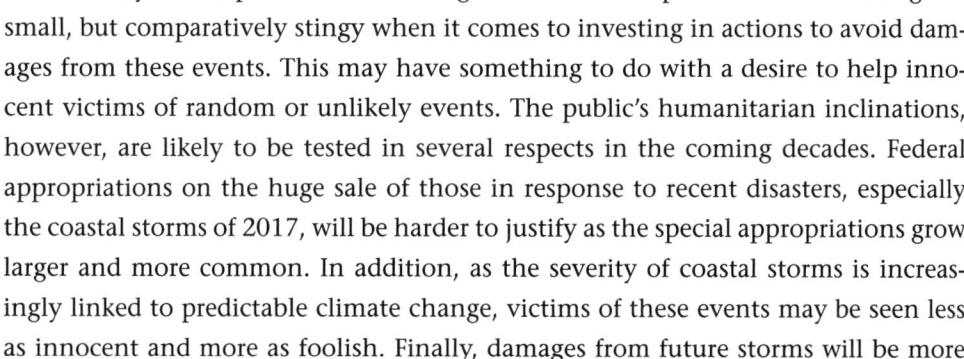

The country is compassionate and magnanimous in response to disasters large or small, but comparatively stingy when it comes to investing in actions to avoid damages from these events. This may have something to do with a desire to help innocent victims of random or unlikely events. The public's humanitarian inclinations, however, are likely to be tested in several respects in the coming decades. Federal appropriations on the huge sale of those in response to recent disasters, especially the coastal storms of 2017, will be harder to justify as the special appropriations grow larger and more common. In addition, as the severity of coastal storms is increasingly linked to predictable climate change, victims of these events may be seen less as innocent and more as foolish. Finally, damages from future storms will be more extensive as sea level rises and may grow to seem less random or unlikely and more inevitable. The best way to sustain generous public support for disaster response is to keep costs from exploding and to more clearly discourage people from taking unwise risks. That will require a new commitment to invest in more up-front disaster mitigation focused on the most serious future risks.

11

Coastal Management Programs: Overcommitted and Underfunded

Revising flood insurance and disaster assistance programs to account for more severe storms and rising seas is an important step toward preparing America to meet these threats. These improvements alone, however, would not deliver a coordinated national program to prepare communities, critical coastal infrastructure, and ecosystems for more damaging storms and rising seas.

Despite lack of a national plan to prepare the country for more severe storms and rising seas, there are some elements of existing national plans or programs that address these challenges. They fit within three broad categories. As discussed in the previous chapter, coastal impacts are among the many topics addressed in disaster and hazard mitigation planning, along with wildfires, earthquakes, and inland flooding. Sea level rise and storms also come up as one element in the national assessment of all climate changes and in some federal agency plans to adapt to a variety of impacts of a changing climate, which are discussed in the next chapter.

The existing national program that comes closest to fitting the challenges that more severe storms and rising seas pose for coastal communities is the planning effort under the Coastal Zone Management Program. A close second was planning under the National Ocean Policy before it was terminated by President Trump in 2018. These coastal and ocean management efforts cover diverse topics ranging from commercial fishing, to recreational access to the coast, to energy development. Although sea level rise and coastal storms are mentioned in each case, they are not the central focus.

Storms and Sea Level Rise in the Context of the National Ocean Policy

Although the country does not have a coastal storm and sea level rise strategy per se, it did have a *National Ocean Policy*. Executive Order 13547, signed by President Obama in 2010, established a broad national policy to "ensure the protection, maintenance, and restoration of the health of ocean, coastal, and Great Lakes ecosystems and resources."[1] The order called for "development of coastal and marine spatial plans that build upon and improve existing Federal, State, tribal, local, and regional decision-making and planning processes,"[2] and created an interagency National Ocean Council to manage its implementation.

Unfortunately, President Trump revoked the executive order in June of 2018, replacing it with a new order, "rolling back excessive bureaucracy created by the previous Administration,"[3] and shifting focus to economic aspects of ocean policy. Although the original order is no longer in effect, work to implement it over the past eight years included attention to coastal storms and rising seas. Assuming for a moment that a future administration might want to restore the ocean planning process along the lines framed in the Obama order, would that be the best way for the country to prepare for more severe storms and rising seas?

The order looked comprehensively at all things marine, but a *National Ocean Policy Implementation Plan* spoke broadly to the need to build "ocean and coastal resilience,"[4] stating that "agencies will also enable and support efforts to understand, minimize, and adapt to the impacts of climate change, ocean acidification, sea-level rise, and extreme weather events, strengthening the resilience of coastal communities."[5] The 2013 appendix supporting the plan provides more details of actions, many of which have occurred, including improving mapping of coastal elevations, providing guidance on sea level rise scenarios, developing tools and training for assessing vulnerability of coastal infrastructure to climate change and providing coastal inundation and sea level change decision-support tools to local, state, tribal, and federal managers.

In 2016, an annual workplan for the *National Ocean Policy* articulated a stronger sense of urgency with respect to sea level rise: "For coastal communities, sea level rise, combined with coastal storms, has increased the risk of erosion, storm surge damage, and flooding. The capacity of marine and coastal ecosystems to buffer the impacts of extreme events is being overwhelmed. . . . The Federal government repeatedly hears from state, local, and tribal leaders that they need consistent, accessible, authoritative, and regionally appropriate projections and scenarios of future climate change, in particular sea level rise, for supporting preparedness planning."[6]

More specifically, the workplan called for developing sea level rise data and scenarios along the lines published in early 2017 by the National Oceanic and Atmospheric Administration (NOAA) and further called for federal agencies to "begin integrating regional sea level rise scenarios with existing coastal risk assessment capabilities in Federal agencies to create an evolving set of informational and planning tools designed to help stakeholders develop improved insight into the additional risks sea level rise over upcoming decades may pose in coastal settings."[7]

The nine regional marine planning bodies created to implement the Obama order had a mixed record with respect to storms and risings seas. Regional ocean plans for the Northeast and mid-Atlantic regions cover a range of ocean-focused topics but do not speak to storms or rising sea levels in detail. The Southeast region got off to a good start, including establishing a Governors South Atlantic Alliance and a technical team on disaster resilient communities. The technical team set an ambitious agenda including to "develop and implement adaptation and mitigation strategies for climate change impacts with plans for retreat of natural and human communities."[8] But, that alliance is no longer operating. However, the Gulf of Mexico Governors Alliance, which dates back to 2004, has developed three action plans, the most recent in 2016, and has established a Coastal Resilience Team. The team has developed a guide to sea level rise models, is working on improving local building codes, and is supporting community resilience projects.

Would a new administration reestablish a national ocean planning process and would this approach provide the national leadership needed to prepare for more severe storms and rising seas? An obvious advantage to putting coastal storms and sea level rise in the context of a wider array of ocean issues is that coordination among the related topics is more likely. The offsetting disadvantage is that sea level rise becomes one of perhaps a dozen major topics, and the available time and energy of a national council and supporting groups is spread across all these issues. The most recent workplan suggests a growing concern for sea level rise, but sea level issues continue to compete with many other ocean and coastal priorities.

A related question is whether the network of regional marine planning bodies, should they be reestablished, could take on the challenge of planning for sea level protection structures and land uses related to relocation of communities and infrastructure assets. Again, this approach would encourage coordination with related work but sea level rise adaptation tasks would be in addition to other coastal and ocean planning work. With the flexibility to develop plans on selective topics of interest, might some regions simply decline to speak to risks related to more severe storms and sea level rise? Integrating the work of new regional entities into existing state and local government processes is also a challenge.

Coastal Zone Management Program: Wide Responsibilities, Limited Funds

The Coastal Zone Management Act, passed by Congress in 1972 and administered by NOAA, is intended to "preserve, protect, develop, and where possible, to restore or enhance, the resources of the Nation's coastal zone for this and succeeding generations."[9] The principal program created by the act is the Coastal Zone Management Program, but it also authorizes smaller programs for estuarine research and conservation.

The "coastal zone" includes the Great Lakes, as well as saltwater coasts, and generally refers to "the coastal waters (including the lands therein and thereunder) and the adjacent shorelands (including the waters therein and thereunder), strongly influenced by each other."[10] Although states specifically define coastal zones, in practice, the zones often include the communities along the coast and rivers to the height of tidal influence. Importantly, the act specifically mentions sea level rise as a consideration in defining the coastal zone: "The zone extends inland from the shorelines only to the extent necessary to control shorelands, the uses of which have a direct and significant impact on the coastal waters, and to control those geographical areas which are likely to be affected by or vulnerable to sea level rise."[11]

Over the years, states worked with the National Oceanic and Atmospheric Administration (NOAA) to develop state programs for protecting the coastal zone in that state, and NOAA approved the programs for thirty-four of thirty-five eligible states and territories (Alaska does not participate, and six of the states are on the Great Lakes, rather than saltwater states). State programs "must include provisions to assure the appropriate protection of those significant resources and areas," but a key feature of the program is that states and territories have "great latitude in both how to participate in the federal program and what topics they emphasize."[12] Some states delegate much of the program to local governments while others manage the program at the state level.

States also can emphasize different topics (e.g., California has emphasized beach protection). Congress and NOAA have from time to time, however, pressed for state action in key areas. For example, in the 1990s Congress amended the Coastal Zone Management Act to add a requirement that state programs "must include a planning process for assessing the effects of, and studying and evaluating ways to control, or lessen the impact of, shoreline erosion, including potential impacts of sea level rise."[13]

The implementing regulations are adamant that action is needed, noting that programs "must include an identification and description of enforceable policies, legal authorities, funding techniques, and other techniques that will be used to manage the effects of erosion, including potential impacts of sea level rise,"[14] but then backs off with the proviso that such steps may be taken only "as the state's planning process indicates is necessary."[15]

States with approved programs are eligible for modest grants. In 2017, NOAA provided almost $70 million, allocated to implement coastal management programs in the thirty-four participating states and territories, and state and local governments matched this with over $57 million of their funds.[16] These funds were divided roughly equally among programs and projects intended to protect water quality, improve access, promote development, protect habitat, manage the program, and engage the public.

Among the areas of state focus in terms of grants is "Mitigating Coastal Hazards," including projects for "minimizing risk from coastal hazards such as storms, flooding, erosion, and sea level rise to make communities more resilient."[17] In 2017, states spent about $13.5 million in federal funds (about 16 percent of the national program grant) and $6.9 million in their own funds for this work.[18] Some of the specific projects supported with this modest funding are laying a foundation to prepare for rising seas. Reported projects include identifying the populations and infrastructure most vulnerable to coastal flooding; adopting best practices for siting buildings and infrastructure out of high-hazard areas; preserving important coastal habitat (e.g., wetlands, coral reefs, and dunes) to buffer communities from storms; and incorporating current and future hazard risk information into community plans and other decision-making processes.

Although these projects help communities begin to address coastal flood risks, this work falls short on several counts. States often look at coastal flood issues only in some specific communities, rather than an entire state coastline. Some of the work looking at coastal flooding may have little or no connection to the more challenging problem of sea level rise inundation. The work to date appears to be framed to address specific topics, such as vulnerable infrastructure, rather than looking at flooding issues more comprehensively. Finally, there is no set path toward a core set of policies states should adopt to prepare for more severe storms and rising seas, and no tracking of progress states are making along such a path.

NOAA has, however, taken small steps to focus state attention on "coastal hazards" generally if not sea level rise more specifically. The act authorizes special "enhancement grants" in nine priority areas that are funded with money set aside from the general grant to states. Among the statutory priority areas are "preventing or significantly reducing threats to life and destruction of property by eliminating development and redevelopment in high-hazard areas, managing development in other hazard areas, and anticipating and managing the effects of potential sea level rise and Great Lakes level rise."[19] In addition, NOAA has issued guidance for the period 2016–2020, making "coastal hazards" an area of national importance, noting that state programs are "strongly encouraged to develop one or more strategies to improve the effectiveness of their program in designated areas of national importance."[20]

NOAA is clearly making an effort to focus states on coastal resilience issues and prompt preparedness planning. Academic appraisals of progress in addressing coastal flooding under the Coastal Zone Management Program, however, are mixed. In a 2017 assessment of coastal managers' response to sea level rise, James London reviewed efforts to address coastal erosion and sea level rise in the program and commented that

> the coastal management community appeared to grasp the severity of the issue and was moving to expand its toolbox accordingly; still, there has been pushback from anti-regulatory, property rights interests, reinforced by the Lucas case [concerning "takings' under the Fifth Amendment] and other court rulings. As such, state policymakers have been reticent to move too quickly with new policy prescriptions. While state leaders have looked to local government, where land use authority resides, to take more of a lead, most local entities have been slow to act given that they are the immediate beneficiaries of coastal development and to date have paid a relatively small share of beach stabilization costs.[21]

London went on to describe the several changes to the Coastal Zone Management Act promoting attention to coastal flooding and erosion and observed that "while the Federal directive was clear, new state initiatives to address increased sea level rise were slow to materialize....Still, the general consensus was that further program development was dependent on clear directives from above. Those directives have been slow to develop due in part to an intergovernmental impasse. State legislators were awaiting pressure from the local level before initiating legislation. Meanwhile, local officials want a push from above but don't like unfunded mandates."[22]

In addition to supporting state coastal programs, NOAA has awarded grants on a competitive basis for both coastal habitat restoration and "community resilience." Grants totaled $35.8 million over the years 2015–2017.[23] A 2016 report by the Government Accountability Office (GAO) pointed to high demand for such grants, noting that the program received "132 qualified applications requesting a total of $105 million during its first application period in fiscal year 2015, when a total of $4.5 million was available for the grants."[24]

Although these grant funds have been used for diverse projects including disaster recovery, dam safety, and habitat restoration, some of this funding is addressing sea level rise. A grant to the San Diego Regional Climate Collaborative, for example, is intended to help "seven cities develop coordinated sea level rise strategies to reduce future economic losses."[25] A grant to the Cape Cod Commission is for "a public planning process to improve community understanding of climate change impacts, sea level rise scenarios, and various adaptation strategies."[26]

Starting in 2018, this coastal resilience grant program evolved to become a National Coastal Resilience Fund managed by the National Fish and Wildlife Foundation with support from NOAA, Shell Oil, and TransRe reinsurance company. The Fund is authorized in the National Oceans and Coastal Security Act developed by Senator Sheldon Whitehouse (see ch. 8) and supports projects "that restore or expand natural features such as coastal marshes and wetlands, dune and beach systems . . . that minimize the impacts of storms and other naturally occurring events on nearby communities." Grants in 2018 totaled $28.9 million and the foundation has requested proposals for $29 million in grants for 2019.

Finally, it is important to note that the federal government has supported acquisition of sensitive coastal land. Starting in 2002, NOAA provided matching grants to state and local governments under the Coastal and Estuarine Land Conservation Program (CELCP) for purchase of over 100,000 acres of ecologically important and threatened coastal and estuarine lands. States developed coastal land acquisition plans identifying critical lands for purchase. Unfortunately, program guidance does not require that purchases be made with sea level rise adaptation in mind. And, funding for CELCP grants for saltwater coastal land acquisition ended in 2012. A similar program, focused on acquiring and restoring coastal wetlands and funded out of the Sport Fish Restoration and Boating Trust Fund, fared better and provided $19 million to twenty-two projects in thirteen coastal states in 2018.

The biggest federal investment in land acquisition comes from the Land and Water Conservation Fund supported with royalties from offshore drilling leases. Royalties transferred to the fund come to nearly $900 million annually, but Congress has been appropriating only about $350 million, although the amount varies. Most of these funds are used by the federal government to purchase land based on Department of the Interior priorities. Some funds have been used to expand coastal parks and wetlands, but only rarely with the idea of adapting to coastal storms or rising seas. States have recently received roughly $40 million per year as matching grants to help meet their acquisition priorities, including park, forestry, wildlife, and recreation projects. State programs for coastal land acquisition are discussed in chapter 14.

Stepping back to look at the larger coastal management program picture, the long cooperation between NOAA and coastal states stands out as a major asset. Some of this cooperation can be traced back to NOAA's willingness to be flexible and give states a major role in setting priorities. On the other hand, the program is literally covering the waterfront of coastal issues with very limited funding. The topics of storm preparedness and building resilience to sea level rise are on the list of things to do, but limited funding and wide discretion at the state level to advance these issues, or not, casts some doubt on any decision to rely on this model to carry an expanded effort to adapt to more damaging storms and rising seas.

Estuary Protection: On the Front Line of Storms and Rising Seas

While the Coastal Zone Management Program engages states and delivers statewide programs, another part of the coastal protection picture is a network of programs focused on protecting individual estuaries around the coast. Originally intended to support coastal research or address conventional water pollution problems in defined coastal areas, these programs are increasingly addressing risks from rising seas and have the potential to draw on strong connections to local leaders to play a constructive role in adapting to more severe storms and rising seas.

One example of a network of estuary programs is the National Estuarine Research Reserve System, consisting of twenty-nine sites and 1.3 million acres of estuaries around the coast. NOAA invested $23 million in these estuaries in fiscal year 2017, and states and others provide some $6.6 million in further funding.[27] About half this funding goes for research and monitoring and the other half is divided among stewardship, management, and education activities. Despite the research focus, the reserves can help advance actions to adapt to storms and rising seas. The *Strategic Plan* for the reserves states somewhat tentatively that reserve managers "are poised to study the impacts on and vulnerabilities of estuaries and coastal communities, support adaptation strategies, and mitigate impacts through greenhouse gas reduction and carbon sequestration."[28] NOAA's Sentinel Site Cooperative Program builds from a base of a reserve or other protected area to bring together networks of people to address sea level rise impacts on coastal communities. Sentinel programs are in place for the Chesapeake Bay, North Carolina, the northern Gulf of Mexico, San Francisco Bay, and Hawaiian Islands.

The National Estuary Program, managed by the EPA, supports state and local government efforts to protect twenty-eight specific estuaries, including development of Comprehensive Conservation and Management Plans (CCMP). Work in each estuary is guided by a management council that "consists of diverse stakeholders and uses a collaborative, consensus-building approach to implement the CCMP."[29] Importantly, four of these local estuary programs are in the state of Florida.

The Environmental Protection Agency (EPA) developed a Climate Ready Estuaries Program that looks at a range of climate-related risks to estuaries, including more severe coastal storm and sea level rise impacts, and offers tools to help define response actions. Several of the estuary programs have identified sea level rise as a priority. The Sarasota Bay Estuary Program, for example, released a video focused on sea level rise and worked with Mote Marine Laboratory to publish a sea level rise planning guide for local governments.

One unfortunate thing these geographically focused coastal programs have in common is the antipathy of the Trump administration that has consistently proposed to eliminate all, or virtually all, funding. Fortunately, congressional support for these programs is strong. Congress has overridden proposed cuts in past years, and it seems likely to continue to do so.

Even a well-informed reader might reasonably suffer some shell shock after bravely marching through the myriad national programs related to flood insurance, disaster relief, coastal zone management, ocean policy, and estuary protection. It is a lot to take in, and gaining any confidence in judgements about the cumulative effect of all this work is difficult. In addition, these programs were conceived before climate change emerged as a major concern and before the risks of rising seas and more severe storms were well understood. Still, all these programs have adapted to the new realities that climate change brings to the coast to varying degrees and might be further adapted to more forcefully respond to these risks. But, before thinking about how to best organize the response to these challenges going forward, it is important to look at one last set of programs and policies—those developed in recent years to specifically address the risks of a changing climate, including risks along the coasts.

12

National Planning for Climate Change: An Answer to Coastal Inundation?

Coastal planning and hazard mitigation planning have long histories. Planning with the specific idea of preparing for a changing climate is a much newer endeavor.

Over the past ten years, the federal government, as well as states and local governments, dramatically scaled up climate change adaptation planning and response projects. This work relied heavily on the continuing assessment of climate change science and impacts provided by the Global Change Research Program and the national climate assessment process. Federal agencies developed climate change adaptation plans and supporting initiatives as part of a coordinated national effort called for in an executive order on climate change adaptation. Coastal storms and rising seas were part of this effort, although often just two among many topics of interest. Unfortunately, much of this work came to an abrupt halt at the federal level in early 2017 with the inauguration of President Trump.

Abandoning almost a decade of steady progress toward preparing for the impacts of a changing climate was tragically shortsighted. When the day comes, as it surely must, that federal agencies reinvest in climate adaptation planning, agency leaders will need to decide how to manage efforts to prepare for more severe storms and rising seas. One option is to deal with coastal storms and rising seas as part of a comprehensive climate change adaptation initiative. As discussed in prior chapters, other options are to treat this as a hazard mitigation, disaster reduction problem, or to encourage states to address storms and rising seas more consistently and completely with coastal zone management programs. This chapter reviews past climate change adaptation initiatives with an eye toward understanding how preparing for more severe storms and rising seas might fit into a revived national effort.

Emergence of a Federal Climate Change Adaptation Program

The National Climate Assessment, developed every four years under the statutory authority of the Global Change Research Act, is intended to inform the nation about the current status of the climate, observed climate changes, and anticipated trends for the future, including presenting the scientific evidence of risks posed to the coast by more severe storms and rising seas. It forms a foundation for climate change adaptation planning and programs.

The Global Change Research Program (GCRP) published the first *National Climate Assessment* in 2000, a second assessment in 2009, and another in 2014. In late 2017, GCRP published a *Climate Science Special Report* as a first element of the fourth *National Climate Assessment* released by the Trump administration in November 2018. The fourth assessment provides the most recent information on relative sea level rise at many places along the United States coast (i.e., from the 2017 National Oceanic and Atmospheric Agency [NOAA] report) and estimates of costs of property lost to sea level rise (see chs. 2 and 4).

The comprehensive process of developing the national assessment reports, involving experts in different fields to draft chapters on diverse topics, works well. Still, it is important to remember that, for all its strengths, the national assessment process focuses on the physical science of climate change and does not address in detail emerging questions related to demographic changes in coastal areas, the diverse economic consequences of more severe storms and rising seas, or the social and psychological impacts of adapting to rising seas. And, the periodic national assessments are not intended to resolve major policy questions or frame a national strategy to adapt to climate change generally or prepare for more damaging coastal storm flooding or sea level rise inundation more specifically.

The scientific assessment provided in the national climate assessment reports is required by law, but no law directs the federal government to take the next step and use this assessment information for national climate adaptation planning broadly or preparing for more severe storms or rising seas. This gap, however, has at times been partly filled by a mix of executive orders and federal agency initiatives.

At the federal level, the Environmental Protection Agency (EPA) released one of the first climate adaptation strategies in early 2008, focused on the National Water

Program. The Department of the Interior (DOI) released findings of its Climate Change Task Force in late 2008 and then required its bureaus to "consider and analyze potential climate change impacts when undertaking long-range planning."[1] The United States Fish and Wildlife Service within the DOI released a *Climate Change Strategic Plan* in September 2009.

The Pew Center for Global Climate Change, now the Center for Climate and Energy Solutions, published *Adapting to Climate Change: A Call for Federal Leadership* in early 2010, arguing that, "while many adaptations will occur at the state and local levels, the federal government is a critical player in an effective and coordinated approach to climate change adaptation in the United States."[2]

President Obama established an interagency Climate Change Adaptation Task Force in 2009 and, in May of 2010, the White House hosted a National Climate Change Adaptation Summit that convened stakeholders to identify challenges and opportunities for collaborative action. The *President's Climate Action Plan,* issued in June 2013, set out an ambitious agenda for cutting release of greenhouse gases and engaging other countries in addressing climate change. In an unusual step, adaptation issues and actions were addressed to the same degree as issues related to greenhouse gas mitigation and international cooperation.

This work culminated in 2013 in Executive Order 13653, which laid a foundation for a federal government effort to plan for the impacts of a changing climate, stating that "the Federal Government must build on recent progress and pursue new strategies to improve the Nation's preparedness and resilience."[3] It created a new interagency Council on Climate Preparedness and Resilience, upgrading an earlier interagency task force, directed federal agencies to develop climate change-adaptation plans (see chs. 5 and 6), and created a task force of state, tribal, and local officials to make recommendations on climate adaptation.

Without a national law demanding action, why did the federal government begin giving serious attention to climate adaptation starting about ten years ago? Until that time, climate change was all about mitigating greenhouse gases. Understanding of impacts was considered useful, but mainly to support the need to forge ahead with steps to limit warming, such as the legislation then pending in Congress to limit emissions using a "cap and trade" model. The worse the climate impacts, the greater the need to deal with the problem. Paying too much attention to adapting to changes was viewed with some skepticism, or even suspicion, as a fatal weakening of resolve to adopt the needed controls on greenhouse gas emissions and a surrender to the idea that, whatever the impacts, society could just adapt.

The emergence of focused federal interest in climate change adaptation can be traced to several factors. A new administration taking office in early 2009 and eager to tackle climate change issues clearly played a major role. The release of the second National Climate Assessment in 2009 had the effect of teeing up the climate change impacts and making the case for actions to adapt to the forecasted changes. States and local governments also began adaptation planning starting in 2007 and 2008. The 2010 release of the *Call for Federal Leadership* by a respected nonprofit organization, the Pew Center for Global Climate Change, put specific, suggested actions on the table and had the effect of inoculating the climate adaptation effort from charges of undermining the greenhouse gas mitigation effort.

Although President Trump revoked the climate change adaptation executive order in March of 2017, it is worth reviewing some of work developed under the order, including the report of a state, local, and tribal leaders task force and federal agency climate adaptation plans. This work was a high water mark in the county's effort to prepare for a changing climate. A grasp of the direction of these efforts, and their attention or inattention to coastal storm and sea level rise challenges, informs thinking about strategies going forward.

State, Local, and Tribal Leaders Task Force on Climate Preparedness and Resilience

The climate adaptation executive order (13653) established a State, Local, and Tribal Leaders Task Force on Climate Preparedness and Resilience, which included governors, mayors, and tribal officials and produced a comprehensive set of recommendations on a range of climate adaptation challenges. The final report of the task force, published in November 2014, is as close to a national plan for adapting to climate change as the country has produced to date.

Although the task force offered many constructive recommendations, sea level rise received limited attention. The report aptly recognized that, "a significant portion of the Nation's population, economic activity, and infrastructure is located near the coast, in floodplains, or in other areas vulnerable to sea level rise, more intense storms, tides, and coastal erosion."[4] But the specific recommendations were for modest steps. For example, the task force encourages the Army Corps of Engineers to conduct coastal climate vulnerability assessments, disseminate this information to communities, and expand regional sediment management programs to address coastal erosion threats in a comprehensive and cost-effective manner. Federal agencies more generally were to provide technical assistance to coastal and island communities as they develop response plans and strategies for sea level rise, increased storm surge, and other climate change related risks.

The task force delivered more pointed recommendations in the specific area of disaster preparedness. Section 5 of the task force report called for modifying "disaster recovery programs to encourage and prioritize projects that are sized and designed to withstand future climate impacts and that are located outside areas vulnerable under current or foreseeable conditions."[5] Although this recommendation is not targeted just at storm surges and rising seas, it speaks to those challenges. The same section of the report identified a need for "mapping based on current and predicted climate impacts to help improve local capacity for effective hazard mitigation planning."[6]

With respect to the National Flood Insurance Program, the task force recommended that "the minimum standards local governments must adopt to participate in the program should be strengthened to prevent the continued degradation of critical floodplains, wetlands, coastal marshes and dune areas that naturally buffer the impacts of storms and rising sea levels."[7]

Federal Agency Climate Change Adaptation Plans

The 2013 climate adaptation executive order also called on federal agencies to update "comprehensive plans that integrate consideration of climate change into agency operations and overall mission objectives."[8] The order did not specifically mention storm surges and rising seas, but the reference to considering "updating agency policies for leasing, building upgrades, relocation of existing facilities and equipment, and construction of new facilities"[9] suggests that coastal inundation risks were of interest.

The Congressional Research Service (CRS) noted in early 2015 that, by the end of 2014, thirty-eight departments and agencies had produced climate change adaptation plans but concluded that "most agencies are in formative stages of their assessments and strategic planning. Some agencies are embarking on more detailed analyses and limited implementation actions. Overall, few examples are apparent of day-to-day agency decisions or actions that are different as a result of their adaptation efforts."[10]

Shifting to a review of progress in implementing agency plans, CRS reported finding "few specific adaptation actions, either planned or taken, that tangibly alter federal vulnerabilities at this point in time."[11] CRS identified half a dozen examples of effective adaptation actions, including the work by the Army Corps of Engineers to evaluate sea level rise impacts, but "was not able to identify widespread changes in federal decision-making, management, or operations associated with adaptation to projected climate change."[12]

In June 2015, the Army Corps of Engineers released a detailed assessment of agency adaptation plans and identified highlights in each plan but reported very few highlights related to storms surge or sea level rise.[13] A detailed crosswalk of agency adaptation activities showed fewer than a dozen related to sea level rise, mostly in the plans by the Corps, NOAA, and the Environmental Protection Agency (EPA). This limited attention to coastal storms and rising seas seems to be the result of several factors. The plans tend to cast a wide net, looking at a full range of possible climate change impacts and giving equal air time to each. Because the direction to develop plans did not come with dedicated money to implement them, agencies may have been reluctant to commit to actions they would then need to fund with their existing resources. Coastal storm and sea level rise challenges often need cooperative attention from multiple agencies, and such cooperation did not fit well in these agency-specific plans. Finally, these challenges are daunting and some agencies may have simply decided to leave them for another day. Although these plans are a bit of a disappointment in the context of preparing for coastal storms and rising seas, they offer some lessons for how to do better in the future (see ch. 18).

Climate Change Adaptation and the National Environmental Policy Act

Another line of attack by the Obama administration to shift the country toward more effective adaptation to climate change was to better account for these risks in development of environmental impact statements under the National Environmental Policy Act. The White House Council on Environmental Quality (CEQ) issued guidance in 2016 to help federal agencies both address the impacts of major federal actions on climate change (e.g., increased emissions of greenhouse gases) and the impacts of climate change (e.g., storms and rising seas) on such projects.

The CEQ developed the guidance in response to the widely differing consideration that federal agencies were giving to climate change in assessing environmental impacts. In 2012, Patrick Woolsey at Columbia Law School published a paper analyzing this topic, finding that "a comparison of agency approaches to EIS [environmental impacts statement] scope and methodology shows widely varying treatment of climate change impacts. . . . While some EISs emphasize scientific uncertainty about the scope and nature of future climate impacts, others provide projections of potential long-term impacts at the national, regional, and project level."[14] In responding to this situation, the guidance highlighted the need to consider future, as well as existing, environmental conditions and encouraged both more resilient designs and consideration of alternatives to an action that have "improved resilience to climate impacts."[15]

Although this guidance document did not amend underlying law or regulations, it was controversial and the 2016 final guidance was preceded by drafts for public comment in 2010 and 2014. Despite this extensive public engagement process, the Trump administration withdrew the guidance in April 2017, shortly after taking office, citing a need for further consideration.

Climate Resilience Toolkit

Although federal agency climate change-adaptation plans got mixed reviews and consideration of climate change impacts on projects in environmental impact statements varied, federal agencies were remarkably adept at developing an array of tools to support climate adaptation and documenting adaptation success stories, including tools and stories related to more severe storms and rising seas.

A common complaint of people trying to make sense out of many different federal agency websites on different aspects of climate change was the lack of a single index to all the tools and information available. In response, federal agencies created a "Climate Resilience Toolkit." The toolkit provides a structured framework for case studies, tools, and reports that anyone working on climate change resilience issues can use to quickly find resources by topic, region, and type of information.

The Toolkit is rich in information on many topics including coastal resilience and sea level rise. There are forty-nine sea level rise resilience case studies from around the country. There are over 140 tools related to coastal resilience, over 30 addressing sea level rise directly. There are over 100 reports presented addressing coastal impacts of climate change and a "Coastal Inundation Dashboard" offering diverse mapping data. As valuable as the Toolkit is, it is so even-handed in its presentation of tools and other resources that it can be difficult to identify the most current and useful products and it does not sort for sea level rise specifically. And, although it includes a brief introduction describing generic steps in resilience planning, it is not intended to be a national preparedness plan for more severe storms and sea level rise (i.e., a pile of bricks is not a wall).

Looking back on the entire federal interagency climate adaptation effort in its 2017 High-risk Series report, the Government Accountability Office (GAO) acknowledged progress but declined to remove this topic from the list of thirty-four high-risk areas across the federal government. GAO concluded that "it is too early to determine the new mechanism's effectiveness at demonstrating progress in implementing corrective measures.[16] GAO called for "leadership commitment to be enhanced,"[17] and a new "focus on implementing this strategy—by developing measurable goals; identifying the roles, responsibilities, and working relation-ships among federal, state, and local entities; identifying how such efforts will be funded and staffed over time; and establishing mechanisms to track and monitor progress."[18] Well said, but not likely in the near future. The 2019 version of the High-risk Series again retained climate risk management as a high-risk area noting that agencies had "regressed" in key areas including assisting state and local gov-ernments and providing national leadership.[19]

The Army Corps of Engineers Engages Coastal Storms and Rising Seas

The Army Corps of Engineers within the Department of Defense deserves special mention for its engagement of climate change and coastal issues. The Corps was an early advocate for addressing the impact of a changing climate on water resources generally and was the first federal agency to adopt policy on recognizing sea level rise in coastal engineering projects. The Corps also works with EPA to implement the Clean Water Act wetlands protection program, including protection of coastal wetlands, and played major roles in designing long-term responses to Hurricanes Katrina and Sandy.

Corps of Engineers engagement on sea level rise can be traced back to the release of a 1987 report by the National Research Council that recommended "feasibility studies of coastal projects (e.g., coastal protection projects of the Corps, and storm surge studies of the Federal Emergency Management Agency) should consider the high probability of accelerated sea level rise."[20]

The Corps first issued formal guidance on incorporating sea level rise into proj-ect feasibility studies in 1989 and has revised and updated it several times. Today, Corps planning for sea level rise is guided by a technical letter issued in 2014. A key concept in this letter calls for recognizing long planning horizons, including up to 100 years: "Infrastructure often stays in place well beyond the economic period of analysis. . . . Many of the SLC [sea level change] projection scenarios include an increased rate of sea level rise further into the future. . . . Using a longer adaptation horizon enables us to improve robustness and resilience compared to planning for shorter time frames."[21]

The technical letter also calls for a shift away from a "most probable future condi-tion" analysis to a new approach, recognizing a wider range of possible conditions: "An example could be the assessment of several potential SLC values in conjunc-tion with different infrastructure development rates in the project area. This is not the traditional singular 'most probable future condition' approach; comparison and selection of alternatives in a multi-scenario setting is . . . a new challenge for planning USACE Civil Works (CW) projects."[22]

To support the complex process of reviewing sea level rise projections and conditions at specific locations, the Corps developed a "Sea Level Change Curve Calculator," shared online to allow users to enter a coastal location and time period and then select from several scientific projections of sea level rise. The calculator generates a graph and supporting data showing estimated changes in sea level at a given place over the time period selected for a range of scenarios or assumptions. The tool includes the latest NOAA 2017 relative sea level rise projections, including local adjustments for land subsidence, allowing planners to gain a site-specific estimate of future relative sea level rise that would be much more difficult to develop without the calculator.

One small element of the $50 billion spending bill Congress enacted to rebuild after Hurricane Sandy hit the northeast coast in October 2012 provided $20 million for the Corps to conduct a comprehensive study of measures needed to avoid such a future catastrophe. The final report, issued in early 2015 after extensive engagement with states, local governments, and the public, includes a preface with an insightful warning of risks resulting from stronger storms and sea level rise.

> Hurricane Sandy made us acutely aware of our vulnerability to coastal storms and the potential for future, more devastating events due to changing sea levels and climate change. Changing sea levels represent an inexorable process causing numerous, significant water resource problems. . . . Addressing these problems requires a paradigm shift in how we work, live, travel, and play in a sustainable manner as the extent of the area at very high risk of coastal storm damage expands.[23]

What is the paradigm shift the Corps proposed? There is a call for innovation, for "expanding from traditional structural risk reduction measures to include more emphasis on nonstructural, natural, and nature-based systems,"[24] and, most significantly, a clear recognition that relocation of coastal communities will become unavoidable.

> Given current and projected sea level and climate change trends, some of our built environment will become unsustainable for the human systems presently located there. Coastal communities face tough choices as they adapt local land use patterns while striving to preserve community values and economic

vitality. In some cases, this may mean that, just as ecosystems migrate and change functions, human systems may have to relocate in a responsible manner to sustain their economic viability and social resilience. Absent improvements to our current planning and development patterns that account for future conditions, the next devastating storm event will result in similar or worse impacts.[25]

Finally, it is important to mention the work that the Corps does with EPA to manage the national program for issuing permits for development in wetlands under section 404 of the Clean Water Act. The wetlands program does not play a direct role in planning to avoid coastal storm or sea level related flooding, or to recover from such flooding. Program managers do, however, make decisions about whether to allow development in coastal wetlands that, if left undeveloped, serve as important buffers for reducing the impacts of coastal storms and storm surges.

Unfortunately, the 404 permit program does not presently have a policy or guidance with respect to coastal wetlands specifically or the long-term impacts of coastal storms or sea level rise on coastal wetlands. However, following release of a report by NOAA and the Fish and Wildlife Service in 2008, showing dramatic losses of wetlands in coastal watersheds, mostly along the Gulf and Atlantic coasts, EPA and other federal agencies organized an interagency Coastal Wetlands Working Group and conducted a series of program reviews around the coast looking at reasons for coastal wetlands losses and possible program improvements.

Some of the reasons for the losses that the working group identified included development projects, sea level rise, and land subsidence. Although the working group did not identify new requirements with respect to permits for development in coastal wetlands, the reviews summarized some of the steps that state coastal managers recommended, including better data and mapping, more enforcement, more funding, and increased use of watershed planning.[26]

Federal Flood Risk Management Standard

There is one additional Obama administration initiative revoked by President Trump worth mentioning. FEMA developed a new Federal Flood Risk Management Standard (FFRMS) in 2015, drawing on ideas developed by the Hurricane Sandy Rebuilding Task Force. It directed federal agencies making investments in infrastructure to take steps to avoid areas at risk of flooding and, if building in an area at risk of flooding or sea level rise, to elevate structures above the base flood elevation. Sadly, but predictably,

President Trump revoked the FFRMS as part of an executive order intended to speed development of infrastructure projects in August of 2017.

Although the FFRMS is no longer in effect, for a brief time it provided an important complement to the National Flood Insurance Program and disaster assistance programs, and it could be a pattern for future efforts to address the magnification of flood risks occurring as a result of climate change. It called on federal agencies investing in new infrastructure, or substantially improving existing infrastructure, to determine whether the project was located in a flood zone or in the wider area at risk of flooding as a result of climate change. Federal agencies had the option of using a "climate informed science approach" relying on climate data and models to identify land areas at risk. Agencies also had the option to default to an area determined by increasing the base flood elevation on floodplain maps by two feet in the case of most projects, or three feet in the case of projects for critical facilities such as hospitals or water plants.

If a planned project was located in this wider flood risk area, the agency was to consider an alternative, safer location outside that area. If relocation to a safer location was not possible, agencies were to protect the structure from damage caused by a flood two feet, or in the case of critical facilities, three feet higher than the base flood elevation. Because most building codes simply call for flood protection to the base flood elevation, this increased elevation was additional protection as well as additional cost.

A key limitation of the FFRMS was that it applied only to investments made directly by a federal agency or indirectly through grants or loans. Some observers hoped, however, that the federal standard would gain wider acceptance and be adopted into state and local codes and other requirements. National organizations interested in infrastructure, such as the American Society of Civil Engineers, supported the FFRMS. A Florida Republican who has called for addressing the threat posed by climate change, former Rep. Carlos Curbelo, criticized President Trump's decision to revoke the order: "It's irresponsible and it will lead to taxpayer dollars being wasted on projects that may not be built to endure the flooding we are already seeing and know is only going to get worse."[27]

To date, the story of national planning to prepare for a changing climate boils down to three steps forward and two steps back. The Obama administration made remarkable progress in a few short years marshalling executive authority and agency capacity. But, the lack of a statutory foundation and buy-in from Congress left this work vulnerable to a new administration willing to wash it away without a second thought. This first major attempt at preparing the country for climate change also highlighted other challenges, such as the difficulty of addressing a specific problem like coastal storms and rising seas within a larger process attempting to cover all the impacts of a changing climate. Fortunately, there will be more innings to play as the country confronts a warming climate and a transition to a new coast.

Part III: Epilogue

A Nation Unprepared for Coastal Storms and Rising Seas

Any objective observer reviewing the diverse array of federal programs and policies related to flooding, disaster assistance, coastal management, and climate adaptation has to be impressed by the good intentions, the sheer diversity of approaches, and the determined efforts to fit together the pieces of a very complex puzzle.

Although each individual program has strengths, each struggles with flaws or shortcomings. The coastal element of the flood insurance program is financially unsustainable and counterproductive. The disaster recovery programs are good at spending money and rebuilding but not so good at planning to prevent the most serious, emerging risks in the first place. The Coastal Zone Management Program has its arms around everything coastal and was sorely underfunded even before the Trump administration began proposing deep reductions. The Obama climate adaptation initiatives made some real headway but never focused directly on coastal storms and sea level rise and have been largely shut down by the Trump administration for now.

Could these programs, individually or collectively, be hammered into a national program with the capacity and resources to prepare the country for more severe storms and rising seas? Some or all of these programs will need to be updated in response to these challenges. And of course, the topics are related and need to be coordinated. But, they all have broader missions that would suffer if preparing for storms and rising seas became the primary mission. And, the significant new programs, authorities, and investments needed to prepare for these risks (see part 5) would be hard to fit into any of the existing frameworks. So, the third argument for a national program to prepare for more severe storms and rising seas is that the existing, related programs are not a good fit for the job.

PART IV

States, Communities, and Businesses Cope with Storms and Rising Seas

It is this spirit of cooperation, the ability to share, trust, and learn from each other, which has led to accelerated action throughout our region—a region so large it accounts for roughly one third of Florida's population. And while all of this gives us great reason to celebrate success, the truth is, we could not have done it without the expertise of our Federal partners.

Kristin Jacobs, county commissioner, Broward County, Florida, commenting on the Regional Climate Action Plan developed by the Southeast Florida Regional Climate Compact

13

Novel Challenges of Storms and Rising Seas

New science and research will continue to focus understanding of the extent to which communities, infrastructure assets, and ecosystems are at risk of more severe storms and rising seas. The federal government will pay flood insurance claims, help with recovery from major coastal storms, and push ahead with coastal management and climate change adaptation programs, to varying degrees. But, a key ingredient for success in preparing for more severe storms and rising seas is having coastal states, local governments, and businesses fully engaged in the effort.

Some state and local governments are already actively engaged in understanding and responding to the challenge of more severe storms and rising seas but many have not yet tackled the topic. In addition, businesses along the coast need to be a major part of responding to storms and rising seas and are generally not yet focused on the risks or engaged in finding solutions. A key step in making better progress toward a coordinated national response to these challenges is to consider why governments and businesses have not made more progress so far.

One key reason for the limited state and local government response is that preparing for rising seas poses novel challenges. Never before have environmental conditions arisen comparable to those causing the gradual and permanent inundation of miles of coastal land and communities that are home to millions of people. There is no textbook outlining options and answers. Although there are some state and local success stories, there is no widely accepted template for states, local governments, or businesses to follow.

Before outlining the choices governments and the private sector will need to make as they cope with coastal storms and rising seas, it is important to better define the novel challenges that must be resolved. One such challenge is that the coastal inundation coming from rising seas is permanent, unlike temporary flooding from storm surges. Another challenge is that the shifting of the coastline inland poses complex issues for property ownership. And, inundated lands can't simply be abandoned; someone needs to be responsible for making them safe. Coastal inundation will reduce property values with implications for owners and for property taxes paid to local governments. And, planning for rising seas is complicated by uncertainty over the appropriate planning horizon and the ultimate inland reach of inundation.

Past Coastal Flooding Experience: Temporary and Random

In looking for ideas for possible tools and policies to prepare for rising seas, it seems like past experience with coastal flooding related to high tides and storm surges would be a good place to start. There is wide experience in preparing for and responding to coastal flooding, but all this experience is shaped by the critical knowledge that the flooding is both temporary and, at any given place along the coast, unpredictable.

Although there is a general understanding that severe storms will occur somewhere and sometime along the United States coast, the uncertainty associated with the specific place and time means that the chance of severe flooding at any given place is often substantially discounted (i.e., 'it won't happen to me . . .''). High tides can be predicted exactly, and there are data about the past frequency of occurrence of severe storms providing some basis to judge future risks on a county or a state basis. Even applying this information, the general perception of coastal flooding events is that they occur randomly at any given place and thus are short-term emergencies, properly treated as disasters to which affected communities and individuals randomly fall victim.

In addition, as damaging as coastal flooding has been, it is understood as a trial to be endured only briefly. Flooding from high tides will recede in a matter of hours. Storm surge flooding commonly peaks for several hours and may take up to several days to recede, allowing the life of a community to shift to recovery and restoration for a time and then return to normal.

The idea that coastal flooding is a temporary disaster falling on any given place randomly has long supported a set of generous government programs that have focused on rebuilding damaged property, communities, and infrastructure. Because flooding is random, an insurance model in which everyone pays a little to provide funds to help the occasional, unlucky flood victim makes sense. Since the water is expected to recede and might not come back to that place for many years, it makes sense to support use of insurance payments and recovery funding for rebuilding of the damaged or destroyed structure, perhaps even paying for some measure of additional protection, such as elevating a structure or hardening construction. In fact, under these circumstances, prompt and complete restoration often becomes a rallying cry for a community and an opportunity to demonstrate resilience in the face of adversity.

Unfortunately, sea level rise changes this equation. Sea level rise applies everywhere (i.e., is not random) and comes to stay (i.e., is not temporary). This creates two problems.

The first problem is that the random and temporary coastal flooding mindset tends to undervalue planning and preparation. With uncertainty of whether a flood will even happen, confidence that impacts will be temporary, and a generous set of national programs to pay for rebuilding, a bias against spending political or financial capital on preparing for coastal flooding is understandable. But, the advent of rising seas, amplified by more severe storms, puts a premium on effective planning and adaptation. State and local governments face the novel problem of educating the public about the changing nature of coastal flooding and gaining support for much more extensive and costly preparations for it.

The second problem is that the random and temporary coastal flooding mindset tends to overvalue rebuilding in the same place and putting homes and communities back to the way they were. With rising seas and more severe storms, rebuilding in place increasingly puts property and lives at risk and is increasingly unsustainable financially. The public has long-held expectations that much-loved homes and communities will be rebuilt as they were. State and local governments face the novel problem of shifting public expectations away from rebuilding in place and toward other solutions. Some communities may decide that paying for expensive coastal structural protection projects or elevation of buildings is justified, at least for a time. Other communities may have more limited options and need to consider relocation.

Shifting public perceptions toward more effective planning to prepare for coastal storms and rising seas and away from expectations of rebuilding in place needs to happen soon. The best case is that communities recognize the need to shift to a new strategy and to do so before sunk costs become too high and the psychological need to double down on past investments becomes insurmountable. Failing to make a timely shift to a new approach will lead to a jarring transition requiring a painful write-off of major past investments and abandonment of physical places on which much toil and treasure has been expended.

Legal Consequences of Adjusting Shorelines and Regulating Coastal Land

Recovery from flood damage to coastal property due to coastal storm surges can be long, demoralizing, and expensive, but once the water recedes damaged property can be either repaired or razed and rebuilt. Better still, the land remains dry ground recorded in a deed and plotted on the local tax map. A novel aspect of planning for sea level rise is that it must account for not just damage to homes and other assets as a result of storm surges that go farther inland as seas rise, but also for the gradual inundation of land and the landward shift of the coastline marked by the high tide line.

Under the public trust doctrine, most states own lands beneath navigable waters up to the high tide line and other states own lands starting at different points. As discussed in chapter 3, the National Oceanic and Atmospheric Administration (NOAA) manages a process to redefine the shoreline for roughly a twenty-year epoch, and new shorelines will move inland over the coming decades putting now dry, private land seaward of the shoreline, thereby shifting ownership to the state.

In addition to the obvious consternation property owners will feel at the loss of property, this raises some vexing legal issues. Would property owners, faced with the loss of land as a result of this government mapping process, seek compensation under the Fifth Amendment to the Constitution that prohibits "taking" of private land by the government without just compensation? Might the federal government be forced to compensate private property owners for adjustments in the coastline to reflect rising seas?

Georgetown University professor Peter Byrne argues that compensation of private landowners for the natural inland movement of the shoreline is unlikely: "The public trust inheres in title to land....Moreover, it will move landward with the tideline. Thus, as the seas rise and the public trust areas move upland, the use rights of owners will either be extinguished or subjected to public property interests that will permit strict regulation. . . . Note that when the public trust applies the private owner . . . has no takings claim at all because the public enjoys a superior property interest."[1]

Writing in the *Duke Environmental Law and Policy Forum*, Michael Haitt made a somewhat different argument but reached a similar conclusion: "The takings clause would seemingly require the states to compensate the private property owners for a possessory taking. Yet finding a taking here is problematic because of the magnitude of the potential sea level rise—thousands of square miles of land and several major cities are at risk of being submerged."[2] Hiatt goes on to argue that sea level rise is a novel case: "This unprecedented threat posed by climate change calls for a narrow exception to the current per se possessory takings rule that it always constitutes a taking when the government physically occupies or appropriates private property. The taking that results from climate change and large-scale sea level rise is distinguished from the typical possessory taking because it is a passive government reaction to an uncontrollable force of nature that is novel and unprecedented."[3]

In addition to legal uncertainty there can be a political issue. For example, in 2015, the town of Nags Head, North Carolina, wanted to remove structures below the high tide line for safety reasons, thinking the lands were public property. But the owners sued arguing that the only the state could move the high tide line and had not done so. The state refused to help, and the town had to buy the property for $1.5 million. Cliff Ogburn, Nags Head town manager, commented, "I don't think there's interest at the state level to remove private property."[4]

A more difficult question raised by the Fifth Amendment is the extent to which it restricts regulation of coastal land by local or state governments that are working to manage storm surge or rising seas (e.g., limiting development on land expected to be inundated). Some of the factors to be considered are whether a regulation prevents all uses of the property, whether the regulation undermines the reasonable financial expectations of the property owners, and the timing of any regulatory impact. In addition, a "passive" taking can arise when a government discontinues service to a risky area.

As Devon Applegate observed in the *Boston College Environmental Affairs Law Review,* as need for coastal land regulations increases, courts "will be forced to handle and analyze these novel, numerous, and essential regulations,"[5] and "as it stands today, Takings Clause jurisprudence lacks both uniformity and clarity. The chaotic state of takings jurisprudence will become even more chaotic as climate change-related regulations emerge, unless the United States Supreme Court provides guidance and a clear legal framework."[6]

State and local officials dealing with coastal management issues will need to learn to be very good at devising response actions that are effective and well coordinated with contemporary court interpretations of the takings clause.

Making Inundated Lands Safe

Buildings and other infrastructure on lands that shift from private or local government ownership to state ownership will need to be demolished and removed prior to inundation by rising seas. Structures need to be removed to prevent them from becoming hazards to navigation or risks to the public who will have access to land below the high tide line when the tide is low. In addition, other infrastructure, including seawalls, docks and piers, roads and paved areas, sewer and water lines, power poles and transformers, and underground storage tanks need to be removed or otherwise decommissioned to address safety and environmental hazard issues. Although the practice of decommissioning land is not novel, managing decommissioning of lands all along the coast, as sea level rises, is a novel challenge.

A further complication of the decommissioning process is the sequence of decommissioning and a formal decision concerning a landward shift in the official coastline. In the normal course of events, NOAA would establish a new coastline based on sea level rise over a period of roughly twenty years, and a state would assume ownership of the inundated land. At that point, however, property seaward of the new coastline would be under water, making any decommissioning action costly and perhaps impractical. A more practical sequence would be for the decommissioning to occur prior to the actual inundation. This would require notice of a likely decision to move the coastline inland years ahead of the actual inundation to allow time for the decommissioning actions to take place.

For example, NOAA might identify both areas officially seaward of a new coastline and those lands that are likely to be shifted to seaward of the new coastline by the time of the next twenty-year National Tidal Data Epoch. With a two-phase process like this in place, state or local planners would be able to identify the decommissioning actions that need to occur and have the chance to take those actions while the land is still accessible and mostly dry.

Some further considerations in this effort are how a state or local government would get access to private lands not yet inundated and who would be responsible for implementing decommissioning actions. In the case of public land, it seems clear that a federal agency, state, or local government owning the land should be responsible for implementing timely decommissioning actions and paying for these actions. Some combination of new federal and state laws will be needed to make these responsibilities binding and create a capacity for decommissioning as a last resort. In the case of private property, binding requirements will need to be established to identify the decommissioning responsibilities of the landowner and the consequences in the event that a landowner is unwilling or unable to implement those actions, including provisions providing for decommissioning as a last resort.

Financial Impacts of Shifting Coastlines on Private Property Owners

As a regular process for the gradual shifting of the coastline inland emerges, the value of newly inundated lands and property will be lost to owners of these lands. As noted in chapter 4, the value of lands and property expected to be inundated by 2100 is estimated to be between $800 billion and three trillion dollars, depending on adaptation actions, and could be much greater as a result of population increases and resulting development. Any process of shifting the shoreline on a rolling twenty-year basis would only involve property that is a small part of all the lands with this roughly trillion-dollar value, but the consequences for individual owners of residential, commercial, and industrial property will still be significant.

Again, there is an important timing question. Today, coastal property values are high. But, growing public awareness of more severe storms and rising seas, and of the potential loss of land or damage to property, may cause the value of these lands to decline. As discussed in chapter 17, prices of property at risk of rising seas and storm surge may come to a tipping point, where prices fall dramatically, putting mortgaged property under water both figuratively and literally.

In 2017, Bloomberg News reported on sea level rise impacts on real estate sales in Florida and found the market still strong, but noted that some homeowners "are deciding to sell now—not necessarily because they want to move, but because they're worried their neighbors will sell first."[7] The article quoted Ross Hancock, who recently sold his home in Coral Gables, describing the Florida real estate market as "pessimists selling to optimists," and said he wanted to cash out while the latter still outnumbered the former. "I was just worried about my life's savings," Hancock said. "You can't fight Mother Nature."[8]

Another consideration is that, at some point, a local government, or water or power utility, will be forced by rising seas to announce plans to discontinue service to groups of properties in a neighborhood. This will be a difficult decision but the loss of property taxes and utility rate payments involved might be seen as preferable to paying the high costs of service to lands at risk of imminent inundation. Nearby property owners may decide to sell, expanding supplies of property and further driving down market prices. This decline may cause yet more homeowners to sell resulting in a downward spiral in market values.

Many coastal property owners have strong attachments to their coastal properties and the communities in which they live. Making a decision to sell a coastal property can be agonizing but holding on to a property as the risk of inundation grows, and public awareness of the risks increases, could prove to be very costly. The Union of Concerned Scientists warned that a tipping point may not be in the far future: "The cliff's edge of a real estate market deflation due to flooding and sea level rise is already visible for many communities if they choose to look."[9] For most coastal property owners, this will be a novel and very disagreeable experience.

Impacts of Shifting Coastlines on Municipal Governments

Property owners will find navigation of the changing market values of coastal properties challenging, but local government officials face the related challenge of sustaining local tax revenues even as the market value of coastal properties declines. The 2018 *National Climate Assessment* pointed to the risk that "coastal property and infrastructure losses cascade into threats to personal wealth and could affect the economic stability of local governments, businesses, and the broader economy."[10] In many coastal communities, coastal properties make up a significant percentage of the local real estate property tax returns, and these taxes make up a significant part of the municipal budget. Even a small decline in local tax revenues can be a serious challenge for coastal communities struggling to keep up with the rising costs of services and employees. For many public officials in coastal towns, declining revenues from coastal property will be a novel and unwelcome development.

As the pace of sea level rise picks up in the years ahead the market value of coastal properties will become increasingly precarious. Both property owners and municipal officials have a strong financial incentive to protect the value of coastal property and delay the reckoning with declining market values and eventual inundation. This interest in protecting coastal property values may make local officials reluctant to initiate planning for sea level rise and make local property owners reluctant to participate in such planning. It can also have the regrettable and costly effect of delaying the shift away from the traditional coastal flood and emergency management model to a strategy that avoids new development in areas at risk of inundation, buys time with limited structural protection measures where cost-effective, and invests in planning for timely relocation of homes and other assets to higher ground.

One unhappy scenario is that a state or local government is jolted into action as property values and property taxes fall, turning first to the most publicly acceptable response of building coastal protection structures, such as seawalls or beach nourishment. With limited funds, the design of a coastal protection structure is likely to tend toward a smaller, more affordable scale. Although such measures can make economic sense in some cases, the broader risk is that small-scale projects consume scarce resources while providing only limited and short-term benefits. These small-scale projects can put a community on a path toward paying for short delays in inundation over and over again as protection structures are abandoned and rebuilt farther inland. At the end of this process, the community might pay a cumulative cost that is far greater than would have been needed to build a more comprehensive protection structure or to relocate to safer ground.

Long Planning Time Horizons

In many coastal places, noticeable impacts of sea level rise are decades away and the most dramatic impacts will not begin occurring until around the turn of the century. For most communities, and many businesses and other governments, these impacts seem a long way off compared to conventional capital investment planning. Most capital planning is in the context of five, ten, or twenty-year planning horizons (e.g., budget for a new roof for the school ten years ahead). Some large communities may plan investments 50 or more years ahead, but only rarely 100 years.

Planning for more severe storms and sea level rise presents to governments and the private sector the challenge of extending planning horizons. North Carolina, for example, opted for a thirty-year planning horizon. Yes, some sea level impacts will occur within a twenty- or thirty-year time horizon and should be addressed. But planning for each small increment of sea level rise without looking at the full effects over a longer term can be needlessly expensive (e.g., making structural protection projects look more cost-effective than they are). It can also make it harder for the public to recognize the seriousness of the problem and understand how it is different from the better-known challenge of random, temporary coastal flooding. When the public does not see the sea level problem as the new and different threat that it is, gaining support for the dramatically different response strategies (e.g., relocating to higher ground or investing in infrastructure at a scale well beyond that needed to meet a twenty-year threat) becomes a challenge.

In addition, because the rate of increase in sea level rise is accelerating, communities looking at just the early stages of the problem can miss major out-year impacts. Investing in measures that fit an initial, small increase might look good at the moment but can result in undersized infrastructure. Planning for sea level rise will involve spending money, but short-term planning that results in repetitive investments in protection measures or a failure to implement timely relocation policies will drive costs higher than they need to be. Convincing communities and others planning for rising seas to adopt longer planning horizons is a novel challenge that will require creative new tools and policies.

How Far Inland Will Rising Seas Ultimately Reach?

Adopting a new, longer-term planning horizon can be a challenge, and it is complicated by the difficulty of identifying an appropriate target date for the new, longer horizon. At what date in the future can it reasonably be hoped that sea level rise will be done and over with? What land areas are totally safe from sea level rise? Uncertainty about this final endpoint, and the rate at which coastal lands will be inundated in the interim, can cause decision makers and the public to see descriptions of risk as inexact and thus less persuasive.

The science tells us that even successful efforts to limit warming to 2°C by 2100 called for in the Paris Climate Agreement will result in continued sea level rise after 2100 for several centuries. Even this 2°C cap on warming is thought to be sufficient only to keep sea level rise from reaching catastrophic levels in the years beyond 2100.

Knowing the date when sea level will finally stop rising would be a big asset to long-term planning. To some extent, however, the end point of a planning horizon for a new investment in infrastructure, such as a bridge or a power plant, is limited by the useful life of the project. If the project has a useful life of 100 years, the sea level rise occurring after that might be of limited significance. On the other hand, some infrastructure, once sited, is difficult to relocate because of supporting facilities (e.g., a water treatment plant has a vast network of underground water lines running to it).

Communities, which citizens expect to exist for an indefinite period, pose a different problem. Citizens may well ask what it will take to preserve the community. The response, asking for how long, seems unsatisfactory. And, in the case of communities that want to relocate to higher ground, it would help to know how much higher this new ground needs to be to avoid all possible sea level rise, or even just sea level rise expected if warming is kept to the 2°C cap in the Paris Climate Agreement. The answers to these questions may become clearer in time as efforts to limit warming either succeed or fail.

Each of these novel challenges presents an obstacle to state and local governments advancing responses to more severe storms and rising seas. Together, they can be overwhelming and delay or defeat efforts to address the problem. What can be done to help state and local governments gain confidence that these challenges can be met and need not be a bar to constructive action? At one level, as experience responding to more severe storms and rising seas increases with the passage of time, these challenges will become less and less novel. Clearly, sharing successful tools and approaches for overcoming them is a step in the right direction. But more affirmative steps, including national-level attention to addressing these challenges from Congress and federal agencies, are likely to lead to more consistent and timely actions.

14

State and Community Choices in Preparing for a Changing Coast

State and local governments along the coast face some big picture choices in preparing for more severe storms and rising seas. The threshold choice is whether to engage the difficult subject at all or simply defer it to some future date. Several factors might cause a state or local government to cross the threshold of putting preparation for these risks on the agenda. For some, a major storm is a deciding factor. For example, interest in coastal inundation in New York State increased after Hurricane Sandy. Often, however, a decision to work on the problem results from the interest of a motivated official, local activist, or local organization, rather than a top-down mandate.

Once a state or community makes the decision to engage storm and sea level rise preparation, the choices of next steps fall into conventional categories of government action. They can educate the public and build awareness of the problem. They can make response plans, commonly involving protecting assets (e.g., build a seawall) or accommodating flooding (e.g., elevate buildings). They can spend taxpayer money or use taxing powers to prompt preparedness actions. And they can regulate to reduce risk. In each of these areas, there are a set of tools that can be mixed and matched to fit the risks a state or community faces and its policy preferences. The final choice, not so conventional and followed as of today by only a small number of coastal communities, is to stage the relocation of homes, businesses, and supporting infrastructure at risk of inundation to higher ground (see ch. 15).

How are coastal states and communities approaching storm and sea level rise risk and what response actions are they implementing? This chapter provides an overview and examples of state and local action. Having a grasp of the progress states and communities are making can inform thinking about how best to design a national response and the appropriate role for the federal government relative to state and local governments, including how federal agencies can support state and local efforts. As discussed at the close of this chapter, existing state and local efforts are commendable but by themselves not a sufficient response to the problem.

Choices: Building Awareness of Coastal Storm and Sea Level Rise Risk

A simple but concrete first step states and local governments can take in adapting to more severe storms and sea level rise risk is to broaden public awareness of these risks and possible response actions. Another step to build awareness is requiring disclosure of flood or sea level rise risk at the time of sale of a property.

Public Education and Information: Spending a period of several years getting the public familiar with flood and inundation risk concepts can increase receptivity to more structured assessment and planning efforts and to more controversial measures that might be proposed at a later date.

States can focus on strengthening delivery of basic information about relative sea level rise and expected inundation consequences using public meetings and information tools such as websites. States can also promote education about basic sea level rise topics, including providing training for state employees, and encouraging school districts to use existing classroom kits on storms and rising seas, such as the grades 4 through 6 materials developed by the National Park Service focusing on Florida,[1] and a curriculum for 9th- through 12th-grade students under development by the Northern Gulf of Mexico Sentinel Site Cooperative Program.

The state of Hawaii, for example, published a *Sea Level Rise Vulnerability and Adaptation Report* in 2017, including strong recommendations for building understanding of the problem. The report called for outreach to schools across the state to engage children in understanding sea level rise and suggested focusing the work of the university-level Sea Grant program on sea level rise.[2] Other actions include maintaining a communications and outreach plan, holding a conference on sea level rise issues in the state every two years, and engaging citizens in monitoring sea level rise. The report also called for strengthening accountability for progress and transparency by adding sea level rise to the "Aloha+ Challenge Dashboard," an online data platform to track the state's progress toward sustainability.

In South Carolina, Governor Henry McMaster issued an executive order in late 2018 creating a Floodwater Commission to review a range of flood issues facing the state. The commission was created in response to "a need for communication and coordination"[3] on the problem and includes state and local officials, academic experts, and business leaders. Ten subcommittees will work on topics ranging from living shorelines, infrastructure, security, and federal funding with the general goal of "identifying potential short-term and long-term mitigation solutions for low lying and coastal areas."[4]

At the local level, public meetings and websites can generate discussion and build awareness of storm surge and sea level rise risks. The community can publish maps showing sea level rise and storm surge risk areas. Communities can also sponsor speakers on coastal flood risk and sea level rise, drawing on experts from area universities, state or federal agencies, or nonprofit organizations. Introductory articles on storm and sea level rise risk can be included in local government newsletters.

Anne Arundel County, Maryland, developed an innovative public engagement process working with George Mason University and others hosting a program called "Future Coast." The process included background information such as issue books, online maps, videos of climate scientists, a Facebook page, and an online "Sea Level Rise Quiz." Citizens were encouraged to host small discussion groups over several months, and project leaders hosted a countywide "Future Coast Citizens' Discussion" meeting at a local high school.

Disclosure of Flood or Sea Level Rise Risk at Time of Sale of Property: Another step that states and local governments can take to build awareness of storm flooding and sea level rise inundation risks is to adopt a law or regulation requiring disclosure of the risk at the time of sale of property located in these risky areas. Although flood risk disclosure at the time of property sale only reaches people who are selling property, over a period of years it can build a wider understanding of local areas at risk of flooding.

Disclosure of various issues with real estate at the time of sale is a widely accepted practice. What legal obligation does a realtor have to disclose the risks of coastal property related to storm surge flooding or future sea level rise? Generally, "common law in the United States mandates that the seller of any real property is required to disclose all facts that are material, that affect price paid and desirability by a potential buyer."[5]

The National Association of Realtors issued flood disclosure guidance in 2014, concluding somewhat ambiguously that, in general, brokers and agents owe buyers duty to disclose adverse material features, conditions, or aspects of property of which they have actual knowledge. Brokers and agents are not, however, generally required to investigate independently whether a property is in a flood zone or otherwise in an area likely to be subject to flooding or flood risks. However, if a broker or agent has <u>actual</u> <u>knowledge</u> [underline in original] that a property being marketed for sale is in an area where flood insurance is required or has specific knowledge that flood insurance has been required for that particular property in the past, those facts should be disclosed to the buyer.[6]

The *New York Times* pointed to some reasons working against disclosure: "Real estate agents risk putting themselves at a competitive disadvantage by overstating threats. Good information is hard to come by. No one knows whether, when or by how much properties will depreciate, seas will encroach, or flood insurance policies will change."[7]

Some states have tried to clarify matters with laws that require disclosure of flood risks, but these laws vary from state to state in terms of what must be disclosed and penalties for noncompliance. The Sabine Center for Climate Change Law at Columbia University, working with the Natural Resources Defense Council, evaluated state flood risk disclosure laws in 2018 and found that nine coastal states had no disclosure laws. Laws in five coastal states were deemed "inadequate," while laws in six states were considered "adequate," and two states—Louisiana and Mississippi—were rated "best."[8]

These laws are often the result of tense battles in state legislatures. Virginia passed a flood disclosure law, leaving the real estate industry "immensely satisfied,"[9] but which some municipal officials felt was incomplete. The *Virginia-Pilot* reported the following: "'It's a nondisclosure disclosure,' said Meg Pittenger, an environmental manager in Portsmouth who works on sea-level rise and flooding issues. 'But it's a step.' She said sellers or realtors should be required to divulge to prospective buyers whether a property lies in a flood zone."[10]

The existence of FEMA floodplain maps makes a mandatory requirement for flood risk disclosure practical, even if inconvenient for a realtor, and interest in disclosure is growing. Legislation to reform the National Flood Insurance Program (NFIP), passed in the US House of Representatives in late 2017 but not enacted, made local adoption of a flood risk disclosure requirement a condition of community participation in the National Flood Insurance Program and required FEMA to notify flood insurance policyholders of property flood risk when a new policy takes effect.

Disclosure of sea level rise risks, rather than conventional flood risks, at the time of sale of a property, is hard to require today because of the lack of official maps of areas considered to be at risk. Several states, however, have considered adopting a state law requiring sea level rise risk disclosure. For example, the sea level adaptation plan for Maryland recommended that the state "develop a Maryland Sea-Level Rise Disclosure and Advisory Statement to inform prospective coastal property purchasers of the potential impacts that climate change and sea-level rise may pose to a particular piece of property."[11]

The problem of defining sea level rise risk areas, however, is not insurmountable. The state of Florida requires that buyers be notified that some coastal property "may be subject to coastal erosion"[12] without mentioning sea level rise. At the local level, communities in the NFIP can get points under the Community Rating System (CRS) program "when prospective buyers of a property are advised of the potential for flooding due to climate changes and/or sea level rise."[13]

Choices: Planning and Risk Assessment

For most states and communities, decisions to implement storm surge and sea level rise programs or invest in protection structures or other measures will be based on a planning process. There are several options for integrating storm and sea level rise risks into planning processes, including amending existing hazard mitigation plans, coastal management plans, or local comprehensive plans. Any of these choices requires thinking about some threshold questions.

Threshold Questions for Storm Surge and Sea Level Rise Plans: Some of the questions states and communities need to consider in coastal storm and sea level rise planning are the following:

- Should a plan focus on just impacts of storm surge, or should it address sea level rise, or should it address both?
- What should be the time horizon for the plan (e.g., 20, 50, or 100 years)?
- How much risk is the state or community willing to accept?
- Should the state or community act independently or develop a plan in cooperation with neighboring jurisdictions?

A state or community might decide to start with questions related to reducing risk from more severe storm surges and hold off on the harder questions of preparing for sea level rise. This might be accomplished, for example, with amendments to strengthen the existing state or local Hazard Mitigation Plan or a report responding to a major storm. The state of Texas, for example, released a report in late 2018 with detailed recommendations to respond to Hurricane Harvey. The report cites research indicating that coastal storms may become more severe and calls for longer planning horizons without mentioning climate change or sea level rise.[14] This "one step at a time" approach can be tempting. But, the science of sea level rise makes it clear that all coastal states and communities will move through these choices for both storm surges and rising seas over the coming decades. States and communities that wait to engage sea level rise until waters lap at the doorstep may find that they have made unsustainable choices (e.g., expensive protection structures) and are likely to pay a far higher cost, in both economic and social terms, over many years.

In the case of a state wanting to address sea level rise, amending an existing Hazard Mitigation Plan or Coastal Zone Management Plan is a good place to start, but developing a freestanding storm surge and sea level rise plan is also a good option. A decision whether to address sea level rise as well as storm surge is a key choice because adding sea level rise will require longer planning horizons and measures to address permanent rather than just temporary flooding (e.g., a policy of elevating structures makes sense if the risk is episodic flooding but does not make sense if the risk is permanent inundation because of sea level rise).

The time horizon for a plan is important in the context of storm surge planning but even more important in the context of plans for sea level rise. A short time horizon of twenty or thirty years is likely to show some early sea level rise while a longer timeframe, such as 2100, will provide a fuller picture of long-term impacts. A full understanding of relative sea level rise can help avoid planning that results in expensive investments that later prove to be short-sighted and help build local support for politically difficult decisions such as early action to limit development in areas at risk of future inundation or staged relocation.

Related to the time horizon question is the question of how much sea level rise risk is a state or community willing to accept. This risk decision requires thinking about the ranges of sea level rise that scientists define for a given location from Low to Extreme (see NOAA sea level rise scenarios in app. 1) and the probability of these outcomes. The lower projections of sea level rise have higher probability of occurring and higher projections have lower probability. Some communities may be willing to accept high risks of inundation while others would rather plan more cautiously, and some projects (e.g., power plants) might warrant a more cautious approach than others.

Several states are helping communities with this assessment work by publishing state-specific sea level rise guidance. The state of California, for example, published sea level rise planning guidance for communities in 2018 simplifying the NOAA scenarios into three "risk aversion" categories of "Low," "Medium High," and "Extreme." The guidance then expresses the sea level rise projected in feet for "Low emissions" and "High emissions" scenarios at twelve coastal locations, allowing a community to make assumptions about future success in reducing global greenhouse gas emissions.[15] With a planning horizon year, a decision on degree of risk, and an assumption on future emissions a state or community can then determine a projected sea level rise number in feet, and use that to map land area at risk of rising seas. In 2018, the state of Florida published a more general adaptation-planning guide outlining a process that communities can follow to assess and respond to coastal flood risks.

Cooperation with neighbors has some costs in terms of added complexity but can also offer some significant benefits. For example, because storm surge risks and relative sea level rise projections often apply to a coastal area larger than a single community, some parts of an assessment can be shared and costs reduced. In addition, some response actions may involve land in more than a single community, and the chance of more effective, cross-community actions increases if actions stem from the same basic assessment. To date, cooperation with neighboring jurisdictions is more common at the local level than state level.

State and Local Hazard Mitigation Plans: As discussed in chapter 10, all states and over 20,000 local jurisdictions have Hazard Mitigation Plans that are required in order to receive federal disaster assistance. Although FEMA requires that new plans assess climate change vulnerability, actions based on the assessment are discretionary. Attention to sea level rise and more severe storms, however, is likely to increase as states gradually revise Hazard Mitigation Plans based on 2015 guidance from Federal Emergency Management Agency (FEMA). In states with limited exposure to sea level rise, addressing these risks and defining response actions in the context of the wide range of other hazards covered in the plan is a reasonable approach. For states with more extensive sea level rise risks, these risks should be mentioned in revised hazard mitigation plans, but the state or local government may want to consider addressing the complex issues relating to adapting to rising seas in another planning effort.

Like state plans, local Hazard Mitigation Plans are thin on climate change and sea level rise assessment and actions. In her 2017 paper reviewing hazard mitigation plans, Dr. Melissa Stults concluded that, "communities are just beginning to integrate climate change into hazards planning."[16] In summarizing these plans, she concluded that they "tend to have a strong emphasis on structural preparedness such as flood defenses, use of culverts, and enhanced building codes"[17] but that "nonstructural actions such as changes in policy and the use of natural systems to lessen the impact of hazards on human systems are beginning to emerge."[18] San Diego

County is an example of a community that has specifically addressed sea level rise in its plan, calling for expanded assessment of vulnerability and better coordination with related plans.

State Coastal Zone Management Plans: States have the option of upgrading existing state Coastal Zone Management Plans to address more severe storm surge and sea level rise. As discussed in chapter 11, the National Oceanic and Atmospheric Administration (NOAA) is encouraging states to address sea level rise in these plans and has made "coastal hazards" an area of focus for "enhancement" grants under section 309 of the Coastal Zone Management Act for the next several years. Most coastal states applied for section 309 funding, and many of these proposals address sea level rise to some degree. Some of the states that are using the Coastal Zone Management Program to address sea level rise include Hawaii (special beach erosion studies), South Carolina (local beach management plans), and Virginia (support for sea level rise preparation in the Hampton Roads area).

Rhode Island is a good example of a state that is using the coastal management program as a vehicle for addressing coastal flood and sea level-rise challenges. In 2008, the State Coastal Resources Management Council adopted a Climate Change and Sea Level Rise section of the Coastal Zone Management Program, describing risks of sea level rise in the state and providing that the council "will integrate climate change and sea level rise scenarios into its programs to prepare Rhode Island for these new, evolving conditions and make our coastal areas more resilient."[19]

After updating the policy several times, the state is now working within the Coastal Zone Management Program framework to develop a "Special Area Management Plan" addressing shoreline change. The plan recognizes "the need for comprehensive planning to address the impacts of storm surge, flooding, sea level rise and erosion"[20] and will include "recommendations for best management practices and adaptation strategies or techniques to be employed at both the state and local level to minimize future risk."[21]

State Environmental Project Review Laws: Many states have adopted environmental review laws similar to the National Environmental Policy Act for assessment of environmental impacts of major projects. These laws are not "plans" in the traditional sense, but they do guide individual major development decisions. Several states have recently amended these laws to consider future environmental conditions, including more severe storms and sea level rise.

In 2018, the state of New York amended its regulations to add climate change and its associated impacts to the list of issues that may be addressed in an environmental impact statement including specific mention of "sea level rise and flooding."[22] This new authority is supported by development of sea level rise projections specific to the state under its 2014 Community Risk and Resilience Act, which also apply to a range of permit and funding programs.

On the other side of the country, Hawaii Governor David Ige signed a bill in 2018 providing for consideration of sea level rise in project reviews noting that "sea level rise is already having an impact on beaches, roadways and homes near the shoreline. As a result, we face difficult land-use decisions, and requiring an analysis of sea level rise before beginning construction is just plain common sense."[23]

Local Comprehensive Plans: For local governments interested in going beyond simply updating local Hazard Mitigation Plans, adding storm surge and sea level assessment and response actions to existing comprehensive plans is a useful step. Comprehensive plans express community goals and are linked to the tools that can accomplish the goals, such as zoning ordinances, building codes, and floodplain ordinances adopted under the NFIP program. As Jessica Grannis of the Georgetown Climate Center noted in her study of sea level rise adaptation tools, "Local governments could include recommendations developed in CMPs [coastal management plans] and HMPs [hazard management plans] into their comprehensive plans to ensure that the recommendations get implemented when land-use decisions are made."[24]

Revision of a local comprehensive plan can also be a first step toward a community entering the Community Rating System (CRS) under the National Flood Insurance Program, or improving its CRS score by adopting additional measures that reduce flood and sea level rise risks. Another consideration is that adding sea level rise to comprehensive plans raises the question of the time horizon for the plan—should an existing local comprehensive plan's 20-year time horizon be extended to 50 or 100 years just to better recognize sea level rise impacts and issues? The answer will likely depend on the degree of sea level risk facing the community.

The state of Virginia adopted a requirement in 2015 that local comprehensive plans include adaptation and mitigation plans and strategies for addressing sea level rise and recurrent flooding. In 2016, the city of Virginia Beach adopted a revised Comprehensive Plan. Part 2 of the plan provides an "Environmental Stewardship Framework" that includes as one element "Sea Level Rise, Recurrent Flooding, and Hazard Mitigation." The city adopted a projection of 1.5 feet of sea level rise as a "short-term" projection and a 3 foot rise "for the long-term planning horizon (50+ years) to be used as a basis for making long-term decisions, such as public infrastructure."[25] The sea level adaptation elements of the plan include actions, such as evaluation of risks to sewer pump stations, and recommendations, such as "concentrate new development at higher elevations,"[26] as well as a call for a process to develop by 2018 a more detailed sea level rise and recurrent flooding response plan.

Freestanding Storm and Sea Level Rise Adaptation Plans: Another approach to planning for storms and sea level rise available to both states and communities is developing a freestanding storm and sea level rise preparedness plan or addressing these challenges as one major element of a larger climate change adaptation plan.

The state of Maryland published a *Comprehensive Strategy for Reducing Maryland's Vulnerability to Climate Change: Phase I: Sea Level Rise and Coastal Storms* in 2008, and the document is still guiding state response efforts today. The strategy starts from the premise that "we must take action now to plan for the impacts of climate change."[27] In addition to protection of communities and infrastructure from inundation, key themes of the document are protection of natural resources and supporting communities in developing local response plans.

Maryland's strategy included a call for "state-wide sea-level rise planning guidance to advise adaptation and response planning at the local level,"[28] an idea that has since been adopted in other states, including California and New York. The strategy included several notable recommendations including a call for a sea level rise disclosure statement to inform prospective coastal property purchasers of the potential impacts that sea level rise may pose for a property, and a clear recognition of the need to retreat from some locations.

Like Maryland, New Hampshire opted for a freestanding planning process focused directly on sea level rise. New Hampshire's plan, published in 2016, included "Guiding Principles" such as "Act Early: By starting now, the normal cycles of reconstruction, replacement and redevelopment can gradually replace vulnerable facilities and construction not designed for future conditions, often at minimal added cost and resulting in long-term cost savings."[29] Some key recommendations address elevating critical infrastructure and acquiring property at risk of flooding. Coastal communities were encouraged to develop their own plans and given flexibility in selecting response actions.

The city of Charleston, South Carolina, was one of the first communities to publish a plan focused on sea level rise with a purpose of providing "an overall strategy and guiding framework to protect lives and property, maintain a thriving economy, and improve quality of life by making the City more resilient to sea level rise and recurrent flooding."[30] The 2015 *Sea Level Rise Strategy* focuses on smart investments and recommends considering "greater than 2.5 ft SLR [sea level rise] for developments with longevities of greater than 50 years"[31] and "building codes that support construction which is more resilient to SLR."[32] Other recommendations are to "encourage green/open space connectivity to marshes and creeks," and "evaluate development policies for low lying areas."[33]

Multijurisdictional Planning for Storms and Rising Seas: A final element of the storm and sea level rise planning picture is the emergence of multijurisdictional, cooperative efforts to assess and respond to storm and sea level rise threats.

In a study conducted in support of the state of California's *Fourth Climate Change Assessment* in 2018, Mark Lubell at the University of California, Davis, identified a "governance gap" related to sea level rise and coastal storms. Professor Lubell reported that local leaders in the San Francisco Bay area identified "the overarching governance challenge as the imperative for multilevel cooperation among all sea level rise adaptation stakeholders. Sealevel rise adaptation entails interdependencies, where the vulnerabilities and adaptation decisions of local actors impose regional costs and benefits. While regional cooperation is beginning to emerge, most stakeholders see a critical need for shared learning, coordination and planning."[34]

Given the importance of regional cooperation as a strategy for managing coastal storms and rising seas, it is worth looking closely at the somewhat different approaches being implemented in the Hampton Roads, Virginia, area, the Los Angeles area, and southeast Florida.

The Hampton Roads Planning District Commission is coordinating with seventeen jurisdictions on sea level rise planning. The commission formed a Recurrent Flooding and Sea Level Rise Committee in 2014 to update sea level rise inundation maps and started work with the US Geological Survey on a land subsidence study. In a related effort, Old Dominion University developed an "Intergovernmental Pilot Project" to design a regional "whole of government" and "whole of community" approach to sea level rise preparedness and resilience planning with extensive participation from military leaders as well as local and state officials. And, in 2016, the Department of Housing and Urban Development awarded $120 million as part of the National Disaster Resilience Competition to support green infrastructure projects in Norfolk, and support a new resilience laboratory to "act as broker between regional assets and innovators to test resilient solutions"[35] and make grants to support resilience projects. The region also agreed to common sea level rise projections in 2018.

Attention is increasingly focused on the question of how to pay for investments in infrastructure or other coastal protection measures. Ben McFarlane, senior planner with the Hampton Roads Planning Commission noted that the military has a major interest in the region and could play a key role in funding both planning and implementation. The Army Corps of Engineers recently completed a report outlining coastal storm and sea level rise measures for the city of Norfolk with a price tag of $1.5 billion, of which 60 percent would be federal funds. Area officials are considering seeking authorizations for Corps studies in surrounding jurisdictions in future water resources bills as well as authorizations for funds to implement studies.

In the Los Angeles area, the *Regional AdaptLA* project is addressing coastal impacts and sea level rise in eleven local municipalities, Los Angeles County, and six supporting organizations. A key goal is to build a "community of practice" by providing leadership, technical assistance, training workshops, and webinars sharing the best available scientific information. These communities "are making significant progress in integrating climate change considerations into their existing planning mechanism."[36]

A key feature of this effort is the leadership provided by the University of Southern California Sea Grant Program. In addition to local governments, the program engaged other partners, including the US Geological Survey, the California State Coastal Conservancy, and the environmental group, Heal the Bay. These organizations help build the capacity of local governments to adapt to changing coastal conditions through workshops and webinars. They also assessed the social vulnerability of communities using US Census data to screen for socioeconomic characteristics translating to higher sensitivity and lower adaptive capacity to coastal risks.

The coastal work by *Regional AdaptLA* is one of five major elements of a *Greater LA Climate Action Framework* (the other elements are energy, health, transportation, and water), managed by the Los Angeles Regional Collaborative for Climate Action and Sustainability (LARC). The ocean and coastal part of the framework defines regional goals and strategies, including support for protecting coastal infrastructure, communities, and natural resources. The last of five major coastal goals, to "begin exploring opportunities and policies to move the built environment back from the shoreline in at-risk areas,"[37] is not, as of yet, supported by specific strategies or actions.

Perhaps the largest and most critical cooperative effort among local jurisdictions to prepare for rising seas is underway in South Florida. In 2010, Broward, Miami–Dade, Monroe, and Palm Beach counties created the Southeast Florida Regional Climate Change Compact that "represents a new form of regional climate governance designed to allow local governments to set the agenda for adaptation while providing an efficient means for state and federal agencies to engage with technical assistance and support."[38] In addition to counties, thirty-five municipalities are partners in the compact, along with other public and nonprofit organizations.

Although the *Regional Climate Action Plan* adopted by the compact calls broadly

Figure 14-1. Intensive coastal development in Palm Beach, Florida, is at risk of storms and rising sea level. Photo by D. Ramey Logan, Palm Beach Shores Florida, D. Ramey Logan.jpg, from Wikipedia Commons, Creative Commons Attribution 4.0. https://creativecommons.org/licenses /by/4.0/deed.en.

for "concerted action in reducing greenhouse gas emissions and adapting to regional and local impacts of a changing climate,"[39] much of the plan focuses on sea level rise assessment and response because "Southeast Florida is considered one of the most vulnerable areas to climate change and sea level rise."[40] Initial work of the compact focused on unified estimates of future sea level and mapping of assets at risk. The plan includes 110 action items, including defining "Adaptation Action Areas" at special risk from sea level rise and flooding, designation of "Restoration Areas" that are a priority for land acquisition, and strategies to reduce sea level rise impacts on transportation, water infrastructure, and natural areas. To support implementation of the plan, the compact worked with area universities to develop climate change indicators to track key impacts, including sea level rise, nuisance flooding, saltwater intrusion to groundwater, and severe storms.

In addition to the plan, the compact manages an annual regional climate leadership summit, holding the tenth such summit late in 2018. The annual summits bring together local officials, business leaders, and citizens to assess progress and identify emerging issues. Business leaders, including four area chambers of commerce, have responded by signing a *Joint Statement on Collaboration for Regional Economic Resilience*, calling for expanded business education on community resilience and coordinated advocacy for state and federal investments in resilience.

Choices: Investments in Coastal Protection Structures and Land

Once a state or community uses a planning process to assess vulnerability to storm surge and sea level rise and sets goals, hard choices arise about how to accomplish the goals. Capital investment in coastal protection structures and acquisition of land designed to reduce storm or sea level rise impacts are key implementation tools.

Investments can take several forms including traditional engineered structures, (e.g., publicly funded seawalls or floodgates), beach nourishment, living shorelines, wetlands restoration, and projects to improve drainage of areas at risk of flooding. Land acquisition can be an outright purchase or purchase of an easement to protect existing natural features that moderate storm surge and inundation impacts.

Traditional Engineered Coastal Protection Structures: Local governments have traditionally invested in structures to reduce coastal erosion on a modest scale, often in the form of a seawall, berm, or armored shoreline to protect a facility, road, or neighborhood. Some communities have made major investments in seawalls, like the Galveston, Texas, seawall, begun in 1902 after a major hurricane, and now extending over ten miles at a height of seventeen feet. Probably the best known recent experience with large-scale coastal flood protection infrastructure is the expansion of levees and pumps around the city of New Orleans at a cost of over $14 billion.

On a smaller scale, a storm surge barrier, including a river gate and 3,000-foot-long barrier on the Fox River, has been successfully protecting Providence, Rhode Island, from storm surges coming up Narragansett Bay since 1966. Charleston, South Carolina, recently announced plans to raise the low battery, a historic seawall that protects much of the historic area of the city by 2.5 feet, at a cost of over $100 million.[41]

Today, communities are considering investing in such structures just to adapt to rising seas. A few larger communities are evaluating very large projects that are intended to offer many years of storm surge and sea level rise protection on a large scale, with costs in the billions of dollars. Some of the projects suggested include the following:

- a project termed the "Big U" designed to protect lower Manhattan behind a ten-mile barrier of berms, walls, and gates, at a cost of $540 million, which is just a down payment on a $10 billion project to elevate land along the East River;
- a project to protect Boston, Massachusetts, with a protective barrier from Deer

Island across the harbor islands to the Hull Peninsula, at an estimated cost of over $10 billion;

- a gate or other control structure across the entrance to San Francisco Bay to protect San Francisco and the many communities and ecosystems in the larger Bay Area;
- the Ike Dike, named after Hurricane Ike, projected to run from Galveston Island to the Bolivar Peninsula and protect the city of Houston and the substantial oil and industrial facilities in the region, initially estimated to cost $3–4 billion but expanded to about $12 billion in the wake of Hurricane Harvey;
- a series of floodwalls, storm surge barriers, and tidal gates for Norfolk, Virginia, estimated by the Army Corps of Engineers to cost $1.5 billion; and
- a gate on the Potomac River south of Washington, DC, designed to protect the metropolitan area, national monuments, and key government buildings, including the Capitol and White House.

The state of Texas has given more attention to coastal protection projects than other states. Following the widespread devastation caused by Hurricane Harvey, Texas Governor Greg Abbott took the proverbial bull by the horns and created a Commission to Rebuild Texas that developed a proposal to "future proof" the state's coastal areas, in late 2017. Demonstrating that everything is bigger in Texas, the "Request for Federal Assistance Critical Infrastructure Projects"[42] seeks $61 billion from a range of federal agencies and includes a major focus on coastal flood protection structures (about 60 percent of funds, including $36 billion from the Army Corps of Engineers). Unfortunately, the proposed projects are focused on reducing the impacts of future natural disasters, and selection criteria make no mention of rising seas.

In 2019, the Texas Land Office released a *Coastal Resiliency Master Plan* calling for a somewhat more diverse set of 123 projects addressing beach sustainability, wetland protection, and shoreline stabilization in addition to infrastructure improvement. The plan is not presented as a comprehensive response to storms and sea level rise, but each project "considers future conditions along the coast, including socially driven changes, such as increased development, and environmentally driven changes, such as relative sea level rise and more frequent and extreme storms."[43]

Internationally, several cities have opted for major infrastructure as a solution to storm surges and rising seas. London has been protected by the Thames River Barrier since 1982. Rotterdam is protected by the "Maeslantkering," a set of door-like structures almost the size of the Eiffel Tower. The MOSE project protects Venice with a series of gates closing inlets to the Venetian Lagoon. These projects tend to be in places where geography is particularly helpful, where options for stepping back from rising waters are limited, or where the area protected is of such high value that pay-ing very high project costs is possible.

In addition to high costs, these projects can have significant environmental impacts and tend to promote denser development behind a seawall, making eventual relocation more complex. Relocation needs arise because protection structures some-times end up providing only a temporary solution. For example, only a year after final completion of the New Orleans flood levee system, the Army Corps of Engineers announced that rising seas and sinking land result in the system failing to provide the intended protection within four years. In a notice seeking comment on a study of alternatives, the Corps sounded almost chastened, asking "if the work necessary to sustain the 1% level of hurricane storm damage risk reduction is technically feasible, environmentally acceptable, and economically justified."[44]

In addition to public investments in massive coastal protection structures, both communities and private citizens build smaller coastal structures, and these projects are discussed later in this chapter in the context of regulating the use and design of structures.

Beach Nourishment, Living Shorelines, and Urban Drainage: Over the past several decades, conventional coastal protection structures have been criticized on several grounds, including cost to build and maintain, harm to natural systems, and failure to perform as expected.[45] A 2007 study by the National Research Council noted a further problem: "Additionally, sea-level rise is chronic and progressive, requiring a response that is correspondingly progressive. Attempts to follow a "hold the line" mitigation strategy against erosion and sea-level rise by coastal armoring will result in a steady escalation in both the costs of maintenance and the consequences of failure."[46] In response to criticisms such as this, communities and property owners have looked to softer structures and alternative strategies.

Beach "nourishment" amounts to adding sand to an area where coastal erosion or storms have removed sand and retains many features of more engineered measures. Adding sand back to a beach can inch the shorefront seaward and give property own-ers on the beach a small measure of reassurance. Coastal communities look to beach nourishment both for coastal protection and to restore or sustain tourism.

A national database of beach nourishment projects maintained by Western Carolina University includes over 2,000 projects, going back to work on Coney Island in 1923, so far involving 1.2 billion cubic yards of sand, with nominal costs of over $5.5 billion on 2,180 miles of beach.[47] The American Shore and Beach Preservation Association main-tains a similar database that reports moving a bit more sand (i.e., over 1.5 billion cubic yards) at about the same cost (i.e., just over $6 billion) on fewer beach miles.[48]

North Carolina is a leading investor in beach nourishment. The state has approximately 160 miles of developed oceanfront shoreline, and communities are planning for approximately 120 miles of nourishment.[49] Prior to storms in the 1990's, only 12 miles of beaches were nourished. There are, however, doubts over whether this investment is sustainable. Stanley Riggs, a geologist at East Carolina University, commented in 2017, that "we have 127 miles of communities in the State of North Carolina that, in order to have an economy next year, they've got to pump sand because there's no beach anymore. . . . This whole system is collapsing."[50]

Critics of beach nourishment point to numerous problems. One concern is the high cost of projects, with federal taxpayers often shouldering more than half the cost. Another concern is that nourishment is often a frustratingly temporary solution as sand washes back to sea in a future storm. NOAA pointed to a range of environmental impacts: "Beach nourishment projects can have serious long and short-term environmental effects at: the beach where the nourishment takes place; the borrow site; and, nearby areas of the water column and the water bottom."[51] In the same report NOAA noted that projects could induce development in high hazard areas by giving landowners and local officials a false sense of security. Finally, there is evidence that beach nourishment projects pose issues of social equity. Federal funding is prioritized based on comparing costs to benefits, and wealthy communities with higher value assets tend to rank higher than poorer communities.

Coastal communities and states are investing in "living shorelines" and sometimes working with federal agencies, like the Army Corps of Engineers, or nonprofit organizations like The Nature Conservancy, to implement marsh and wetland restoration projects. A living shoreline is generally used to stabilize sheltered shorelines and is distinguished by the use of natural vegetation such as grasses. Living shorelines are also friendlier than hard structures to fish, shore birds, and recreation.

The Borough of Fenwick, Connecticut, for example, is implementing a living shoreline project on the coast of Long Island Sound to reduce erosion and restore a marsh. The city of Jacksonville, Florida, is planning a living shoreline to replace failing rip rap in front of the city zoo and garden. The city of New York is working with the Army Corps of Engineers to use dredged material to restore marsh islands in Jamaica Bay.

Another strategy that communities are using to reduce the extent of coastal flooding during high tides and rain events is investment in improved stormwater and drainage infrastructure. This can take the form of "green infrastructure" practices, such as rain gardens, which help flood waters soak into the ground, as well as more traditional storm sewers, retention basins, and pumps.

Figure 14-2. The US Army Corps of Engineers implements a beach nourishment project in Virginia Beach, Virginia. Photo by Pamela Spaugy, US Army.

The city of Miami is already experiencing flooded roads and is making major investments to move water out of the city. The *Miami Herald* reported in 2017 that "the city will embark on a $100 million project to raise roads, install pumps and water mains and redo sewer connections during the next two years across a swath of single-family homes in the La Gorce and Lakeview neighborhoods of Mid-Beach. A sizable chunk of a citywide effort estimated to cost $400 to $500 million, the work is meant to keep streets dry in the face of sea level rise."[52]

Despite much study of a big infrastructure project to protect Boston, the mayor announced a "Resilient Boston Harbor" strategy in October 2018, opting for a diverse set of small steps, including investing in 67 acres of open space, restoration of 122 acres of parks and tidal lands to serve as buffers, and elevating flood prone areas. "Instead of walling off our harbor, we need to work with it,"[53] said Boston Mayor Marty Walsh.

Investing in Land: States and communities, often working with nonprofit organizations and the federal government, are increasingly purchasing land or easements as part of a strategy to adapt to more severe storms and rising seas.

One theory of land purchase is to buy land with the goal of protecting features, such as a coastal forest, which provide a buffer for storm surge and sea level rise and reduce erosion. Another theory is to buy land likely to be inundated to remove buildings or infrastructure ahead of inundation, and to avoid more costly removal after inundation. Still another theory is to buy land to facilitate landward migration of beaches, marshes, or wetlands. Note that these programs are generally focused on open land rather than a "buyout" of a private home. (see ch. 15 for information on buyout programs).

A key challenge in setting priorities for buying land expected to be inundated is when to buy and for what price. Most waterfront land is costly but the price is likely to decline as sea level rises and storm surges demonstrate the extent of lands at risk. Although most land purchases today are from willing sellers, governments may feel compelled to consider purchase using eminent domain powers. Over time, as the federal government formally moves the shoreline inland, inundated lands become the property of the state without a purchase from the original owner, and this will become a consideration for allocation of scarce funds.

The state of Maryland developed an early plan to target state investments in coastal land with sea level rise in mind. In 2012, a report by the Maryland Department of Natural Resources explains that "all of the objectives were in line with the goal of pre-serving the opportunity for coastal habitats, primarily wetlands, to move inland as sea levels rise. By doing so, a corridor network and inland targeting was necessary to help pin point areas most suitable for this adaptive response to rising waters."[54]

Connecticut is another example of a state thinking ahead to rising sea levels when prioritizing land acquisition along its coast. The 2015 *Coastal and Estuarine Land Conservation Program Plan* speaks to the consideration of upland suitability for landward migration of wetlands due to sea level rise in its land acquisition program priority ranking system: "Increased rates of sea level rise will threaten tidal wetlands if upland areas adjacent to tidal marshes do not provide appropriate conditions to support the inland migration of these marshes. Accommodating this phenomena of 'marine transgression' will require support."[55]

In 2016, in the San Francisco Bay Area, voters approved a ballot measure providing for a $12 per parcel tax, expected to raise $25 million a year over twenty years to fund a range of land purchases to "use natural habitats to protect communities along the Bay's shoreline from the risks of severe coastal flooding caused by storms and high water levels . . . and . . . provide nature-based flood protection through wetland and habitat restoration."[56] Although sea level rise is not mentioned in the ballot measure, it was widely described in press reports as a primary issue to be addressed with the funding.[57]

The state of Florida's "Florida Forever" program is an environmental land acqui-sition mechanism funded with an initial $3 billion bond in 1999 and another $3 billion bond in 2008. To date, the program has acquired over 737,000 acres for just under $3 billion.[58] The program funds land acquisitions with a range of environmental and related purposes and is required by state law to "consider lands that help address global climate change by providing opportunity to mitigate and help adapt to sea-level rise effects."[59]

An Acquisition and Restoration Council made up of Florida state officials reviews projects and sets priorities but does not publish a long-term plan or outline an approach to helping coastal ecosystems adapt to rising seas. Although there are climate change related projects on the project list, they make up a surprisingly small part of the total. Future funds for the program will come from a 33 percent share in documentary stamp taxes.

In the Gulf of Mexico, a major windfall of funding is available to states and communities as a result of the Deepwater Horizon oil spill, and some of these funds can be used to conserve and restore coastal lands. This immense resource has the potential, if smartly used, to make possible a successful landward migration of key parts of Gulf Coast ecosystems as sea level rises. Some $8 billion is available from the natural resources damage payment, of which some $4 billion is for coastal restoration, mostly in Louisiana. Over $5 billion in Clean Water Act penalties is provided to a Gulf Coast Restoration Trust Fund and is available to states and others for diverse purposes, including coastal restoration. An additional $6 billion is available to state and local governments under other agreements.[60]

States have some discretion in the use of these funds but there is as yet no overall plan for investment. Roughly $1.6 billion is allocated to a Gulf Cost Ecosystem Restoration Council to implement a "comprehensive plan." The 2016 update to the plan describes goals, including "restore and conserve habitat" and "build and sustain Gulf Coast communities' capacity to adapt to short- and long-term natural and man-made hazards, particularly increased flood risks associated with sea level rise and environmental stressors."[61] After this promising start, however, the plan does not really follow through with a strategy to invest in ecosystem or community resilience with sea level rise in mind. Projects to date address wetland and marsh restoration and living shorelines but do not focus on acquisition or uplands to which ecosystems will need to migrate.

Choices: Tax Incentives, Market Mechanisms, and Litigation

State and local governments are drawing on a variety of tax incentives and market mechanisms to change behavior in order to prepare for more severe storms and rising seas. They are even filing suits against major oil companies seeking funds to adapt to these risks.

Tax Incentives for Adapting to Rising Seas: A key goal of tax policies related to storms and rising seas is to use the incentive of lower property taxes to prompt owners of property at risk to commit to limit new development.

Communities preparing for rising seas face a dilemma with respect to local property tax rates. On one hand, local communities need every tax dollar to support local services, especially if they are coping with costs of flooding and rising seas, and waterfront property with high assessed value often makes a critical contribution to these revenues. On the other hand, allowing the density and value of coastal property at risk of storms and rising seas to keep increasing could lead to high costs to the community in the future. At what point do the future costs of higher coastal density offset the value of the tax revenue coming from these densely developed areas? This can be a dicey question.

Because property is taxed at its fair market value, including its future development potential, the assessed value of waterfront property often increases steadily. But, these increases in assessed values can cause long-standing or lower-income property owners to sell, and a new owner will often upgrade the property to an economically viable use consistent with the assessed value. This upgrade process is good for tax revenues but tends to drive up the density of development along at-risk waterfronts. Ever-rising waterfront density and property value makes any local acquisition of property to accommodate rising seas more expensive and regulatory measures more controversial.

A tax policy solution to this problem is to offer property owners the option of agreeing to limits on future development of the property in exchange for a lower tax rate based on its current use and restricted development potential. The trade-off to the local government is lower revenues now in exchange for cost savings later when sea level rise forces the community to buy or manage inundated properties. Lower taxes might be expressed as a lower rate of tax increase, or as a tax abatement, or one-time credit. The incentives might be offered in exchange for various commitments to limit development. These tax approaches might be applied by a community in which regulatory limits on waterfront development are not politically possible.

Although tax adjustments are a tool commonly cited in academic literature and reports describing sea level rise adaptation options, few communities have actually taken the step of changing tax assessments or tax policies with sea level rise adaptation in mind. Some states, however, have provided for state income tax credits for homeowner investments to reduce storm and flood damage. For example, South Carolina offers a state income tax credit of up to $1,000, or 25 percent of costs incurred, to retrofit a residence to be "more resistant to loss due to hurricane, rising floodwater, or other catastrophic windstorm event."[62]

Market Mechanisms: Some communities have experimented with market mechanisms to counteract the upward spiral in density of waterfront, or near waterfront, properties. This approach has the key feature of not giving up tax revenues. The most common example of a market mechanism is providing for a transfer of future development rights from one set of property owners (e.g., waterfront properties) to another set of property owners in the same community (e.g., properties in an area where development or higher density is desired).

To induce waterfront property owners to give up development rights, they need financial compensation, other than tax reductions. To create a market, communities can impose development limits (e.g., density or building height) in some defined areas where development is desired but provide that limits can be overcome with the purchase of development rights sold by property owners in waterfront areas. Local governments define the areas, but the market sets the price of a given unit of "development right," and money changes hands among buyers and sellers without direct involvement of the local government.

Ocean City, Maryland, is using transferrable development rights as a tool to guide coastal development. The city faced a need to remove properties at the beach as a condition of a federal beach nourishment project. Without the resources to buy the properties outright, the city created a Beach Transfer Program in which developments in a defined inland area received a 25 percent density bonus in exchange for their purchase of development rights to 500 square foot parcels in an area seaward of a "build-to" line. The program "was extensively used by property owners and more than 400 rights were transferred. . . . Some property owners received up to $2.5 million for their credits and the city succeeded in taking control of the beach beyond the build-to line, which allowed it to construct the wider beach and dunes."[63]

Litigation as a Source of Revenue: Finally, starting in July of 2017, about a dozen local governments filed suits against major oil companies seeking financial help to respond to the impacts of climate change, often citing sea level rise impacts.

The cities of Oakland and San Francisco sued five international oil companies claiming they should pay for anticipated harm that will eventually flow from a rise in sea level. In a July 2018 order granting a motion to dismiss the case, Judge William Alsup concluded that "the problem deserves a solution on a more vast scale than can be supplied by a district judge or jury in a public nuisance case. While it remains true that our federal courts have authority to fashion common law remedies for claims based on global warming, courts must also respect and defer to the other co-equal branches of government when the problem at hand clearly deserves a solution best addressed by those branches."[64] No mention is made of the manifest failure of other branches of government to provide such a solution. District Court Judge John Keenan echoed these sentiments in dismissing a similar suit brought by the city of New York, but other communities, ranging from Baltimore, Maryland, to Imperial Beach, California, continue to press ahead with suits seeking funds to address climate change impacts, including rising seas.

In light of setbacks in federal courts, the state of Rhode Island filed suit in state court in July 2018 against major oil companies seeking compensatory and punitive damages, noting that, "as a direct and proximate cause of the Defendants' wrongful conduct described in this complaint, average sea level will rise substantially along the Rhode Island coast."[65] Rhode Island Governor Gina Riamondo remarked that "if the Federal government isn't going to do their job, we'll do it for them."[66] At this writing, defendants have moved to shift the suit to federal courts.

Bravado aside, the chances of litigation against major oil companies providing a financial bonanza to help pay for adaptation to more severe storms and rising seas seems a bit far-fetched. Still, the same was said of state suits against major tobacco companies in the early 1990's, yet the final Master Settlement Agreement reached in that case in1998 provided states with $206 billion over twenty-five years.

Might the narrow view of federal judges evolve, especially if Congress and the president continue to fumble the handoff from the judicial branch? One hint on this point is the surprising persistence of *Juliana v. United States* in which twenty-one plaintiffs between the ages of eleven and twenty-two are suing the federal government for creating a national energy system that causes climate change and fails to protect public trust resources. Considered quixotic by many observers, the case has survived determined efforts to dismiss it by the United States Department of Justice, including an emergency appeal by the government to the Supreme Court.

Regulatory Tools for Managing Storm Surge and Sea Level Rise

Some states and communities will find that a combination of education, invest-ments in land or coastal protection measures, or tax incentives will do the job of preparing for storm flooding and rising seas, at least for a while. Often, however, preparation will require changes in human behavior that are best accomplished with enforceable regulations. The menu of preparedness measures involving the coercive hand of government includes requirements concerning where to build or not to build. Other requirements concern what or how to build or rebuild. Still another regulatory tool involves prohibiting or controlling design of private pro-tection structures, such as seawalls.

In general, these measures are implemented by local governments, rather than states, and have the advantages of being reliably effective and comparatively inexpensive. And, they can often be adopted as amendments to existing codes. The disadvantage of these measures is that they are controversial and can be difficult to adopt or enact. In addition to needing to overcome any local opposition, some of these regulations are vulnerable to challenge as a "taking" without just compensation under the Fifth Amendment to the Constitution (see ch. 15).

Regulating Where to Build or Not to Build: When the Urban Land Institute organized a process in 2007 engaging developers and public officials in drafting principles of coastal development, the resulting report recognized the following: "Conventional coastal development does not protect or conserve natural systems. In conventional practice, houses are built along the shore close to the water where the greatest likelihood exists of interference with coastal dynamics."[67] The authors called for conserving natural coastal systems, arguing that "the best way to protect and conserve natural coastal systems is to allow them full freedom to be dynamic. When the dunes, the channels, the beaches, and their interaction with the wind and the water are not disturbed, they can sustain habitats and fend off erosive forces."[68] Although conventional practice is largely unchanged a decade later, growing understanding of coastal storms and rising seas makes this guidance more valuable than ever.

When thinking specifically about the interaction of the coastal landscape with storm surge flooding and sea level rise inundation, a key idea is limiting development in areas that are likely to be inundated. This is a "don't make matters worse" proposition and applies to both limiting density of existing developed areas and avoiding new development in undeveloped areas at risk of inundation. Managing density in areas at risk of inundation reduces storm and sea level rise damages as well as future costs of decommissioning and removing structures. Projections of significant population growth in coastal areas make this topic especially urgent.

Defining an area where development will be limited or prohibited involves considering the projected rate of sea level rise and the type of uses to be prohibited. Why limit development now on land that is not expected to be inundated by rising seas until 2080? The life of a residential structure might narrowly be thirty to fifty years, but an investment in water and sewer lines, roads, and energy infrastructure is needed to support that new home. Then, as seas rise, both the home and the infrastructure need to be safely removed.

Communities interested in steering new development or redevelopment to specific places have the option of amending existing zoning regulations to include an "overlay" zone designed to manage storm flooding and rising seas. Such an overlay does not change an underlying zone (e.g., low density residential) but does add stricter standards, "such as prohibiting or conditioning: new subdivisions, expansion or major renovations to existing structures."[69] An overlay zone could also impose other limitations, such as building standards for any allowed development (e.g., elevation of buildings to accommodate storm surge) and is thus a mechanism for implementing policies in addition to just the location of different activities.

Although a zoning overlay is a useful way to implement measures to adapt to storms and rising seas, it has not been widely adopted to date. In 2011, the state of Florida enacted new authority giving local governments the option of adopting an "Adaptation Action Area designation for those low-lying coastal zones that are experiencing coastal flooding due to extreme high tides and storm surge and are vulnerable to the impacts of rising sea level."[70] Some Florida communities have identified Adaptation Action Areas, but use of the areas is still focused on vulnerability assessments and infrastructure protection.

The town of Greenwich, Connecticut, adopted a "coastal overlay zone" intended to, among other things "limit the potential impact of coastal flooding and erosion patterns on coastal development so as to minimize damage to and destruction of life and property,"[71] without mentioning sea level rise.

One way to reduce opposition to new zoning requirements is to package them in the context of the Community Rating System (CRS) (see ch. 9), which provides for reductions in homeowner premiums under the National Flood Insurance Program. Most community actions earn scores of 100 or 200 points under a rating system in which 500 points are needed to earn a 5 percent reduction in homeowner flood insurance premiums. But, the single most valuable action, earning 1,450 points, is for "keeping land vacant through ownership or regulations."[72] More points are available for protecting areas at risk of coastal erosion, reducing density of development, and addressing sea level rise on floodplain maps or in a watershed master plan.

Regulating What to Build: Most coastal communities participate in the National Flood Insurance Program (NFIP), and the core requirements they adopt when entering the program build on local building codes and focus on construction standards for new, substantially improved, or substantially damaged buildings in FEMA-mapped 100-year floodplains. In general, any covered project must be at or above the "base flood elevation" and comply with construction standards relating to basements and enclosures and use of flood-resistant materials. In high risk V zones, buildings must be on pilings, not fill, so that water can move under the building. Specific construction standards for wind and water loads apply.

Some communities have stepped beyond the core requirements of the NFIP by adopting additional requirements with respect to buildings, often through the CRS.

This system provides a menu of additional building standards that a community can adopt, some of which relate to sea level rise. Some examples include credit for increasing "freeboard" by requiring elevations above base flood elevation on a sliding scale up to 500 points, and adoption of a required setback on a lot to account for coastal erosion along some or all of the community shoreline for up to 370 points. With respect to sea level rise, a community can get 500 points under the CRS for adopting the V zone standards (e.g., elevation on pilings, not fill) for the part of the floodplain that is subject to wave action (i.e., "A zones"). A community aspiring to the very highest CRS class, which has a 45 percent premium discount, must require flood elevations that reflect future conditions, including sea level rise.

Some coastal states and communities have forged ahead of the core building standards of the NFIP, often selecting from the menu of more stringent standards described in the CRS. The Cape Cod Commission, for example, concluded that "it is prudent coastal floodplain management to elevate buildings to accommodate the relative sea level rise rate for the expected life of the building in the area in which building is taking place,"[73] and recommended in its model bylaw that building in the high hazard V zone be elevated three feet above base flood elevation.[74] Still another approach is extension of various building requirements to areas beyond the FEMA-mapped 100-year floodplain (e.g., to the 500-year floodplain). Larry Larson, a senior policy advisor for the Association of State Floodplain Managers explained, "We now are seeing this trend start because people are getting these extreme events."[75]

Regulating Use or Design of Coastal Protection Structures: Control over installation of coastal protection structures by private parties is shared among federal, state, and local governments. The structure and project types range from seawalls and rip rap to softer "living shorelines" that use plants along protected areas. Although hard structures can temporarily slow coastal erosion, they can also delay the natural migration of coastal ecosystems as sea level rises and degrade wetlands and beaches (see ch. 6). The key regulatory questions are whether any project altering the shoreline should be allowed, and if so, whether engineered structures such as seawalls are necessary or whether a living shoreline approach is appropriate.

At the federal level, the Environmental Protection Agency (EPA) and Army Corps of Engineers implement the Clean Water Act program, requiring permits for projects in waters and wetlands. Large seawalls and related structures require an individual permit but smaller bulkheads and related hard structures have been permitted under a less burdensome nationwide general permit. In March 2017, the Corps issued a new general permit for living shorelines, putting these projects on a more even administrative footing with hard structures. Although the development of a "general permit"

Figure 14-3. Workers elevate a home in Atlantic Highlands, New Jersey, damaged by Hurricane Sandy, to reduce flood risk. Photo by Rosanna Arias, Federal Emergency Management Agency.

may encourage living shoreline projects, in place of hard structures, the flip side is that the easier federal permit process may result in more shoreline alterations, ultimately doing more harm than good.

Even if a federal permit could be approved for a seawall or other hard coastal structure, state or local law may prevent the project from proceeding. Eight states have total or partial bans on shoreline hardening projects. Some coastal states and some local governments, often working through Coastal Zone Management Programs, encourage use of living shorelines, and some have adopted laws giving preference to this method or allowing hard structures only under narrow circumstances.

The state of North Carolina, for example, prohibited most hardened shorelines in 1985, although recent legislation has narrowed the prohibition somewhat. Sarasota County, Florida, prohibits shoreline hardening or the construction of shore protection structures unless it is found to be in the public interest. Some states, such as Texas, are going beyond prohibition of new hard coastal structures and prohibiting maintenance or enlargement of existing structures on beaches.

Despite these government attempts to promote living shorelines over hard structures, it is not yet clear what effect the stated preference for living shorelines is having on new coastal stabilization projects. Even if new preferences discourage future shoreline armoring projects, they will do little to change the existing armoring in place along 14 percent of the coast (see ch. 6).

A 2015 report on coastal armoring in California by the Stanford Law School concluded that living shorelines face an uphill battle: "Among private property owners and regulators alike, there is a lack of attention to nature-based and other non-armoring responses to coastal hazards and sea level rise, in part because existing disaster relief policies, insurance programs, and inadequate mitigation fees for armoring do not sufficiently require property owners to internalize the cost of development in high-risk areas."[76]

Another approach to managing the proliferation of coastal protection structures is to purchase, or impose by regulation, a property-use limitation, prohibiting or limiting use of these protection structures sometimes called a "rolling conservation easement." This tool is often considered in the context of adaptation plans designed to relocate structures, buildings, and communities and is discussed in the next chapter.

Assessing State and Local Preparedness for Coastal Storms and Rising Seas

So, what is a fair evaluation of the diverse efforts of state and local government to respond to more severe coastal storms and rising seas? The roll call of states and communities responding is impressive, ranging from the city of Annapolis, Maryland, to the state of Hawaii. Celebrating and learning from the success that state and local governments are having responding to these challenges is important. But comprehensive assessments of progress to date indicate this work is improvisational and preliminary. As sea level rise expert John Englander observed, "very few are thinking big enough."[77]

A related theme is that federal leadership is a key missing ingredient. In an article for the *Lewis and Clark Law Review* in 2016, examining intergovernmental issues related to climate change and flooding, Dan Tarlock and Deborah Chizewer concluded that "although local governments have taken positive steps toward improved flood management and hold the authority to take more steps, they will not achieve consistent results without federal and state financial and legislative support."[78]

The Surfrider Foundation came to the same conclusion in its assessment of state coastal programs: "The overarching results indicate that the majority of coastal managers and state agencies need to take larger steps to ensure our beaches and nation's coastlines will be protected for future generations. This national trend also denotes a clear need for increased federal leadership."[79] The foundation evaluated coastal states on sixteen different metrics, including those related to sea level rise, and gave report card grades for coastal protection in each state, with average grades of B for Northeast and Western states, C for mid-Atlantic states and the islands, and D for Southeast and Gulf states.

Finally, the 2018 *National Climate Assessment* summed up progress in coastal planning for climate risks: "While many current plans call for risk identification, monitoring, research, and additional planning, there is still little focus on the major investments or immediate implementation actions and cost-dependent tradeoffs required to successfully adapt."[80]

State and local governments are in different places along the spectrum of progress in responding to the risks posed by coastal storms and rising seas and have made different choices among the range of response measures and practices. In theory, this variation is reasonable in that some places face a more pressing risk and are a better fit for some practices than others. A problem arises when little progress in responding to these risks is made in the places where it is badly needed (e.g., the Southeast and Gulf of Mexico states). Another problem is that well-meaning plans and response measures are developed without a full reckoning of the risks of inundation over the long term or the financial feasibility of a preferred approach. With inundation risks fully on the table, communities are more likely to focus limited financial and regulatory investments on stepping back from the coast to safer, higher ground. This key problem of getting relocation on the agenda is the topic of the next chapter.

15

Relocation: Often the Inevitable Choice

Firmness of purpose in planning for severe storms and rising seas is important but there is a risk of selecting and implementing ineffective or counterproductive solutions. The single biggest mistake that states and communities need to avoid is to plan for just more severe, temporary flooding from storm surges rather than for the combined effect of more extensive storm surges followed by permanent inundation of increasing amounts of coastal land.

The consequence of this mistake is to invest limited financial and regulatory capital in a losing battle to fortify the existing shoreline and floodproof or elevate property that will be lost to the sea and then need to be removed at further expense. Because of the gradual, decade-by-decade incremental rise of sea level, there is the potential for states and communities to make this same mistake every several decades, abandoning expensive coastal protection structures and floodproofed property, only to build a new round of structures and properties that will in turn be inundated, abandoned, and removed.

It is important to say that coastline and property protection structures have their place. The mistake is not that these measures are included in a plan, but that the decision to include them is based on a calculation of risk that does not recognize inundation as the inevitable outcome for large amounts of coastal land. This too lightly dismisses inundation as too far in the future to consider in a plan, or commits to structural protection without considering consequences for neighboring jurisdictions, ecosystems, or critical infrastructure. With the inundation outcome clearly on the table, some measures to protect land and property from more severe storms and rising seas may pass muster as affordable but temporary accommodations. In other cases, however, facing the hard facts of eventual inundation will result in decisions to shift investments of scarce financial and regulatory capital away from short-term protection measures and toward relocation of coastal communities and assets.

So far, much of the experience with relocation in coastal communities has been in the context of individual, opportunistic property buyouts, often after storm damage and using federal disaster funding. The good news is that the idea of more proactive, community scale relocation is slowly emerging from the shadows of obscure policy discussions and gaining public attention. In November 2018, the federal agency authors of the fourth *National Climate Assessment,* spoke directly to the relocation idea observing that "in all but the very lowest sea level rise projections, retreat will become an unavoidable option in some areas of the U.S. coastline."[1] The bad news is that relocation has not yet gained a foothold in most coastal planning processes. The assessment authors continued: "but the potential need for millions of people and billions of dollars of coastal infrastructure to be relocated in the future creates challenging legal, financial, and equity issues that have not yet been addressed."[2]

This chapter provides an overview of relocation options focused on measures appropriate for state and local governments, including measures for planning, financing, and regulating in this area. Other challenges relating to takings issues and inland impacts of relocation are addressed. Finally, the choices that communities face in deciding when to confront the relocation question—whether to wait for a storm and use disaster funds to relocate or whether to more proactively plan for relocation ahead of storms and rising seas—are discussed.

Coastal Relocation Planning

Today, most state and local plans for coastal flooding in the United States are intended to prepare for temporary flooding from coastal storms, and a smaller set of plans specifically address preparations for sea level rise. Of the plans that speak to sea level rise directly, very few propose a strategy that includes relocation. Among the plans that address relocation, the proposed relocations are often tactical retreats, sometimes simply proposing relocating buildings landward on the same property. Only a few plans or studies have proposed larger scale, strategic relocation of coastal neighborhoods, infrastructure, communities, or ecosystems to new, safer sites, and just a handful of these plans have been implemented.

International Sea Level Rise Relocation Planning: Relocation as a storm and sea level response strategy gets more attention internationally than it does in the United States. The island Republic of Kiribati in the Pacific Ocean is often cited as a case study in planned relocation. In 2014, Kiribati purchased 6,000 acres on the island of Fiji for a planned resettlement. In 2017, however, a new president changed course and "now plans for his people to stay. He doesn't deny that climate change is happening, but he subscribes to a belief, common here, that only divine will could unmake the islands. 'We are telling the world that climate change impacts Kiribati, it's really happening,' he says. 'But we are not telling people to leave.'"[3]

For many Pacific islanders, migration to another island or country is inevitable. Academic studies have identified options, including migration to Hawaii or Australia, but no comprehensive plan is in place. As Richard Marles, member of Parliament in Australia, commented in 2015, "So Australia being a destination for climate change migrants surely has to be up for discussion. Yet this is a public debate we have barely even begun."[4]

The hundreds of thousands of Pacific Island residents migrating to other places, however, is just a trickle compared to the projected wave of resettlements driven by rising seas around the globe. Research published by Cornell University in 2017 found that "in the year 2100, 2 billion people—about one-fifth of the world's population—could become climate change refugees due to rising ocean levels."[5]

While studies continue, concrete plans for how and where these sea level refugees will relocate are only now emerging, mostly as part of adaptation plans included by some countries as supplementary information in the Nationally Determined Contribution statements developed in response to the Paris Climate Agreement.[6] These plans and related efforts "are a start, but they remain scant and underfunded and are years from application."[7] The United Nations High Commissioner for Refugees concluded ominously in 2012 that, "if climate change continues at its current pace, and if we fail to find sustainable solutions for displaced populations, we will leave a poisoned legacy to future generations and increasing threats to peace and security all over the world."[8]

On the other side of the globe from Kiribati, the community of Fairbourne on the Welsh coast of Great Britain opted for the help of a barrister, rather than relying on "divine will," when faced with proposals for relocation in the face of rising seas. Great Britain has developed Shoreline Management Plans (SMPs) for much of the coast that define for specific areas "how the area is defended today and whether those defenses can continue to be maintained. . . . Many places will continue to be defended, but the shoreline plans admit that in 48 areas around the Welsh coastline some homes may be at risk as a result of the policy decided on. . . . In most of these cases, relocation is one of a number of different options that may be considered."[9]

Fairbourne residents argued that, as a result of the SMP, "house prices in Fairbourne have plummeted and businesses have struggled for long-term investment."[10] Residents are "hoping to claim back the original value of all the properties and businesses in Fairbourne following the enormously damaging claims put forward by the SMP."[11] Despite some controversy, the government has proceeded to advance the plans, including proposals for "managed realignment" that allows "the shoreline to move naturally, but managing the process to direct it in certain areas."[12]

Unfortunately, Great Britain's coastal management problems are much bigger than Fairbourne. After looking comprehensively at the English coastline, an independent Climate Change Committee established by statue issued a wake-up call in a 2018 report concluding that "plans do not reflect the realities of long-term change, are not joined up and are not fully implemented. People who live on the coast are not engaged in the process of planning for future change and are not taking proactive steps to adapt."[13] The report authors estimate that existing SMPs would cost between $23 and $38 billion and that for almost 100 miles of coast the cost of the current plan to "hold the line" exceeds the benefits. For another 9,000 coastal miles, almost 30 percent of the total English coastline, benefits of holding the line are twice the costs. But, because the government is now able to fund only projects where benefits exceed costs by six times, "funding for these locations is unlikely and realistic plans to adapt to the inevitability of change are needed now."[14]

In early 2019, England's Environment Agency proposed a new approach looking more comprehensively at the needs of the entire country and planning for a 4°C rise in global temperature. Project Chair Emma Howard Boyd noted that, "We can't win a war against water by building away climate change with infinitely high flood defences."[15] The draft plan proposes using a range of tools to build resilience including "accepting that some areas will flood and erode and enabling local areas to achieve a managed transition."[16]

Coastal Relocation Planning in the United States: At the national level in the United States, the matrix of coastal planning programs, including the Federal Emergency Management Agency (FEMA) hazard mitigation planning and National Oceanic and Atmospheric Administration (NOAA) Coastal Zone Management Program planning, largely defer to state and local government choices for managing preparations for coastal storms and rising seas. Although federal program guidance speaks to sea level rise, it does not call for attention to relocation as a strategy.

Victoria Herrmann, president of the Arctic Institute, summed up the situation in a 2017 paper for the Atlantic Council: "The reality of internally displaced communities due to sea level rise, flooding, and extreme storm events in the United States has arrived, and is poised to get worse. However, the US federal government remains ill-prepared to deal with the immense and undeniable human security challenge at hand. At present, there is no dedicated funding, dedicated lead agency, or dedicated policy framework to guide communities in need of relocation."[17]

For many Americans, the iconic image of coastal relocation is the Cape Hatteras Lighthouse, moved landward almost 3,000 feet in 1999. This case is instructive in that it demonstrates the inevitability of relocation in some cases. Unfortunately, it also conveys the idea that relocation is a comparatively simple engineering challenge and masks the enormous complexities involved in relocation of more complex infrastructure, neighborhoods, and entire communities.

The 2014 report of the State, Tribal, and Local Leaders Task Force on Climate Preparedness and Resilience recognized a need for federal leadership in addressing coastal relocation issues, boldly recommending that federal agencies "explore . . . addressing climate change-related displacement, needs of affected communities, and institutional barriers to community relocation."[18] The report noted that "the Federal Government has an opportunity to provide international leadership by establishing an institutional framework for responding to the complex challenges associated with climate-related displacement."[19]

Bloomberg News reported in late 2016 that this task force's recommendation was not forgotten, noting that "the Obama administration is quietly trying to accomplish one last big thing on climate change,"[20] and explaining that "the White House has asked 11 federal agencies to sign a memorandum of understanding establishing what it calls 'an interagency working group on community-led managed retreat and voluntary relocation.' The group's goal would be to 'develop a framework for managed retreat'—including deciding which agency should be in charge, identifying obstacles to relocation and how to remove them, and coordinating with communities that already want to move."[21]

Community relocation, of course, is a controversial idea and some states and communities may be tempted to simply avoid these difficult choices. Working independently, without a national program framework or links to neighboring communities or states, makes it more likely that planners put relocation on the back burner. Alternatively, a national program framework alert to relocation, dealing with infrastructure and ecosystems, and encouraging coordination improves the chance that relocation options are considered. Relocation is also more likely to be considered if a larger framework makes clear that structural protection everywhere is not financially feasible.

The Trump administration is unlikely to give relocation the attention that the Obama administration did, but it remains part of the policy discussion. In 2018, Stanford's Hoover Institution convened a forum looking at coastal management challenges and concluded that, although "the policy and practical challenges of climate-induced relocation are enormous . . . deliberate planning and policies that facilitate mobility are crucial for handling the socioeconomic and demographic shifts that accompany displacement and migration flows."[22]

Even in the absence of leadership from the federal level, some states and communities have been taking tentative steps to understand relocation issues and options. Most of these efforts, however, fall into the category of tactical, small-scale relocation, rather than plans for strategic relocation of larger communities and assets.

The state of South Carolina adopted a Beachfront Management Act (BFMA) in 1987, which, among other things, "established strategic retreat as a long-term approach to deal with receding shorelines."[23] After this promising start, and several hundred million dollars in beach nourishment projects over a dozen years, the relocation policy suffered a setback in 2013 when a commission made recommendations for updating the act, including "replace language regarding the policy of retreat with the following: The policy of the state of South Carolina is the preservation of its coastal beachfront and beach/dune system."[24] Although the emphasis on planning for strategic retreat has dimmed, municipal-level plans required by the BFMA still address plans for retreat. These community plans, however, generally focus on beach stabilization and express an intention to address relocation at a later date.

The state of Maryland had a similar promising start in addressing rising seas around Smith Island in the Chesapeake Bay. As the *Baltimore Sun* reported in 2017, "Taking stock after Hurricane Sandy washed over the island, the state proposed using storm relief money to buy out ten homeowners in 2013—a step most of the island's 240 residents viewed as a first toward abandonment. They did more than reject the plan. They organized Smith Island United, a de facto island government to stand up for their interests, and looked toward shoring up their home both economically and physically."[25] The state of Maryland, working with the Army Corps of Engineers, is now planning to spend "tens of millions of dollars to preserve the island's history for at least another generation,"[26] including jetties to prevent erosion and a new sewage treatment plant and economic stimulus projects.

On the other side of Chesapeake Bay, the idea of stepping back from the coast is getting a closer look. Virginia Governor Ralph Northam issued an executive order in October of 2018 calling for a "Coastal Resilience Master Plan," which is to, among other things, "consider potential areas and options for managed coastal retreat when appropriate."[27] The plan is also to include an assessment of needed funding and recommendations for potential funding sources.

The state of Hawaii tackled the issues related to coastal relocation head on in a 2019 project evaluating the feasibility of relocation and defining key issues needing to be addressed, including criteria for setting priorities, costs, legal issues, and public access to coastal land. A report resulting from the project includes a recommendation to establish a "multiprong state leadership committee" [28] to "devise a comprehensive, cohesive managed retreat plan."[29]

Although some community relocation efforts have failed, there is reason to hope for engagement with coastal relocation planning based on a handful of examples of plans for relocation of small communities or parts of communities at special risk of rising seas. One of the best-known relocation initiatives is for the small town of Isle de Jean Charles in southeastern Louisiana. This community of twenty-five homes and about 100 people of the Biloxi-Chitimacha-Choctaw Tribe has lost most of its land area to rising waters. The tribe sought help to relocate from the National Disaster Resilience Competition, a fund of $1 billion that the Department of Housing and Urban Development (HUD) reserved to demonstrate resilience approaches. HUD awarded $48 million to resettle the community in 2016. Officials from HUD, the state of Louisiana, and the community reviewed over a dozen resettlement sites and are now focused on a 500-acre plot of farmland forty miles north of the current site at an elevation of nine feet above sea level.

Several other Native American communities are considering relocation. The Quinault Indian Nation Village of Taholah on the Pacific Ocean in Washington State is vulnerable to sea level rise, storm surge, and river flooding. The village developed a plan providing for relocating 650 residents and vulnerable community facilities a half-mile away from the existing village on higher ground.[30] The communities of Newtok, Kivalina, and Shishmaref, are among a dozen Alaskan communities that have evaluated relocation options.

Relocation planning is also beginning in some major American metropolitan areas, often at a small scale. The Shorecrest neighborhood of Miami is regularly flooded and the city is developing a plan to relocate many of the homes and services. As reported by Reuters in 2017, "properties would be purchased by the government and turned into parks and retention basins to hold back rising water. The plan would rely on owners voluntarily selling their homes."[31]

Yankeetown, Florida, just north of Tampa, is one of the few communities to use the 2011 state law allowing for "Adaptation Action Areas" to manage response to storm flooding and rising seas. After a series of community meetings, Yankeetown voted by referendum to establish an Adaptation Action Area of eighteen square miles of the town with a goal to "preserve natural resources by giving them space to evolve as the world warms, without the impediments of sea walls or development right at the edge. By doing so, natural protections can be fortified, such as oyster reefs that buffer against sea-level rise. The town also is left with room to migrate to higher ground if necessary."[32] Larry Feldhusen, mayor at the time the town established the new adaptation area commented that "we wanted to protect the environment and allow it to evolve as things change and not be in the way of it."[33]

State and community consideration of relocation as a strategy for responding to rising seas has not progressed much beyond these modest efforts to relocate small communities or neighborhoods. One of the few attempts to evaluate larger scale relocation is work by the Massachusetts Institute of Technology (MIT) School of Architecture and Planning in 2015, looking at relocation options for greater Boston. Although not a product of local governments, the MIT project report offers a conceptual framework for relocation planning including relocation principles and a "Relocation Suitability Index" used to identify specific areas for relocation.[34] Relocation principles include the following:

- **Out of Harm's Way**—Relocation sites should be located on ground that will not be affected by future sea level rise.
- **Minimize Stress**—Relocation should aim to preserve, as much as possible, continuity for those affected.
- **Receiving Capacity**—Relocation sites should be selected with attention to the receiving capacity and the preexisting infrastructure of those areas.
- **Build it Back Better**—Relocation policies should favor sites with access to public transportation, in walking distance of necessary amenities, and with underutilized infrastructure.
- **Feasibility of Implementation**—The process of acquisition and the potential obstacles involved in acquisition, such as state-owned versus private lands, should be considered.

The project team applied these principles to scenarios in which residents displaced by sea level rise remain in the community, move to a neighboring community, or move to another location nearby. In looking at options under the scenario involving relocation to a nearby location, for example, the team concluded that the land now occupied by Hanscom Field airport in Bedford, Massachusetts, met many of the suitability criteria for community-scale relocation and would be a prime site for relocation of coastal residents.

Similar forward thinking is included in a plan for the greater New York City area developed by the Regional Planning Association with input from local governments and support from foundations. The *Fourth Regional Plan* addresses a range of social and economic issues and includes recommendations related to climate change, including adapting to a changing coastline. The plan calls for protecting some densely populated communities along the coast from storms and flooding but a "transition away from places that can't be protected,"[35] using buyout programs supported by a moratorium on new development in flood-prone areas. The plan also calls for a new Regional Coastal Commission, establishment of adaptation trust funds with a focus on flooding, and a new national park in the New Jersey meadowlands to promote climate resilience.

Relocation Practices and Measures: Buyouts and Other Financial Tools

Governments that decide to move in the direction of stepping back from the shore have a range of tools and practices to accomplish this result. Most prominently used to date are financial tools, including home buyout programs and purchase of conservation easements.

Home Buyout Programs: Existing home buyout programs, sponsored by FEMA and discussed in chapter 10, are currently small and primarily designed to help the National Flood Insurance Program manage the high costs of rebuilding properties that are repeatedly damaged or destroyed rather than support a communitywide, planned step back from areas at risk of storm surge or sea level rise. These small programs have occasionally been supplemented with larger buyout authorities enacted in response to major hurricanes. For example, the Community Development Block Grant Disaster Assistance program at HUD includes housing assistance.

In their study of relocation after Hurricane Sandy, Anamarie Bukvic and Owen Graham commented that the existing programs "are designed to deal with small-scale or individual cases of repetitively damaged properties and do not have policy framework or financial mechanisms to support extensive implementation, and especially the purchase of high-value homes."[36] In addition, some states, like North Carolina, use the FEMA buyout funding but not for coastal property, partly because of its high cost. As a result, "less than 5 percent has been used to purchase property along the state's 300 miles of coast."[37] Most important, FEMA grant programs generally buy homes after a disaster rather than help people prepare for a situation that is gradually getting worse.

By far the largest investments in buyouts, however, have come from the federal government in special appropriations in response to major hurricanes. These programs spending hundreds of millions of dollars are focused on properties already damaged or destroyed and are limited to purchases from willing sellers. These are "relocation" programs only in the sense that they remove an individual structure, and a homeowner, from a risky location and avoid the obligation to support rebuilding or payments for repetitive damages in the future without dealing with the question of where the homeowner relocates to.

For example, the state of New York received $788 million for housing programs after Hurricane Sandy, including for repair of homes, storm protection (e.g., elevating homes), and for buyouts of homes where damages were more than 50 percent of the value of the home. Participation in the buyout program is voluntary but the state has offered additional incentives, including an additional 5 percent of home value for homeowners who permanently relocate within the same county and an additional 10 percent for homeowners in high risk areas that collectively agree to participate in the buyout.

The state of New Jersey responded to Hurricane Sandy by plowing federal funds into a small existing state buyout program called "Blue Acres," with the goal of using "$300 million in Federal disaster recovery funds to purchase clusters of storm-damaged homes or flooded neighborhoods from willing homeowners at pre-storm value."[38] This funding, along with FEMA grant funds and state funds, was intended to purchase 1,300 properties to be demolished and left as open space "to serve as a natural buffer against future storms and floods."[39] Although there is no overall plan setting priorities for relocation of properties at greatest risk, the state is "seeking clusters of homes or whole neighborhoods that were flooded in Superstorm Sandy."[40]

New Jersey also used $215 million to operate a "resettlement" program providing one-time payments to encourage homeowners suffering Sandy damages to stay in their community or to move only within the same county. The state claims that, "by supporting these 18,500 households with $10,000 grants, the program has been critical in helping families stay in their communities, preserving the character of storm-impacted neighborhoods, helping bring a return to normalcy after Sandy, and stabilizing the municipal tax base."[41] Even taken together, however, the buyout and resettlement investment pales in comparison to the $1.34 billion in Sandy recovery funds that New Jersey allocated to repair or rebuild homes.

Despite these serious and innovative efforts to help homeowners relocate to safer places, the Lincoln Institute of Land Policy, in an assessment of the state of government investment in buyouts and relocation, concluded that "buyout programs were employed in New York, New Jersey, and Connecticut following Irene and Sandy, but they were considered politically unfeasible and thus were available to only a handful of communities. Of the billions of Federal aid spent on resilience and recovery in the New York metropolitan region, at least $750 million has been spent on buyouts, which alleviated the flood risk for more than 1,500 homes. However, the vast majority of recovery efforts focused on other measures of adaptation."[42]

The theme of investments in structural protection and rebuilding being preferred to buyouts also comes through in the case of recovery from Hurricane Katrina. After Hurricane Katrina, the state of Louisiana and city of New Orleans focused on rebuilding and expanding levees and other structural protection, at a cost of some $14 billion of federal assistance. The "Road Home" program was designed to help rebuild damaged structures and elevate structures where necessary. Buyouts got less attention.

The Louisiana *Advocate* evaluated the program in 2015, ten years after Hurricane Katrina, and concluded that "the Road Home paid 130,000 homeowners a total of $9 billion, and 119,000 of those recipients promised to rebuild and reoccupy their homes within three years."[43] Because grants were limited to $150,000, many homeowners felt rebuilding was a better use of the money than sale of the property. In 2018, National Public Radio quoted the head of the state coastal agency at the time as saying that when buyouts were even mentioned in earlier plans "there were some 'very upset people literally threatening us with our lives.'"[44]

Louisiana's 2017 *Comprehensive Master Plan for a Sustainable Coast* describes an ambitious program focused on preparing for future flood events by investing $18 billion for marsh creation using dredged material, $5 billion for sediment diversions, and $19 billion for structural protection projects. Within this largely structural protection program is a $6 billion Flood Risk and Resilience Program that "recommends floodproofing more than 1,400 structures, elevating more than 22,400 structures, and the acquisition of approximately 2,400 structures in areas that are most at risk."[45] The plan outlines a program to pay fair market value for the homes, tear them down, and then fund new homes in safer areas. Unlike disaster driven buyout programs, this program is looking ahead to avoid future storm damage and rising seas, albeit on a very small scale.

Although the plan is a step forward in that it offers prestorm buyouts as part of a planned approach, it is a modest start in two respects. First, the State concluded that 23,000 buildings near the coast would have between three and twelve feet of flood waters in the event of a 100-year storm and called for elevating these structures rather than relocating homeowners with a buyout. Even when elevated, these structures are at risk of storm damage and eventual loss to sea level rise. The second caveat is that "the buyout program is on paper only,"[46] because the funding is uncertain. An April 2019 report focusing on six parishes, known as "LA SAFE," builds on the *Master Plan* but calls for steering development away from high risk areas and a major new buyout program for homes in these areas.

In its 2016 assessment of buyout programs, the Lincoln Land Institute proposed some general improvements including "design buyout programs as long-term adaptations to flood risk, not merely as short-term recovery tools,"[47] as well as to "standardize buyout program requirements at the federal level,"[48] and "pursue housing blocks where neighbors can relocate together."[49] Other studies have cited a need for better transparency in the process, clearer roles for community decision makers, faster action to finalize buyouts, and more determined efforts to avoid inequities in buyout decisions.

Cost is also major obstacle to the buyout idea. Today, faced with a prospect of spending billions of dollars a year in a long-term effort to gradually buy risky coastal property, governments are likely to balk. Yet, past experience demonstrates that billions of dollars will be spent after future major storms to rebuild properties, continue to insure them, and buy them out. Waiting until the damage is done is likely to be far more expensive and disruptive than paying up front (see ch. 19 for proposals in this area).

Purchase of Rolling Conservation Easements: The principal alternative to outright purchase of at-risk coastal properties is purchase of a "conservation easement" that maintains private ownership with constraints preventing some uses of the property, usually at lower cost than a buyout. Payments to landowners might be from a government or nonprofit organization such as a land trust. States have different authority for conservation easements and local authority can vary as well.

A conventional conservation easement becomes a "rolling" conservation easement when owners accepting a payment agree not to build protective structures, such as seawalls or armoring, and to allow the forces of nature to work their will on the property and roll inland. The term "rolling easement" is sometimes used to refer to a set of practices including both purchase of an easement and creation of the easement using regulatory or permit restrictions.

In addition to a conservation easement or property restriction that prevents construction of shore protection structures, a rolling easement agreement might prevent regrading of land to hold back the ocean or recognize "ambulatory" property lines that move inland with the shore. An agreement might restrict the development of undeveloped or agricultural land. It might also provide for the transfer of ownership to a government or nonprofit organization of some or all of a parcel in the event that sea level reaches a given level and for the removal of structures or debris.

A fair question is why a property owner would give up the right to hold back the sea for any amount less than the full value of the part of the property at risk. As the EPA points out, the key to fixing a price is the perception of future risk: "A rolling easement would decrease the property value only slightly, because the eventual submergence is so far in the future. Therefore, a relatively modest near-term inducement can lead a reasonable farmer or developer to agree to a rolling easement— especially if the landowner is more skeptical than the land trust about a large rise in sea level and hence views the eventual submergence as a distant possibility."[50]

Prices for easements vary greatly based on the type of restriction, the land and its potential uses, and the immediacy of the flood or inundation risk. EPA reports rolling easement prices for as little as five percent of the value of the property.[51] A payment might also involve some combination of money upfront, some downward adjustment of property taxes, or a permit for some more immediate action. Conservation easements may also entitle a landowner to a state tax reduction or a charitable deduction on federal taxes.

Unfortunately, a buyout or purchase of a conservation easement requires capital, and scaling a conventional buyout program to the national level would require billions of dollars. States and communities looking to raise capital to finance large-scale relocation have few options beyond relying on federal funds or the occasional windfall of a Deepwater Horizon-type settlement. The success some states have had with bond programs for buying sensitive lands might be difficult to transfer to relocation programs. The dubious proposition put to voters would be to use taxpayer money to buy at risk coastal properties of no special value that are likely to come into state ownership eventually anyway as a result of storm damage and inundation by rising seas.

Innovative Financial Models to Ease Property Transfer: The academic literature includes several ideas to reduce upfront costs to communities seeking to promote transfer of at-risk coastal property implemented either by governments or nonprofit organizations such as land trusts.

For example, sea level rise expert John Englander has proposed creation of non-profit land trusts for defined areas called "Shoreline Adaptation Land Trusts" (SALTs). Property owners would have the option of donating land at risk of rising seas to the trust and are allowed lifetime use of the property. The financial benefit to the owner would be avoidance of future property taxes and perhaps a tax deduction for the gift of the property to the trust at the present value rather than a future much-reduced value. To encourage early action, the percent of the value of property that could be deducted would decline over time. As a further benefit to both the landowner and local government, the trust could commit to be responsible for decommissioning the property. In addition, "as a means of developing further value and working capital, the SALT could rent out properties it acquires, after the donating owner dies or abandons the property."[52]

Writing in the *Vanderbilt Law Review*, Richard Henderson suggested that another option is for land trusts or government to purchase an option to buy coastal property at a specified future event (e.g., the date that sea level rises to a specific point or a storm damages a property to a specific degree, such as loss of more than half of assessed value). These "sea level purchase options"[53] (SLPOs) would give the seller an initial payment and some confidence that the property would retain at least a specified value (e.g., a percentage of assessed value) if the option were exercised. The government or nonprofit organization holding the option would gain some control over future relocation for a small investment, reduce uncertainty over future costs, and avoid takings litigation. This approach could be combined with rolling easements or other measures to limit new development.

As far back as 1990, Joseph Sax, writing in the *UCLA Journal of Environmental Law and Policy*, suggested that the owners of property at risk of inundation be taxed to make annual payments to a fund that would be invested and, on or around the time of inundation and defacto transfer of the property to public ownership, make a payment to the original owner based on the owner's payments to the fund, plus interest. This approach "assumes the owner will experience rising sea levels just as she would experience a fire or a hurricane that destroyed her house. The owner will see herself as an innocent victim, and because—voluntary amortization being unlikely—she will effectively be wiped out, she will feel the need for compensation. If she gets compensation at that point, she is likely to feel that justice has been done, just as she would if an insurance policy paid for the damage she had sustained."[54]

Other Financial Tools for Relocation—Limiting Spending and Taxes: There are financial tools other than purchase of land or easements that a state or local government might use to implement a plan for stepping back from the shore. Financial disincentives, in which some subsidy is reduced or additional financial cost applied, can discourage new development or redevelopment in coastal areas at risk of storm surge and sea level rise.

For example, a community might limit local spending on beach nourishment projects and other publicly funded investments in seawalls or protection structures. Beach nourishment projects sometimes are intended to enhance the recreational and economic value of beaches, thus benefitting a larger community, but often are designed to protect private property and other infrastructure. The federal government usually pays 65 percent of a beach nourishment project. For 2017 and 2018, the annual federal funding for these projects was a bit over $50 million.

Local governments might also send a price signal to property owners in areas at risk of storm surges and sea level rise using tax mechanisms. One approach to tax adjustments is to collect additional taxes that reflect the costs that communities expect to pay to safely remove private structures and supporting public infrastructure damaged by storms or inundation. For example, a community might set up a special taxing district to impose a surcharge on annual property taxes in an amount needed to support a dedicated fund to be used to remove structures or pay related costs of storm damages and rising seas.

Another approach is to offer coastal property owners in risky areas a property tax reduction in exchange for a binding rolling easement on the property in an amount approximately equal to the economic benefit the community anticipates deriving from the easement. The same end might be met by donating property or an easement to a nonprofit land trust with an expectation of a reduced property tax on the property.

Regulatory Tools for Relocation

The primary barrier to use of financial tools to implement relocation plans is scarce dollars. Communities without access to a large tax base or federal storm recovery funds have the less costly option of using regulatory tools to implement a relocation strategy by preventing development of undeveloped areas at risk from rising seas, forcing transfer of property ownership from unwilling sellers using powers of eminent domain, as well as to limiting changes to existing property and structures, including construction of shore protection measures. Any effort in this direction, however, must navigate the changing shoals of judicial decisions concerning the takings provisions of Fifth Amendment to the Constitution.

Communities focusing their storm surge and sea level rise adaptation plans on relocation, rather than structural protection or flood accommodation, can use some of the same regulatory tools discussed in the previous chapter, especially tools concerning where to build and tools limiting use of private coastal protection structures. In addition, the same limits that might be accomplished by purchase of a "rolling conservation easement" can also be accomplished using regulations. As discussed earlier, the primary regulatory tools available to most local government are zoning powers and floodplain ordinances adopted under the National Flood Insurance Program. In effect, communities can use the zoning process to identify where relocation is the preferred response to rising seas and define restrictions for these zones (e.g., prohibition on coastal protection structures or regrading) that allow shorelines to migrate inland as sea level rises. This approach would be less controversial politically and legally if applied to land already zoned for open space or other undeveloped lands and more controversial if the rolling easement zone put homes or other major assets at risk.

Local governments interested in using regulatory authority to implement a relocation approach to sea level rise need to consider the legal basis for regulatory actions. Virtually all coastal states have delegated zoning authority to local government, but the level of government holding these powers varies in different parts of the country and may be different from the government responding to sea level rise. Some states require that zoning be directly linked to a local comprehensive plan. Some states give local governments broad powers while others limit local governments to activities specifically delegated to them by the states.

An example of a rolling easement expressed as a regulation is the Rhode Island Coastal Resources Management Program provision for setbacks of most development activities from an inland boundary with the distance determined based on thirty times the annual erosion rate, or sixty times for larger projects. The county of Kaua'i, Hawaii, has a coastal setback requirement that is based on the average annual erosion rate multiplied by a planning period of 70 to 100 years, providing an even greater rolling buffer.

Finally, although not a regulatory measure per se, some local governments have adopted policies to deny municipal services, such as roads, or utilities such as water, sewer, and power. Services might be withheld to proposed new development in risky areas, terminated to existing properties, or simply not replaced after a storm. This approach can be linked to other strategies and has the admirable characteristic of low cost when compared to land purchase from willing or unwilling sellers. Perhaps more important, because it does not directly impact private land, it is less likely than some other regulatory strategies to takings claims.

Avoiding Takings Claims for Regulatory Rolling Easements

By far the biggest challenge in using regulatory measures to manage sea level rise, however, is crafting a regulatory restriction that is consistent with the Fifth Amendment of the Constitution: " . . . nor shall private property be taken for public use, without just compensation."[55] Like other constitutional amendments, the Fifth Amendment carries with it a complex history of judicial findings and shifts over time as new cases are decided.

For local governments opting for regulatory tools to implement a relocation strategy to save money, an order by a judge to compensate property owners because the regulatory restriction is deemed a taking is not happy news. Although a local government is better off not having to pay a takings claim, it is important to note that a claim for a rolling easement restriction for sea level rise decades in the future may prove to be very small. Still, a critical question is how can regulatory tools limiting development of buildings and coastal protection structures be designed consistent with the prohibition on takings in the Constitution?

A first key issue is whether a regulation denies "all productive use" of the property or simply constrains the use. In *Lucus v. South Carolina Coastal Council*, the Supreme Court "attempted to draw a line in the sand by stating that any restriction on use that leaves an owner with no economic value is the equivalent of expropriation without regard to its purpose or public benefits."[56] Georgetown University professor Peter Byrne concluded that "on its face, *Lucas* presents a formidable barrier to land use regulations implementing a retreat strategy because it mandates compensation for total prohibitions on development even if justified by the need to protect the shoreline ecology."[57]

A local regulation that passes muster under the *Lucus* test, however, might still be deemed to constitute a taking if it fails a further test described in a Supreme Court decision in *Penn Central Transportation Co. v. New York City*. Sometimes referred to as a "three-prong test," the court identified factors that need to be considered in finding that a government regulation was a taking. Was there a major economic impact on the person regulated? Does the regulation upset a person's "distinct investment-backed expectations,"[58] such as an investment in a home with a mortgage? What is the character of the governmental action (e.g., is there a physical invasion by government or does the regulation adjust "the benefits and burdens of economic life to promote the common good).[59]

The good news is that rolling easements applied by regulation are not likely to be found to be a taking. Georgetown University professor Peter Byrne explains:

> The rolling feature, of course, helps retreat regulations pass regulatory takings review. Rolling regulations avoid the *Lucas* rule because they permit development and use now, which should have substantial economic value. . . .
> A court reviewing a rolling development restriction must consider its effect on the whole property for its full duration. For example, a beach house that will eventually become subject even to an extremely strict rolling regulation that would require the house to be abandoned would still have a substantial

current economic value based on the estimate of when sea-level rise would push the restrictive zone upon it. The key to this analysis is that the regulation applies immediately but restricts the property only when necessary to achieve the public purpose.[60]

So a regulatory rolling easement is not a taking under the Lucas rule, but does it also comply with the requirements in the *Penn Central* decision? Most experts in this area of law agree that carefully drafted regulations for rolling easements will not be found to be takings. In the context of the *Penn Central* decision idea that a regulation should not substantially undo a person's "investment-backed expectation," James Titus notes that

> the most likely situation in which a court would find a taking would be when someone buys shorefront property before a regulation to protect tidelands is enacted and then is forced to abandon that property. The more common scenario would involve people who purchase property after the regulation is issued. These people would find it almost impossible to successfully challenge the regulation as a taking, because the regulation will have been factored into their investment-backed expectations.[61]

The takeaway message here is that the process of stepping back from the coast will play out over many years and the sooner communities make their intentions known to property owners the better.

In addition, public education and disclosure of sea level risks can reduce the chance that in the future a court would look back at a risky investment and find that it was reasonable. As Thomas Ruppert observed in the *Journal of Land Use*, "While no one part of the *Penn Central* analysis necessarily trumps, ensuring that coastal property owners have full understanding of the nature of the hazards, the dynamic coastal environment, and existing and potential regulatory limitations should demonstrate that owners' expectations which are drastically out of line with these realities and information are not reasonable."[62]

The squishier prong of the *Penn Central* test is defining what the Supreme Court meant by the "character of the government action" generally and in the context of sea level rise more specifically. Devon Applegate explored this question in some detail in 2016, concluding that the term has remained "remarkably undefined,"[63] but that a 2001 opinion by Justice Sandra Day O'Connor suggests "giving deferential treatment to important regulations that possess strong public purposes. Because of the serious dangers associated with sea level rise, related regulations will necessitate deferential treatment. Regulations related to sea level rise will easily overcome the obstacles posed by the Takings Clause if courts utilize Justice O'Connor's regulatory takings analysis."[64]

Last but not least is the intersection of the takings clause with the idea of government withholding of services or inaction. Can government inaction result in a "passive" taking of private property for which the government is liable? In some cases, courts have found that it does. The circumstances that would lead to a passive taking are not yet well developed by the courts and hotly debated by legal scholars and "the balance of action and inaction required for a 'passive taking' is still in flux."[65] Further complicating the debate, some scholars even argue that liability for inaction could arise from a failure to implement a retreat or protection strategy as well as failure to provide or maintain services.

Making a good match between the circumstances of a coastal state or community and the diverse array of relocation tools and practices, ranging from buyouts to regulations, is an intimidating challenge. Today, states and communities interested in these options are sorting out the issues without much help from the federal government or civil society. Fortunately, this problem is starting to get attention from climate change and coastal management experts. For example, the Georgetown Climate Center held a series of workshops in early 2019 as a step toward developing a "Managed Retreat Toolkit," to help organize practices and case studies in this area.

Looking Inland

Coastal communities will all face different relocation challenges. Some communities will need to relocate only a few homes or neighborhoods while others will need to move larger neighborhoods or the entire community. Some communities may have large numbers of transient, second home properties while others have cultural heritage dating back hundreds of years. Low income and minority communities may have fewer relocation options than wealthier communities. The risk of recurring coastal storms and the elevation of land areas at risk will change the sense of immediacy in addressing flood and inundation issues.

Simply stepping back from the coast may be sufficient in many places. People displaced from inundated property, or property isolated by rising water, can sometimes relocate on the same property or to another property on higher ground in the same community. People displaced from homes in communities along barrier beaches or in areas where the only high ground is in another community face decisions about whether to disperse on an improvisational basis or whether to stay together as a community. In some cases, such as south Florida, coastal Louisiana, and greater Norfolk, Virginia, entire regions will need to develop strategies to step back from the coast.

In a country as vast as the United States, it seems like finding a place for people displaced from coastal areas by storm surges and rising seas would be pretty simple. Recent research, however, indicates that this may not be the case.

A first red flag suggesting that finding places to relocate to could be difficult is the number of people involved. Mathew Hauer and coauthors concluded in a 2016 study that, after adjusting for continued coastal population growth to 2100, between 4.2 and 13.1 million people in the United States will be at risk of displacement as a result of rising seas.[66] The ultimate number will depend on the actual growth in coastal populations and investments in, and success of, adaptation practices and measures.

Another reason managing relocation of migrants away from the coasts will be difficult is that it is hard to know where they will go when left to their own devices, and how to prepare the places they go to. A year after his 2016 study, Hauer published a remarkable paper providing an estimate of both the number and destinations of potential sea level rise migrants in the United States over the coming century, concluding that "unmitigated sea level rise is expected to reshape the U.S. population distribution, potentially stressing landlocked areas unprepared to accommodate this wave of coastal migrants even after accounting for potential adaptation."[67]

Although every state is projected to see some population impact of sea level rise migration the largest shift is from Florida to Texas, "Florida could lose more than 2.5 million residents due to 1.8m [5.9 feet] of sea level rise while Texas could see nearly 1.5 million additional residents."[68] Hauer also found, "CBSAs [core based statistical areas] such as Austin TX, Orlando FL, Atlanta GA, and Houston TX could see more than 250,000 previously unforeseen future sea level rise net migrants each. . . . Thirteen CBSAs could see more than 100,000 sea level rise net migrants by 2100 with 1.8m of sea level rise."[69]

Projected population shifts at this scale are hard to grasp, but there is evidence that some migrations are already underway. Reed College economics Professor Jon Rork reported annual increases in retirement age people leaving Florida for states including Georgia and North Carolina between 2012 and 2017, noting that "there's a hypothesis that those who have left Florida for Georgia and North Carolina have done so to avoid hurricanes and big insurance premium jumps."[70] Some Florida developers are responding by offering housing located inland and designed to be resilient to storms, such as newly developed Babcock Ranch, Florida, located thirty feet above sea level.

This research suggests that large-scale migrations as a result of rising seas will have consequences, not just for people migrating, but for people far from the coast in places the migrants find as their new home. This insight opens the question of whether governments preparing for rising seas should include paths to new homes in specific places and what should be done, if anything, to prepare the social and physical infrastructure to accommodate this process.

Another strand of research is looking at barriers to inland migration and constraints on the land available for the resettlement of millions of people globally. Charles Geisler and Ben Currens published research in 2017 on possible paths of coastal migrants moving inland and found multiple barriers. Some lands are arid, or likely to become so, or otherwise not suitable for habitation. Other lands are already fully occupied by people or by critical ecosystems. And, patterns of private land ownership can result in "exclusion zones" analogous to "gated communities" where coastal migrants may not be welcome.

Geisler and Curren report that the high-end estimate of "total spatial mortgage" globally is "roughly 70% of the Earth's current terrestrial area . . ."[71] Admitting that there are many challenges to "reliable spatial estimation," the authors offer this startling insight, suggesting the inevitability of relocation: "The real choice is not, as some insist, between coastal adaptation and global migration away from global coastal zones but between balkanized versus integrated land use planning across large regions where climate migrants will relocate."[72]

People in the United States can be confident that their "spatial mortgage" is less than that of many other countries, but to be sanguine about large-scale relocation of coastal migrants might be a mistake and issues related to income and race will need careful attention. Geisler and Curren call on countries to prepare for inland migrations and for adoption of "integrated habitability policies"[73] and a strategy that "abates, as much as possible, the barriers to entry."[74]

In the United States, reckoning with the consequences of inland migration is not yet generally recognized as part of the coastal planning model, although the state of California speaks to this idea briefly in its 2018 *Fourth Climate Change Assessment*, noting that "with increasing sea level rise and coastal storms by mid-century, localities may begin to consider retreat strategies, which may require the expansion of inland cities."[75]

Relocation Endgame

As difficult as the initial planning and implementation of a relocation strategy is, there will also be a series of hard choices decades in the future as relocation becomes a reality. For some communities, especially those along the Gulf of Mexico and south Atlantic coast, choices may be forced by a severe storm bringing a decision whether to take a flood insurance loss payment or other buyout offer and abandon property. In places where the fates are kind and no major storm forces the issue, individuals and communities face the question of whether to stay until inundation puts property on the seaward side of the high tide line, transferring it to public ownership, or whether to step back more strategically ahead of inundation. Noted sea level rise expert Orrin Pilkey summed up the problem: "We can plan now and retreat in a strategic and calculated fashion, or we can worry about it later and retreat in tactical disarray in response to devastating storms. In other words, we can walk away methodically, or we can flee in panic."[76]

Storm Damage Forcing Relocation: For many of the places along the coast with the most population at risk, the likely scenario is one or more major storms over the coming decades. These storms can be frightening as they occur and, as sea level continues to rise, will drive farther inland than in the past. As time passes, the repeated example of the huge financial and human costs of these coastal storms, and a growing appreciation of the inevitability of sea level rise, may either drive demand for massive coastal protection structures or prompt a willingness to make plans for relocation of assets at risk.

Making decisions about the future of a community in the aftermath of a storm has downsides. Perhaps most important, decisions after a storm are focused on the temporary surge flooding from the storm rather than the longer-term risks of rising seas. This might result in a decision to rebuild damaged homes in places with the additional protection of elevation or other flood management measures. This strategy can consume huge sums of money, drawn largely from the federal flood insurance funds or special appropriations, for a short-term remedy that is not sustainable as sea level rise brings permanent inundation, making transportation and other infrastructure service to inundated areas impractical.

Decision making in the context of storm recovery is also likely to be colored by a "we will not retreat" outlook and a determination to build back what was lost regardless of cost or future risk. The availability of flood insurance or generous disaster assistance dollars can make a "no retreat" approach seem sustainable.

Finally, in the rush to recover from a storm and get back to life as normal, there is often little time for the deliberative planning that can inform the community about long-term risk of permanent inundation from rising seas or consider other issues such as impacts on the welfare of disadvantaged people or communities.

Relocation Endgame Without Storms: For communities that are spared a major storm, the first impacts of sea level rise on property, public infrastructure, and natural systems may be decades away. Depending on local elevation, rate of land subsidence, and other factors, the speed of property loss and other impacts may be slow or surprisingly fast. A local decision to invest in protection structures can extend the life of some property and communities, but usually it will not be enough to defeat rising seas in the end. State and local governments need to come to terms with an appropriate relocation strategy, the mechanics of transferring inundated private property to state ownership, including whether to compensate these losses, and how to safely remove structures to prevent environmental and human health hazards.

In some cases, shifting ownership of property at risk of sea level rise from private to public ownership may be accomplished by means of a buyout. For most coastal property owners, however, the limited funding available for buyouts means that the end of the line is the time that the property or structure shifts from the landward to the seaward side of the shoreline and becomes the property of the state. The loss of property coupled with lack of a buyout will be a painful and, in some cases, economically devastating outcome for many coastal property owners. Multiplied many times at a community or regional scale, this outcome can lead to social as well as physical disruption.

For property not acquired outright by buyout or another form of acquisition (e.g., donation or purchase option), the community has the choice of waiting until the property is fully inundated and ownership is transferred, allowing the local government to cease providing public services and supporting infrastructure without risk of a takings claim. This is the low-cost approach in the sense that there is no payment to the former owner to acquire the land. It may, however, cost more in the long run in that the local government needs to provide services to marginal property and will need to pay much higher costs for safe decommissioning of structures and infrastructure on land that is already underwater. By being in a strictly reactive position, the community also foregoes the advantages of coordinating with neighboring communities to stage a planned relocation of critical infrastructure and assets.

For communities that do not want to wait for property to simply fall into their hands by the action of a rising sea, the choices are to acquire property at risk from willing sellers, acquire property from unwilling owners using eminent domain authority, or acquire a rolling conservation easement. The timing of these actions is important in that costs are likely to decline the closer the property is to inundation, but the benefits to the community are greater when acquisition occurs far enough ahead of inundation to allow for coordinated management and decommissioning. Communities may be tempted to wait to the last moment in order to sustain property tax revenues. Property owners may be tempted to hold out for a buyout perhaps funded by the deep pockets of the federal government after a storm. Finding the "sweet spot" moment when the community is best served by acquiring property far enough ahead of inundation to allow smart management but close enough to inundation to pay a reduced price, while sustaining property taxes as long as possible, will likely be an art form increasingly appreciated in coastal communities.

Coordination within Neighborhoods and with Neighboring Jurisdictions: With an appreciation of the complications that come with individual property transfer and decommissioning, communities will need to come to grips with the choice of whether to treat each property as a separate case as it presents itself or whether to get ahead of the process by acting preemptively to resolve these issues for multiple properties.

In addition, communities that decide to act ahead of the impacts of rising seas also face choices about whether to simply deal with property adjustments within their jurisdiction or whether to cooperate with neighboring jurisdictions and perhaps the state. Neighboring communities, for example, might agree to stage steps back from the shore along a common line of defense from one jurisdiction to another.

Cooperation might include work with state and federal agencies to develop migration pathways for ecosystems and habitats optimized on a regional basis and protected from alternative uses. Communities might also cooperate with state and federal efforts to identify community relocation sites and the best strategy for protecting or relocating critical coastal infrastructure facilities, such as power plants, water treatment facilities, transportation assets, or coastal ecosystems. Finally, some communities happen to be home to economically significant water-dependent uses, such as commercial fishing infrastructure, which are important to the regional economy, and communities might agree to protect or relocate these uses to the most advantageous sites in any of several communities.

Including options for stepping back from the coast as part of the process of responding to more severe storms and rising seas will be an uphill battle. It raises difficult legal issues. It can appear to be more expensive than short-term solutions offering structural protection from flooding, or temporary accommodation of flooding. In addition, more than other options, relocation faces a psychological hurdle—it is a natural reaction to fight to protect a place you know and love, especially after a devastating storm. And, in many places, relocation options highlight social issues related to income and race. The next chapter explores these psychological and social issues, including recent research into the question of whether relocation strategies can be designed to minimize these concerns.

16

Social and Psychological Dimensions of Coastal Storms and Rising Seas

Supreme Court opinions and arcane financial models will have a lot to do with the success of the country's transition to a new coast. Sorting out the legal and financial issues is important, but it is equally important to come to grips with the consequences for disadvantaged people and communities and for the emotional health of all people in coastal communities.

Unfortunately, public officials struggling to deal with the immediate impacts of a storm or the complex issues around rising seas are sometimes reluctant to acknowledge social inequities or emotional strains. They may simply feel they lack the skills or training to take on these issues. They may feel that their job is limited to storm recovery or defining a sea level strategy and does not extend to correcting long-standing social injustice or the intangible, emotional impacts of these policies. Or, they may feel responsible for these issues but think that they lack the authority and resources to make needed changes.

A first step in addressing the social and emotional dimensions of coastal storms and rising seas is for elected officials and other policymakers to make sure that they understand and acknowledge that storm recovery and sea level rise preparedness have social justice and emotional consequences. A growing body of experts, however, argues that simply acknowledging these problems is not enough. Responses should not perpetuate past inequalities, they say. And, plans should not simply reduce economic costs without also focusing on reducing inequality, emotional stress, and loss of community identity.

Coastal Storms, Sea Level Rise, and Social Justice

There is extensive research on the disproportionate impacts that disasters and a changing climate have on disadvantaged communities and people. The Federal Emergency Management Agency (FEMA) supported a major demonstration project following Hurricane Katrina, looking at disaster impacts on socially vulnerable communities, which concluded that "research shows and experience has underscored the fact that disadvantaged people—children, the elderly, those with low-wealth, the disabled, and those who don't speak English—suffer disproportionately during major disasters. For example, people with low wealth don't have sufficient resources to rebuild or move elsewhere. By virtue of their poverty, many are bound to a piece of land they already own or to a low-tax, low-service locale."[1] The Fourth *National Climate Assessment* notes that response planning can make matters worse: "With the limited and often expensive adaptation opportunities currently under consideration, including elevating properties or constructing seawalls, climate-driven impacts may lead to a great deal of unplanned and undesired community change that is likely to disproportionately impact communities that are already marginalized."[2]

Limited Adaptation Options of Socially Vulnerable Communities: In thinking about coastal storms, sea level rise, and social equity issues, it is helpful to have a sense of the scale of the problem. In a 2013 study, Jeremy Martinich and coauthors developed a Social Vulnerability Index, including factors such as poverty and age, and looked at the coastal populations in areas at risk of sea level rise. They concluded that 22 percent of that population, almost 750,000 people, were in the top two categories of social vulnerability.[3] More than half of this coastal and socially vulnerable population is widespread across the Gulf coast while another 200,000 are in small geographic areas in the Northeast.[4]

Martinich then looked at the costs of structural protection measures, including seawalls and beach nourishment, and the value of property along the coast, estimating that protection measures would be implemented in places where the value of the property exceeded the cost of the protection structures. Places where the cost of structures or beach nourishment exceeded the value of property were expected to be abandoned. When the areas protected or abandoned were compared to areas with high social vulnerability the authors found that, "as social vulnerability increases, the area and population protected from sea level rise risk (armored and nourished) decreases while the area and population abandoned increases, relative to the total area at risk."[5] The authors note, however, that, "benefit-cost criterion may not always be applied to protection decisions,"[6] and "this result highlights the need to consider factors other than just economic efficiency in coastal adaptation decision-making."[7]

Recent studies looking more generally at all disaster impacts, rather than just sea level rise, confirm social inequalities in disaster response. Junia Howell and James Elliott looked at the connection between disaster damages and wealth inequality, concluding in late 2018 that "results indicate that as local hazard damages increase, so does wealth inequality, especially along lines of race, education, and homeownership. At any given level of local damage, the more aid an area receives from the Federal Emergency Management Agency, the more this inequality grows. These findings suggest that two defining social problems of our day—wealth inequality and rising natural hazard damages—are dynamically linked."[8]

National Public Radio released its investigation of this topic in early 2019 finding that "federal aid isn't necessarily allocated to those who need it most; it's allocated according to cost-benefit calculations meant to minimize taxpayer risk. Put another way, after a disaster, rich people get richer and poor people get poorer. And federal disaster spending appears to exacerbate that wealth inequality."[9]

One option for government intervention, though arguably short-sighted, is to invest in equalizing the level of structural protection provided for people in areas where property values would not otherwise justify protection. Another approach is to provide a more far-sighted strategy recognizing the limited value of protection structures and facilitate a coordinated stepping back from the shore for everyone. This would involve finding safe adaptation strategies for entire communities, rather than just wealthy landowners, and a planning framework working on a geographic scale large enough to manage relocation.

Ironically, in a few places, past discrimination has actually steered disadvantaged people away from sea level rise risks. For example, Erika Bolstad reported on sea level rise in Miami, Florida, noting that "neighborhoods, formerly redlined by lenders and in some places bound in by a literal color wall, have an amenity not yet in the real estate listings: They're on higher ground and are less likely to flood as seas rise."[10]

The more common experience, however, is that past patterns of social injustice are magnified in the context of major storm damages. For example, the *Washington Post* reported that food supplies were very limited in Puerto Rico after Hurricane Maria but, "a package of federal programs that could have helped feed thousands—and that channeled hundreds of dollars to needy families in Texas and Florida after hurricanes Harvey and Irma—have not been deployed in Puerto Rico, limiting the reach of the island's emergency food assistance."[11] The *Post* linked this shortfall to an earlier federal government decision to limit funding for the Supplemental Nutrition Assistance Program (SNAP) in Puerto Rico compared to other parts of the United States.

Not all the social justice challenges exposed by Hurricane Maria can be traced back to the federal government. In January 2018, Hilda Lloréns and coauthors commented

> Hurricanes are thought of as 'natural' disasters, but the social and environmental devastation wrought upon Puerto Rico by Hurricane María last September is really an unnatural disaster resulting from a long history of colonial subjugation, economic hardship, environmental injustice, infrastructural neglect, and, at the local level, a broken rule of law. Hurricane María affected all of Puerto Rico to some degree, but in doing so the disaster also exposed the vulnerabilities created by ubiquitous socioeconomic inequality and the differential neglect of the island's rural regions.[12]

Social Justice Consequences of Adaptation to Storms and Rising Seas: Although storm and sea level rise response programs are not likely to be the best path toward righting underlying social injustices, the academic research includes many examples of the potential for government to harm or help disadvantaged communities as it makes decisions on managing storm and sea level rise risks.

Even a government decision to not make an adaptation decision and allow market forces to play out has consequences for disadvantaged communities. Researchers at the Massachusetts Institute of Technology (MIT) looking at adaption options for Boston found that "those areas of Metro Boston that can expect the worst damage are also home to low-income populations with low mobility and scant resources to afford to move. . . . Market-driven relocation, moreover, does not actually force families to leave, but instead leaves them in financial straits after already enduring the losses as a result of disaster."[13]

Alice Kaswan, professor at the University of San Francisco School of Law, points out that communities opting to invest in structural protection must make decisions about what areas to protect: "Using land value as the primary metric would preserve rich neighborhoods and doom poorer neighborhoods. Differences in political power are also likely to determine who receives protection and who must leave."[14] She points out that a policy of buyouts for damaged structures also has consequences for disadvantaged communities noting that "although government purchase would provide essential resources to individual families, it cannot prevent the loss of social capital: the social networks that create 'community' and provide demonstrable economic and social welfare gains."[15] In addition, relocation of entire communities would reduce social and cultural harm, but "even community relocation is no panacea, however; it requires substantial resources, identifying an appropriate relocation site, and, for communities whose cultural identities are tied to a geographic place, the risk of fundamental cultural disruption."[16]

Aptly demonstrating this point are tensions arising in the effort to relocate the Isle de St. Jean Charles in Louisiana. Ted Jackson reported in early 2018 that, despite having $40 million to work with, "islanders have grown impatient and distrustful. Plan administrators, on the other hand, have grown weary and defensive."[17] Other issues have come up, including whether people from outside the Biloxi-Chitimacha-Choctaw Tribe will be relocated to the resettlement site, and whether people relocating to the new site are giving up ownership of property at the old site.

In summing up a study of Hurricane Sandy, Professor Chris Sellers of Stony Brook University concluded that "less affluent groups of people suffered more, both in the initial damage and recovery."[18] He called for governments to recognize and address these inequities: "Now more than ever, we need a nationwide conversation on ways our coastal landscapes have developed so that our most vulnerable citizens are now at greater risk from such massive storms. Officials need to find more reliable ways of illuminating problems faced by the less advantaged, and to ensure these are addressed as quickly and effectively as those of the better-off."[19]

How can decision makers more reliably respond to the needs of disadvantaged communities and people? Professor Kaswan argues that "adaptation policies that treat everyone the same, regardless of underlying demographic characteristics, will result in substantial inequality given underlying differences. Equitable adaptation can be achieved only by explicitly addressing the demographics of affected populations and targeting assistance toward the most vulnerable."[20]

A national commitment to spending the money needed to result in fair treatment of disadvantaged populations in strategies responding to storms and rising seas has important implications for the overall direction of these strategies. For example, a national strategy focused on building protection structures along the existing coast for just select, high-value property might be affordable while protecting all or most property on the coast might not. But a relocation strategy, in which everyone steps back from the coast, puts everyone at the same table working to make the best use of the financial resources that are available, in effect, a fresh start.

Psychology of Storms and Sea Level Rise

Most people appreciate the psychological implications of major disasters such as hurricanes. The American Psychological Association (APA) reports that "disasters carry the potential for immediate and severe psychological trauma from personal injury, death of a loved one, damage to or loss of personal property (e.g., home and pets), and disruption in or loss of livelihood."[21] The APA went on, noting that "well after floodwaters had receded, interviewees . . . noted that they were still experiencing panic attacks, difficulty sleeping, low motivation, and obsessive behavior."[22]

A major 2016 study of health effects of climate change by the EPA came to similar conclusions: "Depression and general anxiety are also common consequences of extreme events (such as hurricanes and floods) that involve a loss of life, resources, or social support and social networks or events that involve extensive relocation and life disruption."[23]

Some strategies for adapting to rising seas, such as building protective structures, are likely to have negligible psychological impacts on people in the community and might even provide some positive reassurance. But, as sea level rises and protection structures are no longer effective or affordable, relocation options will be considered. The emotional distress that comes with relocation, even within a neighborhood or community, can be significant. Relocation to a new or different community can be even more stressful, especially if the relocation is perceived as forced.

Leaving Home: The psychological impacts of relocation in response to rising seas are not as well recognized as the psychological impacts of major storms. In her call for a "Climate Change Relocation Plan," Victoria Herrmann explained the problem:

> The social, psychological, and cultural-heritage loss and damage that come from severing a community's attachment to a place-based identity have been explored in research on development-induced displacement and resettlement. . . . This worsening condition stems not only from the physical stress of being displaced from their homes, but also from the loss of community and social safety nets when relocation is focused on individuals rather than a cohesive, intact community. When people are displaced, they are unlikely to establish new social support systems in their new locations, and when a community is dispersed or co-located to another settlement, those social networks are disrupted.[24]

Another, less analytic perspective on the tensions related to leaving a community came from Anthony Carolina, a resident of New Orleans, in an interview with New Orleans Public Radio:

> But man, the flooding's getting worse and worse and worse and worse. I'm ready, man. I—I—I should have left after Katrina but my wife said 'no, let's—let's just build a house.' It's just a matter—I'm tired. I'm ready to go. The school buses can't turn around on certain days, you know I had to track my

kids through water. I mean I had—the National Guard can't even get back to us, the water gets so high. The Fire Department can't get back there. You're trapped in your house 'cause there's only one way in and out.[25]

Researchers at MIT reported that the mental stress associated with relocation is even more pronounced in cases where there is a perception of forced, as opposed to voluntary, relocation:

> Recent scholarship has articulated the potent and damaging legacy of forced relocation on mental (and physical) health. Fullilove and Wallace (2011) have used the term "root shock" to describe the long-term impacts of dislocation on both individuals and communities. Root shock represents a critical consideration in the context of climate change because it reaffirms the attachment of individuals and communities to physical spaces and the memories associated with those spaces. Keeping families and communities relatively close to their former homes following displacement preserves the valuable social capital of close-knit neighborhoods and must be a critical part of any relocation policy.[26]

Owners of some 600 beach-level homes in Del Mar, California, expressed a mental anguish of another sort when the city proposed a managed retreat strategy. As the *San Diego Union Tribune* reported in 2017, "It is an option that is not sitting well with residents of the small community. . . . Homeowners worry that their property values would plummet, insurance rates would skyrocket, and mortgages would be harder to obtain if word gets out that the city is considering retreat. Instead, they say, the city should find more ways to pump sand onto its eroding beaches."[27] In response, the city "deleted the retreat strategy from the latest version of its sea-level rise adaptation strategy that was released last week."[28]

Homeowners in the communities of Rockaway Park and Oakwood Beach, New York, came to different conclusions when offered a community buyout after Hurricane Sandy, "with residents of Oakwood Beach enthusiastically pursuing the plan, while residents of Rockaway Park are, largely, choosing to remain in their community."[29] Although residents of the two communities were comparable in many ways, the study found statistically significant differences in three areas: "On average, residents of Rockaway Park had higher levels of individual sense of community than residents of Oakwood Beach. Residents of Rockaway Park also had higher levels of sense of place. . . . Residents of Oakwood Beach had, on average, higher levels of trust [in public officials] than residents of Rockaway Park."[30]

Since residents of Oakwood Beach decided to move in 2013, the state of New York has been working to appraise and buy houses, using federal funds, at prestorm values, and removing buildings and infrastructure with the goal of returning the land to open space by 2022. The director of the Governor's Office of Storm Recovery commented that "leaving people in an area that will always flood is just not appropriate anymore, especially as extreme weather becomes more and more prevalent. . . . At some point you have to say, 'We're not going to leave people living in a wetland.'"[31]

Looking more generally at a sample of homeowners affected by Hurricane Sandy, Professor Anamarie Bukvic and Owen Graham from Virginia Polytechnic Institute reported homeowner preferences for a strategy to prevent future storm and sea level rise impacts.

> A significant number of respondents prefer engineering solutions such as levees and seawalls, followed by natural barriers (wetlands, sand dunes, vegetation) in the 'somewhat preferred' category. . . . As for relocation to a safer area, the majority of respondents (63 percent) rank the option as 'somewhat less preferred,' while a marginal number rank this possibility as 'most' or 'somewhat preferred.' This suggests that most have a strong commitment to stay in place and first explore in situ adaptation and hazard mitigation strategies before considering relocation.[32]

A public preference for armoring the coast over relocation, however, is not universal. A study by Karen Akerlof and coauthors evaluating public views of sea level rise options in the mid-Atlantic states found stronger support for retreat and soft protection than hard protection or armoring. In 2017, Miyuki Hino and coauthors looked at twenty-seven experiences with managed retreat from rising seas and other natural hazards all around the world, involving 1.3 million people. Hino found that "managed retreat is often controversial because of the social and psychological difficulties in displacing people from their homes, 'the central reference point of the human existence.'"[33]

Are there circumstances that increase the chance of successful relocation? Hino identified two key factors—whether relocation was sought by residents and whether benefits accrued to broader society, such as reduced spending for disaster relief, termed the "mutual agreement" case. Success declined in cases where government sought relocation for the "greater good" without support of people being relocated and in cases where people asked to be moved without a compelling argument that broader society would benefit.

Hino notes that "mutual agreement" cases are mostly in post-disaster settings but that "enabling pre-disaster managed retreat . . . may boost local input and ownership by eliminating the time pressure of post-disaster settings,"[34] and a key step is for government to "delineate where and when retreat might be encouraged or required."[35] Hino also offers the important observation that, "residents may oppose retreat despite severe natural hazard risk," but that if a relocation destination can preserve valued elements and features, "retreat may become more acceptable."[36]

It is worth noting that some of the factors that seem to reduce public reluctance to relocate are easier to implement as part of a national, rather than more localized, response to coastal storms and rising seas. Defining areas to move from and areas to move to, and especially making a case that relocation is for the good of the country as a whole, are most likely to result from a national effort. In addition, a key way to encourage, rather than require, relocation is with an effective buyout or comparable financial assistance program, which can best be financed nationally (see ch. 19).

Local Decision Makers Cope with Relocation: Given the difficult challenges that coastal adaptation strategies present, it is worth considering the psychology of relocation decisions from the point of view of the government decision maker, often a local elected official, rather than that of a community resident.

Administering a voluntary buyout program responding to the devastation of Hurricane Sandy with large reserves of federal funding is challenging but manageable. Pressing ahead with buyouts or regulatory measures to serve the community's long-term interests, however, can be costly and controversial. Individual property owners can drag out a purchase or eminent domain process, taking up time and resources. Community opinion may be divided over whether to just step back from the coast or relocate to another place. The tools used to meet either goal are likely to result in contested and uncomfortable meetings.

The state of Louisiana coastal plan proposes to relocate some residents, and state official Mathew Sanders clearly recognized the personal impacts of that decision, noting in late 2017 that "not everybody is going to live where they are now and continue their way of life . . . and that is an emotional, and terrible, reality to face."[37]

In a study of the political dimension of local government decision making on sea level rise adaptation, Professor Mark Gibbs at the Queensland University of Technology concluded that "the adaptation approach that provides the most future protection to people and buildings (pre-emptive retreat), is the same one that carries the highest short-term political risk"[38]

Professor Gibbs points to several factors making relocation options politically difficult. Rather than supporting an optimal adaptation approach, "many homeowners at times take a substantially shorter-term view and are often concerned over immediate negative impacts of proposed adaptation plans."[39] He also notes that alternatives to relocation, such as structural measures, can look attractive because they are often funded by "a greater regional tax base."[40]

Recognizing "trepidation among decision-makers to acknowledge relocation as a viable adaptation option and engage host communities in relocation schemes,"[41] Professor Bukvic points out that the occasional coastal storm can have a silver lining, opening the door to conversations about long-term risks such as rising seas: "Coastal disasters, as discernible manifestations of less visible chronic risks, can present an opportunity to increase community awareness of hazards and short- and long-term options to mediate these problems, and introduce the concept and possibility of relocation as potentially the most effective and safest option."[42]

With all the complex issues around responding to coastal storms and rising seas, accounting for the interests of disadvantaged people and communities might end up well down the agenda. That would be a mistake for several reasons. First, it is simply not fair that the interests of disadvantaged people and communities be an afterthought. Second, coming to grips with fairness and equity in the context of responding to these challenges forces some needed discipline in thinking about the limits of financial resources and how limited resources are allocated (e.g., can the country afford structural protection everywhere?). Third, approaching this problem with financial constraints and social justice clearly in mind will tend to focus attention on relocation strategies at a community or larger scale. Any response to storms and rising seas will be stressful, but relocation is especially so. Research fortunately points to ways to make relocation less objectionable, and these steps generally involve approaching the problem at a large geographic scale.

As communities, states, and the national government increasingly debate the choices they face as they respond to coastal storms and rising seas—ranging from planning, to investments, to regulations and relocation—a key voice with significant interest and influence will be that of the business community. The role of the private sector in this process, and the potential for business leaders to strengthen decision making at all levels, is addressed in the next chapter.

17

Business Community Response
to Coastal Risks

Federal, state, and local governments need to take the lead in managing the impacts of more damaging storms and rising seas on coastal communities. But, the private sector also has a critical role in protecting the very substantial residential, commercial, and industrial assets that are at risk. Private sector engagement includes ethical choices by individuals, such as real estate agents disclosing sea level rise risk to buyers, and they extend to the corporate level where business leaders need to consider coastal flood and inundation risks in their management strategies and participate in community choices about investments in flood protection and relocation.

There are cases of constructive action to address more severe storms and rising seas by some leaders in the business community but, on the whole, the private sector has not stepped up to this challenge. Professor Keith Rizzardi of St. Thomas University states the problem well: "Unfortunately, the voluntary conduct of an honorable few will not suffice. In the absence of a well-planned, coordinated, and comprehensive public and private sector response to the real threats of sea level rise, the entire community will remain at risk. Ethical leadership must emerge."[1]

Although the economic and social impacts of more severe storms and higher sea level will ripple through the entire American economy, some economic sectors are especially at risk. The coastal property finance sector—including real estate, mortgage, and insurance businesses—has significant exposure to declining coastal property values. Leaders in this sector can play a critical role in helping coastal communities stay

237

ahead of storm and sea level rise risks and minimizing the economic impacts of a transition to a new coast, but they have done little to take up this challenge.

In 2010, the Securities and Exchange Commission (SEC) released guidance on how to address climate change risks, including sea level rise, in corporate annual reports. In addition, several nonprofit organizations have developed corporate climate risk reporting programs. Providing meaningful reports on climate risks to the SEC, and participating in the growing movement toward reporting of climate risks, are good places for many businesses to start in improving storm and sea level rise preparedness. Unfortunately, comparatively few companies are reporting on climate risk generally or on the flood and inundation risk to physical assets in coastal areas.

Leaders in businesses located in coastal communities often have substantial influence and will need to decide how to react to the impacts of more damaging storms and rising seas on their community. These business leaders may well perceive the significant, interrelated impacts in store for coastal communities before the rest of the population and be better able to make pragmatic decisions in response. Will they use that advantage to follow an "every man for himself" strategy, especially with respect to relocation to optimal inland sites? Or, will they stay in the community and engage coastal risk challenges in cooperation with other businesses and government planning efforts?

Finally, there is a tremendous reservoir of professional expertise among realtors, planners, lawyers, civil engineers, and architects that has the potential to provide important leadership and know-how to the sea level rise adaptation planning efforts by governments at all levels. These professionals, and the organizations that represent them, will need to decide whether they will wait to be told what to do about storms and rising seas or whether they will step up to apply their experience and expertise to help coastal communities, and the country overall, make sound adaptation decisions. So far, their record is mixed, at best.

Coastal Property Finance: A Precarious System

There are three major elements of the coastal property financial sector. The real estate industry markets and sells coastal property. The banking and mortgage sector finances the properties. And, the National Flood Insurance Program (NFIP), supported by private insurance companies, sells insurance to property owners to help them rebuild in the event of flood losses and also, by the way, to assure institutional mortgage holders that their investment is secure.

Voices of Concern: In the past several years, several leading voices in the coastal property sector have suggested that the current finance model is not sustainable. A key reason for the concern is that at multiple points in the financial system the price signal of the risk and cost of coastal property is muted. The lack of a clear price signal results in instability in the entire coastal property finance system with potentially significant consequences for the private sector, coastal property owners, coastal communities, and the American economy. More severe coastal storms and rising seas will make the existing system even more unstable.

In April of 2016, Sean Becketti, the chief economist for the Federal Home Loan Mortgage Corporation (Freddie Mac), the government-backed mortgage agency, reviewed the recent estimates of homes at risk of sea level rise and concluded that "rising sea levels and spreading flood plains nonetheless appear likely to destroy billions of dollars in property and to displace millions of people. The economic losses and social disruption may happen gradually, but they are likely to be greater in total than those experienced in the housing crisis and Great Recession."[2]

Writing in late 2017 in the *Inman Report*, a news source for the housing industry, Bryan Walsh, described a "coastal mortgage time bomb."[3] He summarized the huge impacts of the hurricanes earlier that year and the substantial federal recovery funding that "encourages residents to stay put and rebuild, rather than flee for safer areas."[4] Walsh then reviewed estimates of the number of homes at risk from rising seas, concluding that "eventually insurers could begin to pull out of coastal markets altogether, as could lenders who fear that homes won't be able to retain their value through the lifespan of a 30-year mortgage. Unable to get insurance to repair their repeatedly flooded properties—and tired of navigating the now constant risk of water—homeowners might end up desperate to sell, only to find that no one wants to buy. The result would be a wave of defaults."[5]

Walsh quotes Edward Golding, Fellow at the Urban Institute and former head of the Federal Housing Administration: "All of a sudden we're going to reach a tipping point and no one will touch these mortgages. At some point it becomes undesirable risk and people start pulling out from entire regions."[6]

Muting the Price Signal: It is widely thought that the demand for coastal property is driven by a human desire to be near the water, or even on the waterfront, to appreciate the natural beauty and the reduction in stress provided by the soothing sounds of ocean waves. This innate human attraction to the ocean is clearly a factor in coastal population demographics, but another factor is that the coastal property financial system has muted price signals that would otherwise warn people about the risk and cost of coastal property. Unfortunately, this muting of the price signal is occurring at every stage of the coastal property finance system due partly to a failure of the market to provide full disclosure of risks and partly to a failure of the government to step in to correct the failure to disclose true risks.

The first risk disclosure failure occurs with the sale of coastal property. Most realtors will tell you that the price of any property is set by the invisible hand of the market balancing demand and supply. Coastal property, especially waterfront property, is considered desirable and often has a higher sale price than comparable, noncoastal property. This higher sale price generates a higher profit on the sale for the realtor because the sale commission is a percent of the sale price.

As discussed in chapter 14, realtors are expected to disclose information about a property that is "material," but flood risk is not always reported, and reporting of sea level rise risk is unusual. Albert Slap, owner of a consulting firm that advises on flood risk, called lack of disclosure of flood risk a "dirty little secret" of the coastal property sector that now suffers from "systemic fraudulent nondisclosure."[7] Some states have passed laws governing flood disclosure. Nonprofit organizations have proposed a national requirement for flood risk disclosure. And, flood disclosure requirements were included in a flood insurance reauthorization bill, passed in the US House of Representatives, but not enacted. These steps, however, do not address property at risk of inundation by rising seas rather than flooding by storm surge.

The second major player in the coastal property financial system are the banks and other institutions offering mortgages to make the sale of real estate possible. In a well-functioning market, these institutions would exercise due diligence to assess the risks of storm surge and future sea level rise and price those risks into the interest rate they charge for the mortgage.

Internet-based marketing of mortgages, however, has made the industry highly competitive and measuring uncertain risks like storm surge and sea level rise may not drive a decision to charge a bit more interest. In addition, mortgages are commonly for thirty years and mortgage holders can reasonably hope to be paid off by the time a bad storm hits or sea level rise looms. Even if a big storm hits, the mortgage commonly requires that the property be insured and the insurance will help keep the mortgage viable. The result is lower mortgage interest rates not reflecting true risk. Buyers miss another potential price signal.

Not only may a mortgage institution hope to dodge the storm or sea level rise bullet over a thirty-year mortgage, the mortgage institution likely will not hold the mortgage anywhere near that long. A common practice in the mortgage industry is to bundle home loans into "mortgage-backed securities" that are then sold to investors, such as large pension funds. Because the bundle includes many mortgages with different financial characteristics and risk elements, investors rely on ratings agencies, such as Moody's and Standard and Poor's, to evaluate the underlying risks and rate the security. The mortgage institution wants as high a rating as possible to get the most from the investor so is not prompted to disclose to the rating agency more risk information than required, and flood and climate related risk can be well down the list.

Climate risk disclosure is a tool to help investors be informed about the risk exposure of a company, bond, stock, or other investment and is getting more attention recently. Climate risk reporting by all kinds of private companies is expanding, and investment ratings agencies like Moody's have announced new initiatives to improve climate analysis of state and municipal bonds. Unfortunately, in the field of mortgage-backed securities, there is less progress on climate risk disclosure. The 2006 enactment of the Credit Ratings Agency Reform Act gave the Securities and Exchange Commission (SEC) authority to improve transparency and disclosure. In 2011, the Dodd-Frank Act added requirements to disclose ratings models and authority to hold agencies accountable for ratings.[8] None of the new authorities, however, empowered the US Securities and Exchange Commission (SEC) to require rating agencies to consider climate change risks generally, much less coastal storm or sea level rise risks more specifically.

The Federal National Mortgage Association, known as Fannie Mae, has a "Green MBS" (mortgage-backed securities) program but focuses on green buildings and property improvements to reduce energy and water use, with no reference to climate risk or sea level rise. Nonprofit organizations working to build awareness of climate change risk into bonds and other investments, such as the Climate Bond Initiative, have not addressed mortgage-backed securities.

The net result is that, if you were running a mortgage company, you might be inclined to think that, despite doom and gloom forecasts, your industry does not have much to fear from more severe storms and rising seas. You are able to stay competitive by writing low-interest mortgages for coastal properties at risk and bundling them with other mortgages for sale to large-scale investors. Mortgage backed security rating agencies are not looking for these risks or reporting them in ratings and you are thus able to disperse risks across large segments of the economy. Even in the event of storm damage to a property, most people keep paying their mortgage in order to protect the home value they have left. With most mortgages written for thirty years or less, there is still plenty of time to sort out sea level rise risks. At some future date, interest rates on new mortgages can be raised or mortgages simply denied as too risky.

Two events could darken the picture for the industry. First, the losses along the coast will increasingly shift from repairable damages due to storm surges to full loss of property due to rising seas. With a total loss, there is a greater chance that homeowners will stop paying the mortgage, and the value of the property that reverts to the mortgage holder is negligible or even a liability. Second, the rate of losses due to sea level rise will increase faster in the future than it has in the past. Rather than just the occasional sea level rise loss, sea level rise will start affecting a larger percentage of mortgages and driving defaults. A worst-case scenario is that the losses and defaults mount at an unsustainable rate, and mortgage companies, and the larger economy, pay a high price.

The third failure of the financial system to convey the true cost of coastal property comes in the pricing of insurance for the property. Today, the vast majority of coastal flood insurance is provided under the National Flood Insurance Program (NFIP) (see ch. 9). The operation of the NFIP mutes the price signal in several ways. First, the very existence of a massive, federally operated flood insurance program suggests that the government has an interest in helping people stay in risky coastal areas or to move to these areas. In addition, the NFIP's long-standing subsidy of some premiums below market rates has the effect of improving affordability but also suggests that flooding costs are manageable. The cap of $250,000 on federal insurance for a single-family home suggests that damages are unlikely to exceed this amount, even though many insured homes have much higher values and as seas rise will be total losses. And, the decision to map flood zones based on historical information, without accounting for more severe storms or rising seas, has the effect of leaving some homes that are in risky places out of the program, suggesting that the scale of risk is less than it is.

A fourth muting of the price signal results from past experience with federal disaster recovery policy and funding. To date, most people's experience with government policy concerning coastal risk is to see the staggering sums delivered by special appropriations for recovery from coastal storms. As described in chapter 10, the federal government response to disasters is generous, and its response to major storms commonly runs into tens or hundreds of billions of dollars. This humanitarian policy can be misread as a "we have your back" endorsement of choices to locate in a risky coastal area or stay there even after suffering major damages.

The final capping failure to provide a price signal about coastal property risks can be ascribed to climate skeptics, including senior government officials and President Trump, who dismiss the science supporting climate change and sea level rise. It is hard to know how much impact the government's discounting of risk of more severe storms and rising seas has in encouraging people to stay in risky coastal places or to move to these places. As noted in chapter 8, however, the denial of climate science by the current administration is pervasive and strong.

Consequences of Muting the Price Signal: What is the harm in muting the price signal concerning the risks of coastal property? Under a business as usual scenario, the sales, mortgage, and insurance processes all function. Low-cost insurance backs up affordable mortgages that keep sale prices high. Each segment in the industry makes money and property owners see their investment appreciate. Local government gets higher property values and tax revenues while the federal government gets to provide a popular insurance service that improves affordability of homes and thus strengthens communities. In the event of a major disaster, the federal government swoops in with generous recovery funding.

Do all these muted price signals encourage people to stay in, or move to, risky coastal areas? It is hard to be sure, but without price warning bells ringing loudly, people are doing just that. Even the scenario of rapid population growth in risky coastal areas might be financially sustainable if climate change were not causing coastal storms to be more severe and sea level to rise. These new realities of a changing climate make the coastal property financial system less stable and move it closer to the "tipping point" described by Edward Golding, where property values spiral downward, owners default on mortgages, and mortgage-backed securities are jeopardized.

The key question becomes when the combined effect of storm surges and rising seas push the coastal property finance system to a breaking point. Sean Becketti of Freddie Mac puts his finger on the critical problem of predicting the timing of a change in property values:

Consider an expensive beachfront house that is highly likely to be submerged eventually, although "eventually" is difficult to pin down and may be a long way off. Will the value of the house decline gradually as the expected life of the house becomes shorter? Or, alternatively, will the value of the house—and all the houses around it—plunge the first time a lender refuses to make a mortgage on a nearby house or an insurer refuses to issue a homeowner's policy? Or will the trigger be one or two homeowners who decide to sell defensively?[9]

The consequences of a breakdown in coastal property finance will be wide ranging. For the individual property owner, the investment in the property is likely to stop appreciating and start depreciating, perhaps dramatically. Becketti notes that home equity represents a large share of the wealth of many families and "if those homes become uninsurable and unmarketable, the values of the homes will plummet, perhaps to zero."[10] He points out that, in the 2008 housing crisis, many homeowners continued to make mortgage payments, but "it is less likely that borrowers will continue to make mortgage payments if their homes are literally underwater. As a result, lenders, servicers and mortgage insurers are likely to suffer large losses."[11]

There is already evidence that the value of some coastal properties is sinking. In late 2017, Asaf Bernstein, University of Colorado at Boulder, and his coauthors, looked at sales of property rights along the coast between 2007 and 2016 and found an average discount of seven percent with properties projected to be inundated later (i.e., by the end of the century) having a four percent discount. Sea level rise information played a role in pricing, especially for investors not living on a property: "Our evidence further suggests that this discount is driven by non-owner occupiers, who we argue and provide evidence for are more sophisticated investors. Within this market segment [non-owner investors], the average SLR exposure discount is approximately 10% and has increased over time, coinciding with the release of new scientific evidence on the extent and timing of ocean encroachment."[12]

In February 2019, First Street Foundation reported its analysis of declines in coastal home value in seventeen states totalling $15.8 billion lost in some 3.7 million homes since 2005.[13] The greatest losses occurred in Florida, some $5.4 billon, followed by New Jersey with $4.5 billion, New York with $1.3 billion, and smaller totals in other states. A "FloodiQ" website developed by First Street Foundation allows homeowners in these states to enter an address and see an estimate of a change in value.

Florida and New Jersey topped the list again in a GOBankingRates study of declining coastal property value looking at the forty communities expected to see the greatest declines. Eighteen of these communities are in Florida and fourteen in New Jersey. Looking forward, the authors estimated lost homes and value by 2033 and 2100 for each of these forty communities. Miami, Florida leads this group with 37,000 homes (seventy six percent) worth $33 billion underwater by 2100.[14]

This evidence of recent decline in value of existing coastal homes does not seem to have steered new home construction out of risky areas. The nonprofit organization Climate Central and the real estate website Zillow issued a report in 2018 focusing on rates of new construction of coastal homes between 2009 and 2017, finding that "more than half of the country's coastal states...have recently seen higher housing growth rates inside the flood-risk zone than outside of it."[15] *Governing* magazine came to a similar conclusion, finding that population increased faster inside Federal Emergency Management Agency (FEMA) 100-year flood zones than outside of flood zones between 2000 and 2016 (i.e., 14 percent compared to 13 percent). The report did not look specifically at coastal counties but, ominously, flood zone population increased the most in Miami–Dade County, Florida, (i.e., 130,000 or 24 percent), and seven of the ten counties with the highest flood zone population growth are coastal counties, five of them in Florida.[16]

Looking beyond just coastal communities, the *New York Times* pointed out that declining home prices lead to increased coastal mortgage defaults that have the potential to harm the wider economy because of the way bundled mortgage-backed securities spread risk widely.

> Like a game of hot potato, builders, homeowners, banks, flood insurers and buyers of securitized mortgages try to hand off risky properties before getting burned. Developers erect houses and sell them typically within a couple of years, long before their investments depreciate. Banks earn commissions even on risky home loans before bundling these mortgages into securities and selling them to large pension funds, insurers or other buyers.[17]

Sean Becketti argues that these financial losses have the potential to morph into societal losses: "Some homeowners outside the impacted areas will nonetheless suffer losses as businesses are forced to relocate, taking employment opportunities with them....Non-economic losses may be substantial as some communities disappear or unravel. Social unrest may increase in the affected areas."[18]

Risks to mortgages for commercial property in coastal areas—such as office buildings, retail properties, and lodging—are less studied than those to residential property, but there is new evidence of rising concern here as well. The Blackrock Investment Institute reported in 2019 its finding that many assets underpinning commercial mortgage backed securities (CMBS) "are located in regions that are vulnerable to the increasing incidence of severe storms."[19] In addition, Blackrock is concerned that "FEMA flood maps understate true risks"[20] and that under-insurance of flood risk is a growing problem. Although only six percent of commercial properties in BlackRock's database of 60,000 properties are in a FEMA flood zone, their analysis indicates that "the number of properties subject to 1% or more storm surge risk per annum would rise by 1800 percent by 2060–2080 under a "no climate action"[21] scenario.

Declining coastal property values also are bad news for communities that rely on property taxes to fund government services. Bryan Walsh cites the potential for communities to "enter a death spiral, as property taxes vanish even as the cost associated with responding to ever more frequent floods rises...,"[22] and quotes South Miami Mayor Philip Stoddard: "You don't need to be too smart to figure out how this affects your tax base. . . . No one is going to buy or invest in the community after that. This is not going to be pretty."[23]

Making matters worse, eroding tax revenues and spending to manage increasing coastal flooding undermines the ability of local governments to issue bonds to finance capital projects. The BlackRock Investment Institute concluded that climate-related risks "are underappreciated in the U.S. municipal bond market . . ."[24] noting that "hurricanes, floods and other extreme weather pose a host of financial challenges for state and local issuers."[25] Some of the reasons for undervalued risk are a lack of risk information, long time horizons, and a perception that risks are insured or that disaster aid will replace losses. BlackRock found that "within a decade, more than 15% of the current S&P National Municipal Bond Index (by market value) would be issued by MSAs [metropolitan statistical areas] suffering likely average annualized economic losses [due to climate change] of up to 0.5% to 1% of GDP. This would have big implications for the creditworthiness of MSAs—and their ability to fund adaptation projects. The impacts are set to grow more severe in the decades ahead . . ."[26]

Credit-rating agencies have only just begun to consider climate risks in bond ratings and it may take some time to develop sensitivity to coastal storm and sea level rise risks more specifically. For example, Moody's announced in late 2017 that their analysts will "weigh the impact of climate risks with states and municipalities' preparedness and planning for these changes when we are analyzing credit ratings."[27] In 2018, Breckinridge Capital Advisors announced development of a "quantitative approach for assessing risks associated with coastal flooding and sea level rise."[28]

Transition Toward a Stable Coastal Property Sector: The coastal property financial system is cruising along now with all flags flying and everyone making money. So far there is little recognition of the coming storm and the potential consequences for the profitability of the financial system, as well as the financial health of property owners, communities, and investors. There is also little evidence of initiative from this financial community to accept government regulation with respect to disclosure of risks or to monitor the financial health of the system and stay ahead of the destabilization that could lead to a more damaging breakdown, requiring expensive government intervention. Will the coastal property financial community heed the voices of concern and rally in time to meet this challenge? Or will it cling to the old financial model as the waters rise?

On the public sector side of this equation, Congress and FEMA are working to put the NFIP on a sounder footing that reflects risk by gradually removing premium subsidies and improving flood risk disclosure at time of property sale. By making rates more risk based, Congress is hoping to encourage more private companies to enter the flood insurance market, providing either original policies or policies that supplement the NFIP coverage. This will send a price signal that better reflects the true costs of living in risky coastal areas.

But, one person's market signal can be another person's financial disaster. Even the more moderate rate of increase in flood insurance premiums is proving to be unaffordable for some coastal property owners. Writing in the *New York Times Magazine*, Brooke Jarvis reported that people in Norfolk, Virginia, are seeing NFIP premium increases of 18–26 percent per year, and that each $500 annual increase in flood insurance lowers a home's value by $10,000. She quoted a local insurance provider: "People are getting killed. . . . To an appraiser it's still worth $300,000, but to the real world it ain't worth nothing, because it's not going to sell."[29]

Although governments benefit from a stable coastal property price system, it is a mistake to protect a system that does not reflect the risks of storms and rising seas. The goal of government policy should be to recalibrate coastal property values to reflect values based on current and future risks and clear price signals. The tricky part is making the transition to a new, stable, risk-informed coastal property financial system in an orderly manner. The question of what governments might do to transition to such a system is discussed in the next two chapters. The best outcome, of course, would be for the private sector to take responsibility and fix the problem.

Corporate Responsibility, Coastal Storms, and Sea Level Rise

Corporate America has an important role in helping the country respond to a changing climate and can play a critical part in building awareness of the need to respond to more severe storms and rising seas as well as crafting plans and measures to meet these challenges. To date, however, corporate reporting to the SEC on climate change risks has been limited. Some business leaders are supporting voluntary climate reporting mechanisms led by nonprofit organizations, but many others are not participating. Some investors are responding to this mixed record by pressing corporations for more disclosure of climate risks, but climate risk reporting so far has little to say about sea level rise.

Reluctance in corporate boardrooms to engage climate risks may be due to skepticism about the scale of climate change impacts or doubt that the economic risks reach beyond limited economic sectors to threaten wider economic stability. But evidence of risks to the broader economy is mounting. In March 2019, Glenn D. Rudebusch, executive vice president in the Economic Research Department of the Federal Reserve Bank of San Francisco, reviewed potential economic losses related to climate change including "storms, droughts, wildfires, and other extreme events . . ."[30] as well as risk exposure "through loans to affected businesses or mortgages on coastal real estate."[31] He concluded, "If such exposures were broadly correlated across regions or industries, the resulting climate-based risk could threaten the stability of the financial system as a whole and be of macroprudential concern."[32]

SEC Climate Reporting Guidance: In 2010, the SEC voted 3–2 to release *Guidance Regarding Disclosure Related to Climate Change* to "provide guidance to public companies regarding the Commission's existing disclosure requirements as they apply to climate change matters."[33] SEC regulations require publicly traded companies to file annual financial information and "such further material information," [34] as needed to avoid misleading investors. Reporting of material environmental information is well established but the nonbinding guidance was intended to outline SEC thinking about how climate risk should be treated in reports.

The SEC document speaks to the potential impacts on a business of climate regulations and international agreements as well as the physical impacts of climate change, and it specifically mentions coastal storms and sea level rise: "Significant physical effects of climate change, such as effects on the severity of weather (for example, floods or hurricanes), sea levels, the arability of farmland, and water availability and quality, have the potential to affect a registrant's operations and results."[35] It also notes that, "for some registrants, financial risks associated with climate change may arise from physical risks to entities other than the registrant itself. For example, climate change-related physical changes and hazards to coastal property can pose credit risks for banks whose borrowers are located in at-risk areas."[36]

Since it was issued, interested parties have debated whether companies were complying, whether enforcement of the underlying "materiality" requirement should be pressed, or whether the guidance should be made mandatory. The *New York Times* reported that, "in the two years after the interpretive guidance, the S.E.C. issued 49 comment letters to companies addressing the adequacy of their climate change disclosures,"[37] but has issued fewer in subsequent years. The Congress, as well as the state of New York and nonprofit organizations, have pressed for a firmer hand on climate reporting. Senator Brian Schatz, a Democrat from Hawaii, commented that "the S.E.C. has been underreacting in the extreme."[38]

Investor engagement on sustainability and climate change has long been a focus of the nonprofit organization CERES. In early 2018, CERES released the latest in a series of assessments of progress toward corporate focus on climate change, including both management of greenhouse gas emissions and climate change risks that threaten corporate profitability. CERES concluded that, although the percentage of companies mentioning climate risks in SEC filings has grown from 42 percent in 2014 to 51 percent in 2017, "boilerplate language about future climate change regulation has become commonplace, but this language does not give investors the information they need to understand the complexities of company-specific climate-related risks."[39]

In 2016, the SEC issued a concept paper outlining ideas for future reporting and asking for comment, including this question: "Are existing disclosure requirements adequate to elicit the information that would permit investors to evaluate material climate change risk?"[40] In the view of Linda Lowson, editor-in-chief of an American Bar Association publication on climate change, "the preponderance of public comment letters expressed a clear preference for mandatory CCSFR [Climate Change and Sustainability Financial Reporting] disclosure."[41]

Although mandatory disclosure would likely increase the percentage of companies reporting climate change risks above the current 51 percent, it is less clear that it would correct the boilerplate problem or prompt more serious attention to report risks related to storm surges and rising seas. Of course, the premise of the reporting is that disclosure of risk will prompt a company to address the risk (i.e., a company reporting a major facility at risk of sea level rise will make plans to protect or relocate the facility). Unfortunately, companies are not specifically asked in the guidance to speak to responses to identified physical risks, and there is little evidence that this is occurring.

Sea Level Rise and the Task Force on Climate-related Financial Disclosures: Beyond the annual filings to the SEC, some states impose reporting requirements, and corporations have the option of participating in several sustainability and climate reporting mechanisms managed by nonprofit organizations.

For example, the Global Reporting Initiative (GRI) works with governments and businesses to report performance in a comprehensive range of management, economic, and environmental risks, but not including sea level rise. The Climate Registry works with companies, governments, and others focusing on reporting of releases of greenhouse gases and carbon footprints. CDP, formerly the Climate Disclosure Project, works with over 800 investor organizations and requests environmental and climate related information from over 6,000 companies and 500 cities, with a focus on greenhouse gas emissions and energy use, including identification of physical risks such as those related to rising seas.

The most recent, most ambitious, and most promising advance in corporate climate reporting is the announcement in late 2017 of a new reporting mechanism by the Financial Stability Board's Task Force on Climate-related Financial Disclosures (TCFD). Sponsored by the G20 group of nations, the TCFD developed "recommendations for disclosing clear, comparable and consistent information about the risks and opportunities presented by climate change."[42]

The TCFD recommendations address key topics including governance, strategy, risk, and metrics. Climate risks include both "transition risks" associated with the transition to a low-carbon economy and physical risks. In describing physical risks, the report specifically mentions both more severe coastal storms and sea level rise: "Acute physical risks refer to those that are event-driven, including increased severity of extreme weather events, such as cyclones, hurricanes, or floods....Chronic physical risks refer to longer-term shifts in climate patterns (e.g., sustained higher temperatures) that may cause sea level rise or chronic heat waves."[43]

Importantly, the TCFD approach goes beyond simply acknowledging a risk and encourages a company to come to grips with framing a strategy for responding to the risk, finding the financial and other support for this work within the governance structure of the company, and tracking progress with clear metrics. Although the task force encourages companies to use the new format as they report under existing requirements, such as to the SEC climate guidance, the international sponsorship of the framework provides comparable information among companies in different countries and sets a common standard that investors can use in making investment choices.

The goal of the TCFD disclosure process is to "ensure that the effects of climate change become routinely considered in business and investment decisions,"[44] but the success of the approach "depends on near-term, widespread adoption by organizations in the financial and non-financial sectors."[45] To that end, the report outlines a five year "implementation path" but notes, "advocacy for these standards will be necessary for widespread adoption."[46] Early reports on implementation of the framework are positive. As of September 2018 457 companies worldwide with a combined market capitalization of more than $7.9 trillion publicly had committed to supporting the TCFD recommendations."[47]

Today, the jury is still out on whether most companies around the world will use the TCFD framework and whether it will succeed in going beyond just reporting to engage corporate management in framing the strategies needed to both reduce greenhouse gas releases as well as prepare for physical risks such as coastal storms and sea level rise.

Investor Climate Responsibility Advocacy: Some American corporations have been leaders in disclosing climate information but a key part of the climate disclosure story is the increasing demands of investors for more information about greenhouse gas emissions and, to a lesser degree, climate risks such as rising seas.

There is still a long way to go in engaging corporate America in disclosing climate change risks. In 2017, analysts at the management firm KPMG studied financial reports of the 250 largest companies that "can reasonably be expected to lead the way when it comes to acknowledging and disclosing the financial risks of climate change to their business,"[48] finding that 48 percent currently recognize climate risks.

Of the largest companies, those headquartered in France, Germany, and the UK were most likely to recognize risks (90 percent, 61 percent, and 60 percent, respectively) while only 49 percent of United States headquartered companies did so. In addition, only about 2 percent of companies "are currently quantifying the potential impact of those risks in financial terms."[49]

Given the state of current corporate climate reporting, investors clearly have a role to play in advancing climate risk disclosure. TCFD expert Mike Krzus observed in 2017 that "if investors make the case that climate disclosure will be used to drive capital market decisions, companies will see that it's important."[50] The Institute for Energy Economics and Financial Analysis reported some recent progress: "2017 marked a turning point for the climate disclosure push. Not only did shareholders of publicly traded companies pass three climate disclosure proposals that year, but the average percentage of votes cast in support of environmental measures climbed to 39.2%, from 27.5% in 2016 and 16.7% in 2015."[51]

Although these vote percentages seem slim, they represent a significant improvement. Nick Dawson, managing director at Proxy Insight, noted that "a few years ago, climate change was a fringe issue. Ignored by mainstream investors, environmental resolutions were lucky to receive 5 percent support. Now, issuers who try to ignore the associated risks could face serious financial consequences."[52]

Investor demands for climate risk disclosure have been controversial at Exxon-Mobile. In May of 2017, ExxonMobil shareholders voted to require the company to report on the impacts of climate change to its business, "defying management, and marking a milestone in a 28-year effort by activist investors."[53] The 62 percent vote in favor of reporting was a major improvement over the 38 percent a comparable measure received in 2016. Andrew Logan, director of the oil and gas program at CERES, observed "we are witnessing a truly historic shift in shareholder support for these resolutions."[54] When published in early 2018, however, ExxonMobile's climate report received mixed reviews. Patrick Doherty, codirector for Corporate Governance for the New York State Comptroller, concluded that the report, "offers too many generalizations and too few specifics."[55] Then in 2019, the SEC denied a shareholder request for a ballot proposal calling for the company to adopt greenhouse gas reduction targets, calling it "micromanaging."

Corporate Engagement with Communities Responding to Rising Seas: Corporations have a financial interest and an ethical responsibility to understand climate change risks to their facilities and to take action to protect the facilities and the surrounding community. In addition, as governments in the state and community in which a facility or business is located begin to cope with storm surge and rising seas, leaders of businesses large and small will face the ethical choice of whether to sell assets at risk of sea level rise while market values remain high and whether to devote resources to participating in communitywide planning for storms and rising seas.

Business leaders bring several important perspectives to the planning process. When they have assessed physical risks to facilities they can share that information with the community. They can help the community understand the risks they consider acceptable and the risks that would prompt them to invest in protection structures or to relocate. If they are considering relocating, they can discuss timing and alternative locations. In some coastal communities, a decision by a large employer to relocate will have a major influence on the community's plans.

Professor Rizzardi summed up the case for corporate engagement with the community: "While public sector representatives wrestle with decisions to adapt to, mitigate for, or retreat from sea level rise, the private sector has a role to play. Corporations, by law, have rights and privileges; with them must come corporate social responsibility. . . . Ethical behavior by the real estate professions and corporations means informing the people, partnering with the public sector leaders, protecting the public interest, and ensuring a resilient community with a sustainable future."[56]

State, local, and tribal leaders echoed this view in their 2014 report to President Obama on climate preparedness and resilience noting that "the private sector is responsible for much of the infrastructure of physical plants, supply chains, and retail, commercial, and industrial facilities that local and regional economies rely upon."[57] They called on federal agencies to "support regional, state, tribal, territorial, and local efforts to engage the private sector in community resilience and hazard mitigation planning and related projects, including Chambers of Commerce and major employers, as well as architects, engineers, and other designers and the professional organizations that represent them."[58]

Professional Responsibility and Sea Level Rise

Professions where an understanding of coastal storms and sea level rise and their implications will become increasingly important include real estate sales and development, law, civil engineering, architecture, and insurance. People in these professions are guided by ethical codes of conduct that will increasingly intersect coastal projects. Professional leaders also face ethical choices as they make decisions about whether to maintain connection with a community facing rising seas.

Professional Ethical Codes and Sea Level Rise: As indicated in the discussion of the coastal property financial sector earlier in this chapter, there is already wide discussion of possible legal requirements for realtors to disclose information about flood risk and, to a lesser extent, sea level rise risks. Realtors, as well as other classes of professionals that support coastal property transactions, work under codes of professional ethics that express obligations with respect to consideration or disclosure of future flood and sea level rise conditions. This raises the question of what ethical obligations realtors and other professionals have with respect to disclosure of potential future storm flooding or sea level inundation of coastal property.

The real estate industry faces several ethical challenges related to coastal flooding and sea level rise. First, the Code of Ethics and Standards of Practice of the National Association of Realtors calls on its members to "avoid exaggeration, misrepresentation, or concealment of pertinent facts."[59] The supporting guidance on flood risk, however, only applies in cases of "actual knowledge" of a risk. Because realtors are not encouraged to investigate these risks, and to do so would reduce the chance of a sale or reduce the sale price, adopting a policy of ignorance can make business sense and does not directly defy industry guidance (see ch. 14).

Second, even when a flood risk is known, industry guidance only indicates that this risk "should" be disclosed. Given that the guidance plainly indicates that there is a legal obligation to disclose "material" risks, the ambiguous "should" seems to suggest that flood risk might not be material, leaving the decision up to the individual realtor.

Finally, the National Association of Realtors guidance speaks to flood risk but is silent on sea level rise. Although flood risk is linked to location of a property in a FEMA-mapped flood zone, which can be determined easily with an Internet search, risk of eventual sea level rise inundation is more speculative. It is speculative with respect to a specific property, and the timing of inundation, and perhaps can even be denied as a legitimate risk in the mind of an individual realtor. As uncertainties related to sea level rise give way to wider understanding and high confidence of impacts, however, disclosure of the risks will become an undeniable ethical obligation.

Codes of ethics for other professions related to coastal property transactions do not speak directly to flooding or sea level rise but do contain the notion of an ethical obligation to disclose risks that are "material." The American Bar Association (ABA) publishes Model Rules of Professional Conduct, and state bars adopt supporting codes. Rule 4.1 provides that "a lawyer shall not knowingly: (a) make a false statement of material fact or law."[60] Other codes of ethics have similar provisions, including those for the American Institute of Certified Planners and the American Institute of Architects.

A sometimes-overlooked point is that these ethical codes apply not just to clients but to other parties. As Professor Rizzardi observed, "Again and again, the various ethical codes all make it clear that honesty—to everyone— is expected. Material facts must be disclosed and professionals cannot participate in the concealment of truth, particularly if it leads to fraud or misconduct."[61]

There seems to be a good case that ethical codes apply to risks of storm flooding, but do they also apply to risks of rising seas? Professor Rizzardi says yes: "The potentially transformative nature of sea level rise, and the magnitude of the potential problems, renders these traditional land use development laws and measures insufficient. Given the material facts of sea level rise, mere disclosure seems insufficient, too. Taking a consequential view, is it ethical for a planner, architect, engineer or lawyer to include a small print disclosure in a document, informing a buyer that their land is likely to be flooded by rising seas within the life of the buyer's mortgage, and then to sell the property anyway?"[62]

Insurance, Ethics, and Sea Level Rise: Several ethical dilemmas loom for the insurance industry connected by questions of how to interact with the NFIP.

Today, private insurance companies have lucrative contracts to sell and administer NFIP policies. In 2016, the National Public Radio and the Public Broadcasting System program *Frontline* released an investigative report describing high profits and questionable practices by these companies: "Each year, the companies take about a third of the premiums they collect as fees for running the program. The rest goes to settling claims. Those fees come to about $1 billion a year, according to the investigation."[63]

Determining the profit from these fees is difficult, but Robert Hunter of the Consumer Federation of America, and a former administrator of the NFIP, concluded that "we calculated that the pre-tax profits for the WYO [write your own] companies, for 2011 to 2014, averaged about 29% of the fees they received, a whopping $317 million a year! Even worse, when Sandy claims were paid, since the WYOs receive income from handling claims, the profit for 2013 soared to $406 million!"[64]

Are these estimated profits reasonable? Most business enterprises would be happy with a 29 percent profit and some would argue that if the government is willing to pay that it is just good business for companies to take it. From an ethical point of view, however, it is important to note that this is a federal government contract providing a service to people, some of whom are facing life-changing losses, and common decency should prevent price gouging. In addition, the profit is taken with little risk as the federal government is paying the claims. Under these circumstances, are insurance companies acting ethically?

Somewhat more ominously, *Frontline* and National Public Radio also reported that many NFIP policyholders felt their claims following Hurricane Sandy were not fairly paid by the private companies making the payment determinations. As Hunter explains, "Three years after Sandy many claims still remain unresolved and there have been 1,700 lawsuits filed by storm survivors who believe the insurance companies handling their claims have unfairly and fraudulently lowballed, delayed, or denied them. Because of the outcry, FEMA offered reconsideration to 144,000 homeowners with claims problems; 19,000 have appealed. Almost 80 percent of the appeals resulted in more money for the homeowners."[65]

Why would private insurance companies determining claims lowball claims that are being paid by the NFIP? Some theories are that cost constraint is simply habit in insurance companies, and that companies wanted to limit losses to help maintain the financial stability of the program and avoid calls for radical restructuring, which would undercut the handsome profits.

A final ethical question arises for private flood insurance providers operating outside of the NFIP structure. Congress is increasingly encouraging the private sector to help share the large financial risks of flood insurance. Some private insurers have been reluctant to enter this market in the past, but the advent of new mapping and data systems that allow for property-by-property analysis of risks allows private firms to target—some might say cherry pick—properties where lower risks justify a premium that is marketable.

This strategy may make money for an insurance company, but it leaves just the higher risk properties to be covered by the NFIP. Over time, the consolidation of high-risk properties in the NFIP either drives premiums higher or, with capped premiums, enlarges the gap between premium income and claim outlays and forces the federal taxpayer to cover the difference.

Governments at federal and state levels have an obligation to manage a transition from federal to private flood insurance in a way that minimizes harm to consumers, but what ethical obligations do private insurance companies have to support these policies? They can be expected to look to peers in the private sector for signals as to appropriate ethical behavior.

Looking at the big picture of business engagement with the problem of coastal storms and rising seas the key question is not what the private sector should do but what will prompt them to step up to take needed actions. In the case of professional conduct, extensive codes of ethics exist for multiple professions and the challenge is to encourage these professionals to reach for the aspirational elements of these codes. In the case of corporations, domestic and international reporting standards speak to coastal storms and rising seas, and the challenge is prompting more widespread and sincere compliance with these mechanisms. Even in the case of the coastal property financial system, key steps such as improved disclosure of risks at time of sale and in mortgage backed securities clearly would better account for risk, but the challenge is to motivate consistent and meaningful disclosure. Ideally, the private sector will make improved engagement with government on coastal storms and rising seas a priority. Failing that, governments should prompt these actions, and suggestions along these lines are offered in part 5.

PART IV: EPILOGUE

States, Communities, and Businesses Cope with Coastal Storms and Rising Seas

In *The Rising Sea,* authors Orrin Pilkey and Rob Young make a number of insightful recommendations including, "Take the reins from local government,"[1] arguing that "local governments of towns (understandably) follow the self-interests of coastal property owners and developers."[2] This is a provocative idea that runs counter to the widely held view that local governments, holding land use powers, are central to responding to coastal storms and rising seas.

States and communities face novel challenges and hard choices as they plan for more severe coastal storms and rising seas. Tools for managing the problem, be they financial or regulatory, are complex and often controversial. Relocating coastal communities and assets to higher ground is a complex and difficult process, made more challenging by the social and psychological aspects of storms and rising seas. The private sector has the potential to provide leadership in sorting out options and prompting timely local action but has not stepped up to the challenge. And, yes, local politics can result in short-sighted decisions. With so much at stake, are local governments in over their heads?

My conclusion is that, in some places, where state and local commitment is strong, local financial resources are deep, and the scale of impacts comparatively small, local governments can manage. But for most coastal states and communities, getting ahead of the one-two punch of coastal storms and rising seas without substantial help is an unrealistic expectation. Help needs to come from the federal government in the form of leadership, technical expertise, policy guidance, and significant funding.

So, a fourth and final argument for a new national program to prepare for more severe storms and rising seas is that it can best deliver the expanded technical support and resources that state and local governments need to meet this challenge. State and local governments are irreplaceable and need to continue to manage coastal flooding and rising seas place to place. At the same time, the federal government needs to provide a national frame of reference that looks to timely solutions that consider the full range of societal interests, with special care for disadvantaged people and communities, in the context of the money that the country is willing to make available to address these challenges.

What were those first three arguments again? The science predicting more severe storms and rising seas is strong. The impacts on the American coast—communities, infrastructure, ecosystems, and private assets—are significant and need national attention. And, existing national programs—flood insurance, disaster recovery, coastal management, and climate adaptation—can contribute to addressing these risks but are not set up to provide the focused attention and resources that are needed.

What might a national program to prepare for more severe storms and sea level rise look like? Can it strike the right balance between federal resources and state and local capacity? How might a campaign to adopt a national program be organized and funded? What can citizens do to help? These questions are addressed in the next and final part.

PART V

Campaign for a New Coast

Just because we cannot see clearly the end of the road,
that is no reason for not setting out on the essential journey.

John F. Kennedy

18

Framework for a National Storm and Sea Level Rise Program

After hearing testimony on the impacts of sea level rise at a hearing of the Senate Committee on Energy and Natural Resources in 2012, Senator Maria Cantwell (D-WA) commented that "we just can't sustain this level of sea level rise without a plan. We need a plan."[1] Over six years later, the country lacks a plan for responding to more severe coastal storms and sea level rise and might even be said to be moving in the wrong direction in some respects.

There is evidence, however, of growing recognition of the calamity that is coming for America's coast, and an understanding of the need for national leadership to manage these challenges. Some of the strongest voices speaking to the need for action are from South Florida. For example, in June 2018, in an op-ed in the *Palm Beach Post*, Mitchell Chester, CEO of Climate Monitor Media, offered this assessment: "Through myopic indifference, society is making adaptation to sea-level rise exponentially riskier than it needs to be. There is a noticeable absence of meaningful federal and state legislation to create practical and innovative financial tools necessary to help ordinary people and small businesses cope with the projected harm of rising waters. Each day that passes, reluctant lawmakers are unnecessarily escalating the destabilizing physical and emotional effects of swelling oceans with an unprepared and risky fiscal future."[2]

So much to be done. Where to start? A first step in framing a national program to prepare America for more severe storms and rising seas is to define its goals. Ideally, goals are defined in a process involving diverse stakeholders and subject experts. For the purposes of this framework, a working goal is to reduce future loss of life and property, and the costs to the government for disaster relief, resulting from more extensive coastal flooding due to more severe storms and inundation of coastal land by rising seas.

A national program working toward these goals should have several attributes. First, the program should be complete (i.e., it addresses risks to coastal communities but also ecosystems, critical infrastructure, and military assets, and assures that plans are coordinated). The program should be proportional to the financial resources reasonably available. And the program should be fair (i.e., it protects the interests of both wealthy and disadvantaged people). A final, less tangible attribute of a program is that it has the integrity to make hard, complex decisions, especially related to financing and relocation, and to capitalize on bold ideas to transition to a new coast.

Accomplishing a program meeting these goals will require new authority and action in five key areas:

1. Establish new national coastal storm and sea level rise leadership capacity.
2. Revise existing flood and disaster response programs to account for rising seas.
3. Define new coastal storm and sea level rise preparedness policies.
4. Develop national sea level rise plans for critical infrastructure and ecosystems.
5. Provide new authority and funding for state and community sea level rise inundation and storm flooding preparedness planning.

The critical challenges related to creating new funding authorizations and financial mechanisms for sea level rise and coastal storm preparedness are discussed in the next chapter.

Together, these actions make up a framework for a national program to prepare for more severe storms and rising seas. This framework is intended to suggest the scope and scale of the response needed and some general directions. New information, experience, and discussion will surely suggest revisions to this framework.

Key Area 1: Establish New Coastal Storm and Sea Level Rise Leadership Capacity

States and communities are doing good, creative work to prepare for more severe storms and rising seas, but viewed from a national level, this effort is improvisational and lacks the capacity to prepare the entire country to manage the risks and impacts. The country needs a new entity able to provide national leadership in defining and addressing the problem of more severe storms and rising seas and the authority and resources to coordinate a national response. The art of defining this new governmental capacity is to provide leadership—the ability to engage key players to move consistently in a common direction—without creating a heavy-handed government bureaucracy.

Coastal Storm and Sea Level Rise Preparedness Council: With caution about bureaucracy fresh in mind, it must be said that multiple federal agencies play a part in preparing for more severe storms and rising seas, and keeping them moving in the same general direction requires some formal coordination. To provide this navigation function, federal agencies should establish a federal interagency Coastal Storm and Sea Level Rise Preparedness Council. The new council should be led by three agencies—the Federal Emergency Management Agency (FEMA), the National Oceanic and Atmospheric Administration (NOAA), and the US Army Corps of Engineers—and include diverse other federal agencies.

One mission of the council should be to provide leadership for federal agencies as they respond to coastal storm flooding and sea level rise inundation. Another mission should be to provide direction to preparedness work by state and local governments and the private sector. The council also should develop innovative and promising policy and program responses to more severe storms and sea level rise at the national, state, and local levels. In addition, the council should advise the president on approval of national, state, and large metropolitan area response plans and recommend funding levels for these efforts.

A federal council would greatly benefit from input from a committee made up of representatives of state and local governments, academia, nonprofit organizations, the private sector, and the public. To provide this input, a Coastal Storm and Sea Level Rise Preparedness Advisory Committee should be created under the Federal Advisory Committee Act and chartered to provide an interface between the council and the many constituencies that have interests in preparing for coastal storms and sea level rise. Because of the potential for cost-benefit considerations to tilt funding toward wealthy areas, and the evidence of past inequities in disaster spending, special care should be given to include representatives of disadvantaged communities and low-income people on the advisory committee.

The council should be charged with working with the advisory committee to develop periodic reports to Congress and the president evaluating the status of coastal storm and sea level rise preparedness and making recommendations for needed actions. Reports should be timed to follow by a year the quadrennial updates of the *National Climate Assessment* so as to be guided by the most recent science.

Coastal Storm and Sea Level Rise Research and Assessment: Leadership and coordination is also critical for research and assessment relating to more severe storms and sea level rise. Many federal agencies now develop or fund research related to these events and their likely impacts. The existing interagency Global Change Research Program (GCRP) periodically synthesizes a vast array of climate change science into the *National Climate Assessment.* The good news is that the process of integrating climate research and impact assessment into the periodic assessment is well established and respected, and this building block of a national program is in good shape. In addition, in April 2019, an independent Advisory Committee on Applied Climate Assessment recommended steps to further strengthen this process, including engaging a broader range of civil society in the assessment work and increasing "support for practitioners to apply climate-relevant science in multiple ways,"[3] including formation of a new Science for Climate Action Network.

As presently constituted, however, the GCRP does not frame a plan for federal coastal storm and sea level rise related research in coordination with other scientists and in consideration of the information needs of those responding. Another concern is that, today, most science related to sea level is focused on projections of global mean sea level rather than its more localized impacts, and less work is being done on issues like coastal storm frequency and the demographic, economic, and social impacts of rising seas. The World Climate Research Programme has identified "Regional Sea-Level Change and Coastal Impacts" as one of seven climate research "Grand Challenges" and has called for "development of sea level predictions and projections that are of increasing benefit for coastal zone management."[4] Dr. Judith Curry, considered a skeptic of some climate science, has pointed out the importance of improved understanding of local factors, such as vertical land movement and ocean currents, in coastal decision making.[5]

To meet the need for a broader research agenda, a new Coastal Storm and Sea Level Rise Research Program, within the GCRP structure, should work with scientists from federal agencies and academia to define physical and social science research needs over time and guide resources toward these topics. As part of this work, scientists should generate consensus projections of the probability of future sea level rise along the American coastline over time (e.g., at twenty-year intervals). In addition to strengthening research of the domestic impacts of more severe storms and rising seas, this effort should better define goals and funding needs for basic research, including long-term changes in ice sheets in Antarctica and Greenland, and strengthen mechanisms for cooperation with other countries in this work.

Chartering authority for a new research program should also provide for defining and addressing current and future financial implications of coastal storm flooding and inundations expected to result from sea level rise and developing mechanisms to reduce costs and protect low income people and communities. Research demonstrating use of social science tools and measures to reduce barriers to responding to these risks should also be a priority. Special attention is also needed to removing barriers to relocation and developing models that promote sustainable communities.

Expansion and Coordination of Coastal Mapping: A final element of a national leadership and coordination framework needed to prepare for more severe storms and rising seas is expansion and coordination of mapping of these risks. Today, there is something of an embarrassment of riches in coastal mapping but room for improvement in the type of mapping and coordination among agencies.

The most critical new part of a mapping program is to expand the existing work that NOAA does to revise the official American shoreline to reflect the National Tidal Data Epoch. NOAA needs more formal statutory authority for collecting sea level data and revising shoreline delineations on a regular schedule (e.g., twenty years).

In addition, NOAA needs new direction to develop not just new shorelines, but maps defining lands expected to be inundated by rising seas within the near and far future (i.e., 20-year periods, 2100, and 2200). Mapping lands soon to be inundated is essential to support local sea level rise planning, especially actions to decommission lands prior to inundation. Many decommissioning actions, including removal of power and water utilities and buildings or other structures that can be hazards to the public or navigation is far less expensive prior to inundation. Mapping sea level rise in the far future is an essential complement to the near-term mapping, making clear that sea level rise will continue for decades and centuries and is critical information that is needed to avoid expensive, incremental solutions to rising seas.

Producing maps of future inundation will require complex decisions concerning the degree of risk that should be accepted and the appropriate Representative Concentration Pathway (RCP). For example, drawing on the NOAA 2017 sea level rise scenario report, a decision to keep inundation risks low in the context of RCP 4.5 might result in use of the Intermediate-High scenario and mapping the projected area of relative sea level rise along the coast for that scenario. These decisions should include input from other federal agencies, such as FEMA, be informed by input from the advisory committee and be finalized by the interagency council.

Another needed mapping product is a set of maps that combine storm surge maps, now developed by both FEMA and NOAA, with those of projected sea level rise to show the landward extent of storm surges that will ride on top of higher sea level. These maps, produced with the authority of federal agencies, would provide states and communities with a full picture of the geographic extent of both inundation and storm flooding.

Finally, there is room for improved coordination among federal agencies responsible for coastal mapping. Today, NOAA maps shorelines and coastal waters and offers maps of hurricane-driven flooding. FEMA maps floodplains and identifies V zones at risk of storm impacts. The United States Geological Survey maps coastal erosion and identifies coastal vulnerability based on geology and rising seas, in addition to mapping dry land elevations. Both the US Fish and Wildlife Service in the Department of the Interior and the Natural Resources Conservation Service in the Department of Agriculture map coastal wetlands. The Army Corps of Engineers develops precise elevation data for its coastal engineering work.

As the country comes to terms with more severe storms and rising seas, basic and reliable mapping information will be essential for effective planning. It seems unrealistic to expect federal agencies to give up existing coastal mapping responsibilities, but more affirmative cooperation in development of mapping products seems attainable. To that end, the agencies should use their own initiative to create a Coastal Mapping Board to share information and work to improve the collective mapping output. A more formal authority for this work by executive order or statute should be considered at a future date.

Key Area 2: Revise Flood Insurance and Disaster Response Programs

Looking back at all the ideas for changes to the National Flood Insurance Program (NFIP), including adjusting the path to risk-based rates, promoting the role of private insurers, or improving operational efficiencies and fairness, a direct response to the problem of more severe coastal storms and rising seas is notably absent. Admittedly, proposals to generally prohibit insurance for any new construction and reduce repetitive losses by buying risky properties would have some benefit in terms of rising sea levels. And, a FEMA advisory committee has called for better mapping of areas at risk of rising seas. For the most part, however, the interest groups looking at the NFIP, the Trump administration, and Congress are not focused on sea level rise risks.

Although FEMA has studied climate change and sea level rise, neither FEMA nor other organizations have proposed a strategy for the NFIP to account for sea level rise or to survive the expected dramatic increases in program losses. Looking forward, it must be said that the existing network of flood insurance and disaster response programs is wholly inadequate to respond to the immense destruction and relocation that will occur as a result of more severe coastal storms and rising sea level expected as the planet warms. The programs were simply not designed to face the one-two punch of more severe coastal storms and rising seas. Just stretching and tweaking the existing programs will not be sufficient to meet these new challenges. Four major reforms are needed:

- The existing NFIP needs to be split into separately financed and managed programs serving inland areas and those coastal areas that NOAA identifies as at risk of inundation by rising seas by 2100.
- The new flood insurance program applying to areas at risk of inundation by rising seas needs to be phased out over a long time period of roughly thirty years, in conjunction with development of new programs to help owners of at-risk coastal property avoid major financial losses (see ch. 19).
- In the period prior to phase-out of the new insurance program for sea level rise risk areas, key reforms to better recognize more severe storms and sea level rise need to be adopted.
- Reforms recognizing storm surge and sea level rise need to be adopted in disaster assistance programs, including more attention to predisaster planning.

Separate Flood Insurance Programs for Inland and Sea Level Risk Properties: A key reason to separate areas at risk of sea level rise from inland areas is that the inland part of the program is financially sustainable, and the part applying to properties at risk of sea level rise (i.e., most coastal properties) is not. Continuing to insure properties at risk of rising seas is not financially sustainable for several reasons. First, extremely costly coastal storms will drive very high annual losses of coastal properties. As in the case of past major storms, income from premiums is not likely to cover these costs. Second, premiums from inland policyholders are covering losses and some of these payments are subsidizing coastal property claims. This inland subsidy of coastal properties will become politically untenable over time.

More important, the current coastal premiums do not reflect projected gradual increases in sea level rise that will inundate all of these properties, rather than just a few places hit with a flood or storm. The current program assumes that flooding is temporary and, while it occurs in known flood zones, the timing of any given flood is unknown and most policyholders are not filing claims at the same time or even the same year. An insurance model in which a large group pays annual premiums to support assistance to a small number of parties making claims can be made to work. Sea level rise, however, makes this model unsustainable because it brings permanent inundation to most coastal areas to varying degrees, resulting in a very high ratio of claims to policyholders. Premium increases are not likely to keep pace with claims.

The conventional insurance model basically works for inland communities. Separating inland policies from sea level risk policies will allow the program to continue to serve inland policyholders and provide the valuable incentive of encouraging communities to implement flood control practices to reduce flood damages.

In addition, inland policyholders will not be asked to weaken a working insurance model by covering increasing losses incurred by coastal policyholders.

As part of the divorce of the inland and sea level risk elements of the program, there needs to be an agreement for a long-term plan to retire the existing debt the overall program is carrying (see funding proposal in ch. 19).

Long-term Phase-out of Flood Insurance for Sea Level Risk Properties: The dramatic differences between properties at risk of rising seas and other properties facing conventional flood risks make a case for separating management of the properties. But, why is a long-term, thirty-year phase-out of insurance for sea level risk properties necessary?

First, as noted above, the flood insurance model will become increasingly unsustainable financially for sea level risk properties. As sea level inundation gradually imposes losses on a higher and higher percentage of covered properties, and rate subsidies are reduced, actuarial rates will become unaffordable for more and more property owners and lead policyholders who have the option to back out of the program. This will create a downward spiral in the pool of these coastal policyholders. The occasional major coastal storm will generate large one-time payments that will generate a new standing debt for this part of the insurance program.

Second, the NFIP is designed primarily to provide a financial mechanism to restore or replace property lost to a temporary flood at the original site that was flooded, providing security to both homeowners and institutional mortgage holders. This financial support, whether subsidized or an actuarial rate, is an incentive to live in a place that might be flooded again and to rebuild at the same location. In a future with more severe coastal storms and rising seas, rebuilding in the same place is no longer a game of chance but a losing proposition for both a homeowner and the government. The NFIP will increasingly create a costly lag in the time that a community will be willing to come to grips with the true costs of remaining in place rather than engaging safer and less-costly alternatives, including relocating to higher ground. On top of everything else, advancing the NFIP indefinitely in areas at risk of sea level rise puts lives at risk and is morally indefensible.

The problem arises of how to ramp down the NFIP in a way that results in the minimum adverse financial impacts on homeowners at risk of rising seas now relying on the program. Fortunately, sea level rise is slow and there is still time to phase out the program over a transition period. A first step is to promptly set a future date for terminating flood insurance for these properties. A phase-out date should be set to provide current property owners reasonable notice and the chance to make financial plans with a phase-out in mind. A phase-out date of thirty years would allow for even the most recent mortgages to be paid on property. This would also have the advantage of discouraging new development or substantial redevelopment of at-risk coastal property.

Even with a phase-out schedule, annual appropriations will still be needed to make up the shortfall between premiums and actuarial and administrative costs. In some years, major coastal storms will require supplemental appropriations and generate a new NFIP debt that will need to be paid on a schedule that will likely extend beyond the coastal flood insurance phase-out date. The chance of Congress providing this long-term funding commitment is increased if there is a clear plan to end the program and the continuing costs, running as far as the eye can see, at a specific date in the future.

Of course, terminating the federal insurance program for property at risk of rising seas at a date that is years in the future is unlikely to mean that these homeowners will be uninsured for flood risks after that date. Private insurance companies are already playing a larger role in the coastal flood insurance market, either directly as primary insurers or through policies that supplement the NFIP policies for high-value properties. As the NFIP premium subsidies phase out under the current law, owners of risky coastal property will be faced with paying higher rates anyway, and the federal program will increasingly be left to insure a dwindling subset of the most-risky properties as private insurers use new data and tools to focus on lowest risk properties, further destabilizing the federal program.

The most significant downside to a phase-out of the coastal element of the NFIP is the impact on low- to moderate-income homeowners. They have fewer resources to make a move well ahead of program termination and are more likely than wealthy people to be caught in a downward spiral of coastal property prices. This part of the population has the most to gain from adoption of a Coastal Property Price Stabilization Fund along the lines described in the next chapter.

Reforms Needed Prior to Phase-out: As the existing flood insurance program for sea level risk properties ramps down, some ideas for program reforms that either speak directly to more severe storms and sea level rise, or at least clearly fit within a model of an NFIP well adapted to these risks, should be adopted.

For example, the Trump administration's proposal to prohibit issuance of federal flood insurance policies for new construction needs to be adopted for sea level rise risk properties. This policy, which builds on the model of the Coastal Barrier Resources Act, has two major benefits. First, it discourages new development in areas likely to eventually be flooded by storm surges and then inundated by rising seas. Second, prohibiting insurance for new projects also puts the sea level risk element of the flood insurance program on a more sustainable footing in that it avoids expanding commitments to cover property under a financial model that is not self-sustaining even before accounting for rising seas. It is one thing for the federal program to cover existing properties bought before a full understanding of the risks of more severe storms and sea level rise with a long phase-out period. It makes little sense to continue to provide an insurance subsidy for persons choosing to build on property that is known to be at substantial risk that will increase over time.

Shifting to the case of homes and buildings already in place in sea level risk areas, FEMA should minimize the incentives for remaining in risky coastal areas and rebuilding in these areas after a storm. A first key step in this direction is for FEMA, in cooperation with Congress, to revise the calculation of premiums to reflect the risk of rising seas and more severe storms, rather than simply the risks to a property based on past storms. Properties in risky V zones already pay higher rates, but these rates do not reflect rising seas. The development of new maps of areas at risk would support these rate adjustments.

Repetitive losses of property are a major cost to the NFIP, and the chances of repeated loss of a property in an area at risk of sea level rise inundation is higher than in other cases. FEMA and Congress should acknowledge this risk and amend the NFIP so that properties in areas at risk of rising seas that suffer damage valued at more than 50 percent of the value of the property are not eligible for future federal flood insurance. As a complementary policy, FEMA and Congress should significantly expand funding for buyouts and develop new procedures to make up-front, prestorm agreements for market price buyouts of property in areas identified as at risk of future sea level rise and that suffers extensive damage. A new Coastal Property Price Stabilization Fund (described in ch. 19) would complement this policy.

Finally, the core principal of the NFIP—using insurance as an incentive to get communities to adopt local practices that reduce flooding—should be strengthened. This step would be a good improvement to NFIP operation generally, but also helps adapt to more severe storms and rising seas. Coastal communities facing rising seas have more limited options for managing flooding than inland communities, which can work to improve flood control upstream on a watershed basis. Coastal communities, however, can make some progress in this direction, and FEMA should take steps to encourage preparedness.

A key place to start this effort is to strengthen the Community Rating System (CRS) by offering significant incentives in the form of insurance premium reductions for local adoption of stronger local storm surge and sea level rise management practices. For example, the existing offer of 1,450 points, or over a 10 percent reduction in premiums, for prohibition of any new construction in flood zones, should be increased. Additional points should also be offered for community action to define areas intended for relocation after a buyout and adopting into local ordinances protection of natural features that can minimize storm surge and sea level rise, including coastal wetlands, mangroves, and dune structures.

These reforms, together with the additional improvements to operational efficiency and fairness described in chapter 9 (e.g., flood-risk disclosure, affordability, financial soundness), would strengthen the sea level risk element of the flood insurance program and carry it through the next several decades toward a long-term phase-out date.

Adapt Disaster Assistance Programs to More Severe Storms and Sea Level Rise: A key step that would make disaster assistance programs better able to cope with more severe coastal storms and rising sea level is shifting investment from postdisaster planning (i.e., planning to avoid a future disaster just like the last disaster) to predisaster planning that looks more comprehensively at future risk, including rising seas.

As the Congressional Budget Office has found, most of the resources now available for disaster planning are focused on looking backward to avoid past disasters rather than looking forward to anticipate and avoid a likely future disaster. Aside from being a costly way to plan, this backward-looking approach is less likely to recognize emerging and slow-moving threats like rising sea level and increasingly severe coastal storms.

The good news here is that Congress has taken an important step toward increasing funding for predisaster planning in enacting the Disaster Recovery Reform Act in late 2018, but this commitment needs to be made mandatory and supported with guidance. Funding for the small Flood Mitigation Assistance Grant Program should also be significantly increased from the current $160 million level to at least double that amount, as proposed in Senate legislation.

In addition, FEMA should amend guidance for state and local hazard mitigation plans to more specifically address sea level rise and revise the CRS to give substantially more points, and thus greater flood insurance premium discounts, to policyholders in communities that address sea level rise in hazard mitigation plans. For states and communities facing minimal risk, these plans along with Coastal Zone Management Plans, may be sufficient for the time being. As noted in chapter 10, however, most coastal states and communities will need more focused planning and response actions looking at long-term storm and sea level rise risks rather than all hazards.

Another needed reform to disaster assistance is to rebalance the relative contributions of federal and state governments so that states have an incentive to take greater responsibility. A key action FEMA and Congress should take is to develop new metrics for declaring a federal disaster that set the bar somewhat higher. For example, the existing metric of estimated disaster costs exceeding about $1.40 per capita in a state should be raised or be replaced by a sliding scale adjusted for state population, or be replaced with a metric comparing disaster cost to the population of the area affected rather than the population of the state.

In addition, Congress should adopt the proposal by the National Infrastructure Advisory Council for a "disaster deductible"[6] for each state. A state or region would start with a fixed federal share of say 60 percent and could increase that share by adopting various actions to strengthen state capability. Rather than setting just one national ratio of federal-to-state cost sharing, this approach encourages each state to find the right balance considering its circumstances and capabilities. And, FEMA should implement recommendations outlined in its "after action" report on the 2017 hurricane season. Although these recommendations do not speak to the risks of more severe storms and rising seas, they address operational issues identified in the "stress test" FEMA endured as it responded to three major coastal storms.

Finally, the biggest federal spending for disasters is in the form of supplemental appropriations following major coastal storms. This funding is generously applied (i.e., tens or hundreds of billions of dollars for a major disaster) often with adjustments to reduce state or local cost share. The track record of mega spending for mega coastal storms can dim the determination of both governments and property owners to prepare for these risks. Although capping this spending does not make sense, Congress should adopt a requirement for higher state and local match for this supplemental funding in places that have not adopted, or are not implementing, approved coastal storm and sea level rise response plans (see key area 5).

Key Area 3: Define New Coastal Storm and Sea Level Rise Preparedness Policies

In addition to the "old business" of updating federal flood insurance and disaster relief programs, new business on the agenda is to consider new policies and programs to support effective coastal storm and sea level rise planning and response actions. Some key new policies are the following:

- expand disclosure of information about storm surge and sea level rise risks;
- adopt policies to discourage development of coastal lands likely to be inundated by rising seas;
- provide for safe decommissioning of inundated lands; and
- clarify existing law regarding ownership of lands inundated by rising seas.

Expand Disclosure of Coastal Storm and Sea Level Rise Risks: Disclosure of storm surge and sea level rise risks is an essential tool for building awareness and understanding of the risks. Opportunities to improve disclosure exist in the context of floodplain mapping, real estate transactions, and corporate climate change risk reporting.

The first step in improving floodplain mapping is to invest in updating existing floodplain maps used in the National Flood Insurance Program (NFIP). Over half these maps are not current, and an investment of roughly twice the current spending of about $200 million per year is needed. Floodplain maps are seen by millions of Americans, and a simple, next step toward building awareness of sea level rise risk is to identify, for informational purposes, areas at risk of rising seas over the coming 100 years. FEMA's Technical Mapping Advisory Committee (TMAC) has recommended providing separate maps of sea level risks to communities, but such maps are likely to be seen by far fewer people.

A second key opportunity for disclosure of coastal flooding and sea level rise risks is during real estate transactions. Today, the ethical standards regarding disclosure by realtors of flood risk are ambiguous and the legal standards vary among states. A single national standard of disclosure is needed. The national disclosure standard should require that a seller disclose that a property is located in a floodplain, and the owner is subject to paying flood insurance premiums. More important, a national disclosure requirement should require disclosure by the seller of a property's 100-year risk from rising seas. Congress has considered legislation to reauthorize the NFIP that provides for flood risk disclosure but not disclosure of sea level rise risk.

The new 100-year sea level rise maps proposed to be developed by NOAA would provide a basis for this sea level rise risk disclosure. This real estate transaction could be modeled after the requirement for disclosure of lead paint in the Residential Lead-Based Paint Hazard Reduction Act of 1992. Failing a national statute, states should adopt more effective disclosure laws, and coastal property professionals should develop clearer risk disclosure standards in the context of ethical codes. FEMA should also amend the Community Rating System to provide more generous points for communities that adopt flood and sea level risk disclosure requirements in local ordinances.

Finally, the Securities and Exchange Commission (SEC) should make sure that investors have meaningful information about the physical risks that companies face from more severe storms and rising seas. A key step toward this goal is for the SEC to make clear that the existing guidance regarding corporate disclosure of climate change risks is a clarification of underlying mandatory reporting requirements, rather than description of optional reporting. The SEC should also update and strengthen the current guidance and improve enforcement.

Adopt Policies to Discourage Development of Coastal Lands at Risk of Inundation: One of the most urgent questions the country needs to consider is how to limit new development on lands that are likely to be inundated by rising seas by the year 2100.

Coastal population is growing faster than the national average and current projections are that population at the very edge of the coast may as much as double by 2060. Allowing this rapid and intensive development of lands at risk of inundation to occur unchecked will result in dramatic increases in the damages due to eventual inundation and storm surges that occur in the interim. These costs will be devastating for property owners and, under current flood insurance and disaster policies, very costly to taxpayers more generally. It also means that costs of decommissioning these more intensively developed lands will grow. The old admonition, "when you find yourself in a hole, the first thing to do is stop digging," applies here. How might this be accomplished?

One of the most widely recognized incentives for coastal development is the National Flood Insurance Program (NFIP) and the subsidy it provides to help property owners manage costs of storm damage and rising seas. Proposals for changes to the NFIP in this chapter would narrow insurance-based incentives for some coastal development and begin to phase out the program for properties at risk of sea level inundation.

A second approach to discouraging development in coastal areas at risk of rising seas is to adopt policies that strengthen consideration of future environmental conditions, such as more severe storms and rising seas, in review of major projects and federal investments. Two good examples of such policies were developed by the Obama administration but revoked by the Trump administration. Guidance for consideration of climate change in environmental impact statements under the National Environmental Policy Act promoted consideration of future conditions, such as rising seas, in review of major projects. And, the Federal Flood Risk Management Standard provided for shifting federal investments away from places at risk of flooding, effectively discouraging federal investments in risky coastal areas. These policies, which were subject to extensive public input, should be reinstated.

Other ways to encourage local governments to steer development away from risky coastal areas are to provide more points for such local controls in the CRS and making adoption of statewide controls on development in these areas a major element of a "disaster deductible" model in which states taking preparedness actions get a higher ratio of federal to state disaster assistance.

Of all the categories of coastal lands that will be inundated by rising seas, wetlands are especially valuable, providing storm surge buffers and essential fish habitat, among other services. Coastal wetlands, however, are being lost at a rapid rate (see ch. 6). The wetlands permit program, under section 404 of the Clean Water Act, is intended to avoid development in wetlands or minimize impacts and compensate for damages when this is not possible. The initial avoidance review looks at the purpose of a project, including whether the project is water dependent and whether an alternative location is "practicable" in terms of costs and other factors. The Army Corps of Engineers reports, however, that fewer than one percent of permits are denied.[7]

This avoidance policy dates to 1990 and assumes that floods may come and go but wetlands and other waters are basically stationary. Climate change and rising seas add a new element to the 404 permit process. With the knowledge that seas are rising, and that the land areas to be inundated as seas rise by 2100 are mapped by NOAA, Congress should set a higher bar for approval of development of wetlands in these areas at risk of rising seas. Instead of simply allowing most development in these wetlands to proceed if no practicable alternative site is available, development should also be evaluated in the context of both current and future conditions at that site. Factors to consider include timing of future inundation, project life expectancy, public safety issues, and damage to the evolving aquatic environment. In addition, Congress should give the Corps new authority to establish a 404 permit condition requiring an applicant to post a bond in an amount sufficient to pay costs of decommissioning any new development as needed in the future (see discussion of Coastal Hazards Removal Funding in ch. 19).

Finally, rising seas pose a timing problem. As seas rise, some lands that are dry now will become wetlands or open water. Today, the 404 permit program only applies to existing wetlands and does not apply to existing dry lands that are transitioning to wetlands as a result of rising seas. Some of these dry lands will be developed in response to population growth pressure. This development, and supporting infrastructure, will be inundated in due time and pose costs to the government in the form of insurance losses, disaster relief, and decommissioning. A better approach is to add lands expected to be inundated by rising seas by 2100 to the 404 program as these lands are mapped by NOAA. Standards for permits on these soon-to-be wetlands should be adjusted to account for the timing of inundation and the permanence of a proposed development. A national permit program for these areas has the critical advantage over a similar state or local permit programs of providing a single, consistent approach thus protecting state and local governments from charges that any measures in this area confer a competitive disadvantage.

Any proposal to expand the scope of lands covered by the 404 program will be controversial and bitterly fought. Although there would surely be cries that such an anticipatory approach to defining wetlands is jumping the gun, there is some precedent in environmental law. The Endangered Species Act, for example, provides for designation of public and private lands as "critical habitat," where federal actions will be reviewed to avoid harming species, and it even provides for designating as critical habitat lands outside the geographic area of the species at the time it is listed as critical habitat (i.e., is "unoccupied" by the species), if that is essential for the conservation of the species. Not surprisingly, the Trump administration recently proposed regulatory changes to narrow this authority.

Most decisions about development of coastal land at risk of inundation are in the hands of local and in some cases, state governments. As described in chapter 14, some of the tools available are land purchase and easements, zoning, and other regulations, and tax incentives. Each has pros and cons, including varying exposure to a constitutional takings claim. From a national perspective, the goal should be to support effective state and local preparedness planning that applies these tools appropriately. Proposals to support state and local planning for coastal storms and rising seas are provided later in this chapter.

Provide for Safe Decommissioning of Inundated Lands: As the extent of sea level rise inundation of coastal lands grows in the coming decades, the areas inundated will increasingly include developed areas with buildings and supporting infrastructure. A threshold policy question is whether the country will take responsibility for making sure these inundated lands, both publicly and privately owned, are safe and do not pose a risk to people, the environment, or navigation.

Assuming that the country wants to see these lands safely decommissioned, several questions arise, including whether there should be a national minimum decommissioning standard, how such a standard should be developed and implemented, and how this expensive undertaking should be funded.

Decommissioning standards for inundated lands will need to address topics such as whether standing buildings need to be torn down and removed or whether they can remain in place. Although removal of some infrastructure, such as power lines and transformers, can be accomplished relatively easily, removal of water and sewer lines and pump stations can require more complex and expensive projects. Removal of roads might not be needed in some cases but removal of related structures such as guard rails and parts of bridges may be necessary. Structures posing special risks to the environment, such as underground gasoline storage tanks, will raise challenges. EPA is a logical agency to lead development of initial decommissioning guidelines with the support of other federal agencies such as the Army Corps of Engineers and FEMA.

Once national decommissioning guidelines or standards are published, who will be responsible for implementing them? A model already in place in several national environmental laws provides national authority while allowing for delegation of that authority to a state that wants to take on the work. For example, most states now implement the national pollution discharge permit program under the Clean Water Act under delegated authority from the Environmental Protection Agency (EPA). In the case of national decommissioning guidelines or standards, the Corps has wide experience conducting coastal projects and is well qualified to implement decommissioning standards. States wanting to do the work themselves, or delegate the work to a local government, could seek delegated authority from the Corps.

A final, critical question is how these expensive decommissioning projects can be funded. Government should pay for decommissioning on public land but funding for decommissioning of private property should come from a combination of federal, state and local sources, including in the case of new developments, a bond posted by a developer to pay these costs (see Coastal Hazards Removal Funding in ch. 19).

Clarify Law Regarding Ownership of Lands Inundated by Rising Seas: As sea level rises, and NOAA shifts the shoreline landward, property owners and states will increasingly be faced with the need for a clear legal process for the transfer of land ownership from a private party to a state. Because of the constitutional questions that arise with respect to whether the transfer of ownership is a taking, all parties would benefit from action by Congress to set standards for such transfers. Federal agencies and other stakeholders, but especially state and local governments, need to work closely with Congress on these questions.

An initial question Congress needs to address is what specific action triggers transfer of ownership from a private party to the state. For example, Congress needs to speak to whether the simple publication of new shoreline maps triggers transfer of ownership and whether NOAA should be required to provide for notice and comment, or more formal treatment under the Administrative Procedures Act, as it revises maps.

Perhaps the most important contribution Congress can make in this area is to clarify expectations as to any financial obligations the federal government (as the party taking an action to cause an ownership change) or a state (as the party taking ownership of land) may have to pay private owners of inundated lands compensation for the loss of the land. Current law seems clear on the point that land lost to rising seas is not a taking and that government action recognizing this fact is not a taking. Unlike inundation from an action like building a dam that floods a valley, sea level rise inundation is not the result of any specific government project or action.

Still, future court decisions might change this situation and the immense value of coastal lands at risk of inundation suggests that any compensation obligations could be at least several trillion dollars. Even a remote chance of having to pay compensation for inundated land would have a chilling effect on any government action to declare formerly dry land to be inundated and accurate and timely mapping of retreating shorelines.

Although Congress has no authority to change a fundamental constitutional obligation, there are steps that it can take to frame this issue. Congress should, for example, frame a program that provides long notice to property owners of the risk of inundation to inform their "investment backed expectations." And, Congress should protect local regulations to manage rising seas by explicitly recognizing the critical public purpose they serve, reinforcing Justice O'Connor's 2001 opinion (see ch. 15).

This is a subject for which congressional hearings and discussions with interested parties might yield a policy direction that would allow planning for the inundation of these coastal lands to proceed without the debilitating uncertainty of a huge financial liability hanging over the planning process.

Key Area 4: Develop National Plans for Critical Coastal Infrastructure and Ecosystems

Coming to terms with challenges posed by more severe coastal storms and rising seas requires a national assessment of risks to critical coastal infrastructure and ecosystems and plans to protect or relocate these assets.

A Policy Foundation for National Storm and Sea Level Rise Plans: Before launching a major planning effort, it is important to review the question of whether any existing plans or related efforts now provide, or could provide, the guidance needed to protect critical coastal infrastructure and ecosystems as sea level rises. As discussed in chapter 5, there is no national storm and sea level rise plan or sector-specific plans for critical infrastructure. Federal agencies have made some studies and preliminary assessments of risks, but these have not been expanded into more action focused plans. State Coastal Zone Management Plans include varying sea level rise elements, but do not focus on infrastructure at risk to sea level rise and do not look at a national picture. State hazard mitigation plans address storm risks and, in some cases, sea level rise, but do not focus on key infrastructure sectors or look at a national picture.

Infrastructure is a big topic and another important question is where to start in developing national plans to protect critical coastal infrastructure from storms and rising seas. Although the list of infrastructure types is long, infrastructure providing basic human needs should be a priority. On this criterion, transportation and water infrastructure rise to the top. Energy facilities are also a priority but something of a special case because of a high degree of private ownership. Military facilities do not so neatly fit the "basic human need" criterion, but the continuing operation of the military in the face of rising seas is clearly in the paramount interest of the nation. These four infrastructure sectors are a starting point, and other sectors, such as health care facilities, telecommunications assets, and local government and emergency service facilities, should be addressed at a future date.

Advancing a national assessment of sea level rise risks to coastal ecosystems and a plan for minimizing these risks does not neatly fit the basic human need criterion, but effective management of coastal ecosystems as seas rise will be critical to the long-term viability of coastal communities and economies. More localized state or regional plans can define steps to protect coastal ecosystems but can't replace a national plan looking at high priority areas and ecosystem types needing protection and providing for protection of space needed for inland migration of beaches and wetlands.

Questions arise with respect to how best to coordinate among national infrastructure sector plans and how such national plans should be coordinated with state and local sea level rise plans. Would a single national coastal infrastructure storm surge and sea level rise adaptation plan be a better approach than sector specific plans? A key argument for four distinct coastal infrastructure sector plans is that the federal agencies that will lead these planning efforts operate separately and work with different constituencies. Forcing an integrated planning process would generate bureaucratic friction and numerous complications. Coordination can be advanced, however, by pressing lead agencies to work with other federal agencies and states and to share findings and proposed actions early in the planning process. The proposed Coastal Storm and Sea Level Rise Preparedness Council and Federal Advisory Committee can also promote coordination.

National Planning for Critical Coastal Infrastructure Integrity as Sea Level Rises: New authority is needed to direct federal agencies having expertise with respect to critical coastal infrastructure sectors to take the lead in developing plans for adapting to more severe storms and rising seas. Sectors and lead agencies are suggested here:

- transportation, including roads, bridges, ports, railroads, and airports (Department of Transportation);
- energy, including refineries, energy production and generation, and transmission (Department of Energy);
- water, including drinking water and wastewater infrastructure (Environmental Protection Agency); and
- military facilities, with special attention to naval facilities (Department of Defense).

Congress or the president should direct federal agencies to develop a national plan addressing sea level rise and storm surge risks to infrastructure in the sector. The Coastal Storm and Sea Level Rise Preparedness Council can provide general guidance to agencies on framework and elements of plans. Agencies will need to engage other federal agencies, states, and a broad range of constituencies in the planning process, giving special consideration to private-sector owners of infrastructure. Agencies should release a draft plan for public comment within two years, and the council should submit proposed final plans to the president for approval within three years of the start of the process.

National plans will need to address existing and new facilities. In the case of existing facilities, agencies will need to identify and assess existing facilities and assets at risk of more severe storm surges and rising seas by 2050, 2100, and 2200, including definition of geographic areas along the coast where the density and importance of infrastructure makes implementation of preparedness actions a top priority. For these facilities, agencies should propose measures for protecting each facility, including elevating structures, building berms or seawalls, or preserving natural infrastructure such as wetlands or mangroves, where such measures are feasible and cost-effective. Agencies need to address costs and timing of such measures as well as site-specific options for relocating specific existing assets, including timing, siting, likely cost of such relocation, and decommissioning.

In the case of proposed new facilities, agencies need to define policies or other steps necessary to avoid siting of new infrastructure assets in areas at risk of inundation as sea level rises, giving consideration to the importance of the asset and the probability of inundation within its expected operational life at the site. Planning horizons should consider long-term risks and avoid defaulting to a protection strategy, in lieu of a relocation strategy, simply because the former has lower short-term costs.

Agencies developing national plans also need to consider interdependencies among critical infrastructure plans and state and local sea level rise preparedness plans, with special attention to coordination for sharing of preferred sites for potential relocation. Planners should explore opportunities for operational and economic efficiencies that may arise from coordinated relocation of infrastructure across sectors. Planners should also work with disadvantaged communities to identify environmental justice issues (e.g., location of infrastructure in disproportionally low income or minority communities) and avoid such issues in relocation proposals. Other topics these plans should cover include estimates of costs and timing of safe decommissioning and environmental remediation of infrastructure, financing strategies for protection and relocation, and a process to amend and update the plan in coming years.

Planning for military facilities along the coast is a special case in several respects. The operational integrity of these facilities is in the paramount interest of the country and relocation of military facilities needs to be given priority within local and regional sea level rise adaptation plans. Existing military bases also pose special challenges for safe decommissioning.

In addition, military facilities are not isolated islands and rely on civilian infrastructure for water, power, housing, and related services. Retired General Ron Keys, United States Air Force, summed up this concern in a 2016 speech describing challenges of rising seas: "We need to start considering, what can we do? Now I can build a moat, or a barrier around Langley Air Force Base, but the problem is a lot of my people live in Newport News, live in Hampton. A lot of my electricity comes in from outside. My fuel comes in from outside. So at some point we get to the point: 'I've got to move to higher ground.'"[8] Given these interconnections, local sea level rise planning needs to keep up with military planning, and military planners need to engage and support local preparedness efforts as well as national planning for critical coastal infrastructure.

National Planning to Sustain Coastal Ecosystems as Sea Level Rises: Unlike critical infrastructure, which can be defined in terms of discrete facilities and assets, coastal ecosystems are ubiquitous along the coast. While infrastructure sectors provide a narrow set of services or benefits, coastal ecosystems provide more diverse services ranging from storm buffering, to fisheries support, recreation, and carbon sequestration. In addition, the inland migration of ecosystems as sea level rises is comparatively fixed by geography, and options for relocation are often very limited. For these reasons, the approach to national planning to sustain these coastal ecosystems as storms intensify and sea level rises needs to be somewhat different from the approach to critical infrastructure sectors.

The Department of the Interior (DOI) is the federal agency best prepared to lead a national planning effort to sustain coastal ecosystems. DOI's expertise and authority ranges from coastal geology at the United States Geological Survey (USGS), to wildlife at the US Fish and Wildlife Service, to critical resource protection at the National Park Service. The existing USGS Landscape Conservation Cooperatives and Climate Science Centers can play an important role in supporting this effort. At the same time, other federal agencies need to be part of a team to tackle this challenge, including NOAA programs at the national level and the NOAA Regional Integrated Sciences and Assessment organizations.

Because of the complexity and diversity of coastal ecosystem services, the risk assessment phase of a national planning effort needs to be more extensive than is required for critical infrastructure sectors. As a key first step, the Department of the Interior should lead a cross-agency effort and work with states and other stakeholders, including tribal governments, to develop a national atlas of coastal ecosystems.

This effort should review the existing inventories of critical coastal ecosystems—ranging from wetlands, to beaches, to salt marshes—with special attention to defining with precision the elevation of these resources above current sea level and other risk factors. Drawing on new, localized estimates of relative sea level rise, the atlas should map the expected landward migration of coastal ecosystems and identify barriers to landward migration, including natural features and existing human developments. Fish, wildlife, and plant resources, including Essential Fish Habitat and carbon sequestration potential, should be addressed in the atlas drawing on existing information.

The atlas should also map property ownership of ecosystems (e.g., federal, state, tribal, local government, private, nonprofit organization), including identification of existing national parks, marine research reserves, and related protected areas. Specific ecosystems judged to be of exceptional importance on a national level should also be identified along with the potential for reestablishment of the ecosystem landward of its present location. Upland areas of importance for landward migration of ecosystems should be identified. A shoreline atlas recently developed for San Francisco Bay offers a useful model for this approach.

An initial draft of an atlas should be completed in a year and be provided to other federal agencies and to states for their use as they develop storm and sea level rise preparedness plans. A key goal of the early sharing of ecosystem information is to promote coordination of planned future uses of upland areas. Because the landward migration of ecosystems is so constrained, plans for use of upland areas for relocation of critical infrastructure, where location options are less limited, need to be coordinated with planning to protect ecosystems. After providing for public comment, the department should finalize the atlas within an additional year.

As the coastal ecosystem atlas is being developed, the Department of the Interior should begin drafting a national storm and sea level rise plan for coastal ecosystems. This effort should track the process and schedule outlined for the critical coastal infrastructure plans (i.e., draft plan within two years and final in three years). Drawing on the coastal ecosystem atlas, the department should develop a plan that identifies measures and processes to promote the successful landward migration of coastal ecosystems, including identification of areas important for migration of coastal ecosystems and measures to avoid relocation of infrastructure or communities to these areas. DOI should work with state and local governments and others to consider interdependencies among elements of coastal ecosystems, including land, water, fish, plants, and wildlife, with special attention to coordination of preferred sites for relocation.

The national plan should also identify existing authority or resources for protecting or acquiring lands critical to landward migration of coastal ecosystems and identify any new authority or resources needed. Karen Thorne put her finger on the problem in her study of wetland loss to rising seas along the Pacific coast, arguing that "our results suggest that mitigating wetland loss from SLR [sea level rise] may be limited in some regions of the Pacific coast due to urban development and steep topography. . . . Therefore, innovative ideas and the public and political support for future land-use planning and, possibly, land reallocation for wetland expansion may be needed if these systems are to persist."[9]

In the plan, DOI should describe the cumulative costs and timing of steps to protect coastal ecosystems or to facilitate their landward migration and steps needed to finance these measures. Opportunities to use natural infrastructure, including beaches, dunes, wetlands, and "living shorelines" as buffers to protect existing infrastructure and communities from storm surge and rising seas should be described. And, the plan should account for coastal wetland carbon sequestration and promote increased sequestration as wetlands migrate inland.

Finally, once the plan is complete, federal permits for coastal structures, such as seawalls and living shorelines, should only be approved if they are consistent with the plan and any applicable state or local plans. As noted in chapter 6, federal "general permits" under section 404 of the Clean Water Act make proliferation of hardened coastal structures possible, and regulations should be revised to raise the bar for approval of these structures and assure that permits are a tool to protect coastal ecosystems as well as property and communities.

Key Area 5: Provide New Authority for State and Community Storm and Sea Level Rise Planning

A national approach to respond to storm and sea level rise risks facing critical coastal infrastructure and ecosystems is essential, but the most difficult decisions driven by more severe storms and rising seas involve communities ranging from large metropolitan areas to villages and neighborhoods. The country needs to build on the commendable efforts already underway by coastal states and communities and implement a national program able to help all states and communities meet this challenge. Philip Stoddard, mayor of South Miami, spoke to the scale of the challenge in 2018: "We need a plan as to what will be defended because at the moment the approach is that some kid in a garage will come with a solution. There isn't going to be a mop and bucket big enough for this problem."[10]

A key first step to prepare coastal communities for more severe storms and rising seas is for the federal government to provide coastal states with information and financial support so that they can lead work with coastal communities in the state to develop storm and sea level rise preparedness plans for each state.

A second step is to recognize that, for roughly a dozen large, metropolitan areas along the coast, potential storm surge and sea level rise impacts are exceptionally serious, and the resources available to these communities to respond to the risks are well beyond those of coastal communities more generally. Many of these large coastal areas are already well along in planning to adapt to a changing climate, including more severe storms and rising seas. These regions need to coordinate with state efforts to plan for rising seas in smaller communities but should develop preparedness plans that fit their needs and response capacity.

State Storm and Sea Level Rise Preparedness Plans: New authority is needed to provide grants to states to develop and implement preparedness plans with coastal communities within the state that are located, in whole or in part, in areas expected to be at risk by 2100, as identified by NOAA. In addition to areas directly at risk, plans should cover areas expected to be affected by relocation and migration.

NOAA should develop guidance for state preparedness plans. A core element of a state plan should be an assessment of communities at risk of more severe storm surges and rising seas, including definition of communities in the state facing the most significant inundation or related risks and for which implementation of preparedness actions is a top priority. Socially vulnerable parts of communities should be identified as well as important historical, cultural, or natural landmarks that need special consideration for protection or relocation. NOAA guidance should require states to engage disadvantaged populations and to define impacts of protection and relocation investments on these populations and reduce impacts to the extent possible.

State plans should clearly identify policies that the state or local government will take to limit new development in areas identified by NOAA as at risk of inundation as sea level rises, and thus minimize future costs of protection, property buyouts, or relocation. State or local initiatives to steer new development or significant redevelopment away from these risky areas are essential to reducing loss of life and minimizing property losses and government costs of disaster relief. These efforts need to be coordinated with any new regulatory authority at the national level.

The most critical function of state plans is to set statewide direction for how existing community infrastructure and private assets in areas at risk of storms and rising seas will make the transition to a new coast. NOAA guidance should outline the two major options—protection along the line of the existing coast or stepping back to higher, safer ground. Although conditions are different in every community, states need to consider these options with the interests of all communities in mind. For example, a decision by one community to step back from the current coastline might undercut a decision by a neighboring community to protect the current coastline. NOAA guidance should also speak to the importance of defining a common planning horizon for the state, and encourage adoption of long time horizons that fully reflect risks (e.g., a shorter planning horizon might make investment in a seawall or beach nourishment look prudent while a longer planning horizon might shift thinking toward relocation options). A related consideration is that an investment in structural protection might induce new or more dense development behind the structure, making eventual relocation more challenging.

Finally, NOAA guidance should encourage states to think through the best use of limited funds for implementing response actions. Some communities may be able to pay the costs of adapting to rising seas, but most will need financial assistance from state or federal governments. States should make clear that the plan will consider the needs of all communities, rather than those that are first to the statehouse seeking funding or are best represented in the legislature.

States also need to think about their overall capacity to assist all coastal communities over many years and describe any fiscal constraints, including limits on paying for multiple, short-term solutions, early in the planning process. Taking a lesson from recent experience in England, state plans should avoid calling for measures where high cost of individual structural protection measures makes coastwide implementation of such measures unlikely. The topics of the right amount of federal financial assistance for implementing state plans and new federal funding to stabilize coastal property values are discussed in the next chapter.

With statewide direction on protection and relocation approaches generally defined, states should work with coastal communities to identify response measures, timing, and funding that best fit the situation of the community. Communities choosing to protect assets along the line of the existing coast are likely to describe measures such as seawalls, beach nourishment, coastal armoring, and use of living shorelines. Communities choosing to step back from the coast need to consider topics such as property buyouts, coordinated termination of utility and other services to areas at risk, decommissioning of infrastructure and buildings, and whether to encourage relocation to a specific, alternate site.

Another critical function of state plans is to provide an interface with national

planning for infrastructure and ecosystems and any plans developed by large metropolitan areas. For example, state plans can be a forum for reconciling the competing
interests that may arise for relocation of communities, critical infrastructure, or ecosystems to a particularly favored upland site (i.e., is an upland site best used for an airport,
or a factory, or a water treatment plant?). Where competition exists for a site, the state
or local government may have the option of zoning the site for a specific use or, if a
public use is anticipated, using eminent domain authority. A concern in this allocation
of favored sites to best uses is that places that are critical for landward migration of
ecosystems, that have no other place to go, may be committed to other uses.

Although the focus of state plans should be on preparing communities for more
severe storms and rising seas, plans might also include topics of special importance
to a particular state. For example, California might want to add attention to assessing
sea level risks to beaches and facilitating the landward migration of beach systems.
Louisiana might want to focus on the special needs of the oil and gas industry to prepare energy infrastructure for rising seas and more severe storms. Florida might want
to focus on saltwater intrusion as well as inundation. New Jersey might want to add
special attention to the needs of ports and harbors.

Finally, state plans are the best forum to prepare the tourism and fishing industries for the impacts of more severe storms and rising seas. Rather than have a federal
agency lead a national planning process, states can draw on existing working rela-
tionships with their tourism and fishing industries and reduce concerns that a federal
agency is deciding the future of coastal tourism or fishing in the states. In addition,
these industries are highly decentralized and would be difficult to effectively engage at
the federal level. Some states may be reluctant to appear to jeopardize the economic
benefits of tourism and fishing, but putting potential losses in the larger context of
a state plan describing the risks is the best way to demonstrate the seriousness of the
problems. The Department of Commerce, however, should develop general guidance
for the tourism and fishing elements of state plans and support state work in this area.

State plans should be developed within five years and a state's governor should
have the option of submitting the plan to the president for approval. States with
plans approved by the president become eligible to implementation grants (see ch.
19). State plans should be revised not less often than every twenty years.

A key question is how new state plans to prepare coastal communities for
more severe storms and rising seas should be coordinated with existing state plans
approved by NOAA under the Coastal Zone Management Program. In some states,
the existing coastal program plans address sea level rise to some degree, and a state
might prefer to amend these plans to more specifically address sea level rise and

storm surge risks. Other states might prefer to have a separate plan to manage these risks. Governors should be allowed to decide how to coordinate planning for storms and sea level rise with existing coastal program plans. States should also take steps to coordinate this planning with state hazard mitigation plans developed in cooperation with FEMA and with any more geographically focused sea level preparedness plans for large urban areas.

Large Metropolitan Area Sea Level Rise Preparedness Plans: Large coastal metropolitan areas have made some of the most dramatic progress to date in preparing for more severe storms and sea level rise. These often multijurisdictional planning efforts should be supported by federal grants and coordinated with storm and sea level rise planning by states and at the national level.

Coastal cities facing significant risks from storms and rising seas that are in some stage of developing sea level preparedness plans in cooperation with surrounding communities include Boston, New York City, Norfolk, South Florida, New Orleans, Los Angeles, and San Francisco. Other areas where a storm and sea level rise preparedness plan might be developed separately from a state plan include Washington, DC, Charleston, Tampa, Houston, and Seattle.

In addition to making grants to states, NOAA should make grants to large metropolitan areas that have the approval of the governor to develop a plan to complement the one developed by the state. Large metropolitan areas should have five years to develop plans and should, jointly with the governor, submit the plan to the president for approval. Approved plans would give metropolitan areas access to the grants to pay costs of implementation.

NOAA should be charged with reviewing the status of existing plans in these areas and developing guidance that identifies innovative ideas and approaches and outlines key elements that plans should address, including those elements described above in the context of state plans. As in the case of those state plans, policies to limit development in areas at risk of inundation as sea level rises and minimize future costs of protection or relocation should be a priority, especially in the near term. Metropolitan areas should be asked to make special efforts to cooperate with surrounding jurisdictions and the state or states in which they are located. As with state plans, NOAA should require that disadvantaged populations be actively engaged in this process and that impacts on this population be evaluated and reduced to the extent possible.

Another important factor distinguishing these large urban areas from most other coastal communities is that they have significant economic capacity that makes possible both sophisticated planning efforts and implementation of high-cost, capital-intensive protection measures. Because they are able to implement major protection projects, they may be able to delay, reduce, or avoid the need to relocate parts of the community.

At the same time, this capacity to fund major projects can be a mixed blessing and could result in investment in very expensive protection structures without a full assessment of potentially cheaper and better relocation options. As the Association of State Floodplain Managers noted in 2016, "past attempts to master the environment have too often yielded narrowly focused, protective solutions that encourage more at-risk building and behaviors, and create the potential for catastrophe when structures and systems fail or are overwhelmed."[11] With this thought in mind, NOAA guidance should require metropolitan areas to carefully consider the limits of protection structures and to review options for relocation.

The country needs to update flood insurance and disaster assistance programs and adopt new policies in key areas like risk disclosure. And, plans to minimize losses from coastal storms and rising seas are needed at national, state, and local levels. But of the five elements of a national program to transition to a new coast, the one that is hardest to get right, and needs the most attention, is leadership. The individual parts of a national program need to be well coordinated. The program needs to be able to set new directions and goals as conditions change over time. This strategic thinking is not possible without a rigorous commitment to public engagement and program evaluation. Adjustments to account for changing conditions will require political sensitivity. The leadership needed to add these values to a national program is not an easy download. Much will depend on the care taken in creating this leadership authority and finding people with the skills and dedication to make a national program a success.

19

Funding Coastal Storm and Sea Level Rise Preparedness

Building confidence across levels of government and among interested parties that laudable plans will actually be implemented is critical, and nothing can strengthen this confidence better than a reliable funding framework. National funding approaches to addressing more severe coastal storms and rising sea level will evolve over the many years the country will need to manage these challenges. As a starting point, a funding approach needs to define appropriations of general revenues for existing and new federal agency operations related to coastal storms and sea level rise. Adjustments in federal spending to support flood insurance and disaster assistance will also be needed. Entirely new funding mechanisms able to support state and local governments as they develop and implement plans to respond to rising seas need to be created. And, state and local governments need to be relieved of the burden of financing the transition of coastal property at risk of storm surges and rising seas from private to public hands by establishment of new federal capacity able to offer nationally consistent assistance.

Appropriations for Federal Operations

The cost of expanded work by federal agencies to prepare for more severe storms and sea level rise described in the program framework in the previous chapter is comparatively minor but essential to success of the larger effort. Although this work clearly benefits coastal states and communities, reducing the financial and social costs of major storms and rising seas also has benefits for the entire country, and the costs of federal agency response can fairly be charged to American taxpayers.

Some of the work agencies need to do to respond to storms and rising seas can be paid from their existing funds but some new responsibilities will need small increased appropriations. Some key examples of areas needing some new funding are the work of the National Oceanic and Atmospheric Administration (NOAA), the Federal Emergency Management Agency (FEMA), and other agencies to improve projections of relative sea level, expand mapping of areas likely to be inundated, and update coastal floodplain maps. The Environmental Protection Agency (EPA) will likely need new funds for its work with other agencies to develop guidelines and standards for decommissioning inundated lands.

More significant new funding is needed for work by the Departments of Transportation, Energy, and Defense, and the EPA, to develop national sea level preparedness plans in key infrastructure sectors. In addition, the Department of the Interior will need new funding to develop national plans for facilitating the landward migration of coastal ecosystems. A ballpark estimate of new annual costs of these mapping and planning efforts is $400 million per year.

Once these national infrastructure and ecosystem plans are finalized, funding will be needed for implementation. Some implementation can come in the form of program and policy revisions and some costs can be paid from existing accounts. Additional funds will be needed but the amount is difficult to know without a better idea of proposed work schedules. Costs will also vary depending on whether existing facilities can be safely retired ahead of rising seas in favor of new facilities located on higher ground, or whether existing facilities, or poorly sited new facilities, face significant risk during their expected life and need to be relocated.

Another variable is the federal share of costs of implementing national infrastructure plans. For example, in the case of military facilities, the Department of Defense will bear most costs of implementing plans that can be billions of dollars for a single facility, such as Naval Station Norfolk. Energy facilities, on the other hand, are mostly privately owned and can be expected to pay costs of adapting to coastal storms and rising seas from utility rate bases. Most water treatment facilities throughout the country are owned by local governments that are eligible for low-interest loans capitalized by annual federal appropriations of between $2 and$2.5 billion annually, and this funding should be increased to cover costs of protecting or relocating existing facilities. Transportation project funding is a mixed bag coming from many sources and the federal share of costs and total amounts will vary by type of transportation. A ballpark estimate of these costs to the government, strictly as an initial placeholder, is $2 billion annually, mostly for military and transportation facilities.

Funding the implementation of national plans to support coastal ecosystems, including acquiring land to allow inland migration of ecosystems as seas rise, poses a special challenge. Land acquisition is one element of a coastal ecosystem strategy likely to include regulation and easements, and funding needed for land acquisition will vary as needed to complement these other measures. In addition, lands should be acquired on a cost-shared basis with states and local governments, perhaps in cooperation with nonprofit organizations. Federal funding of $200 million annually is an initial estimate but this figure should be adjusted based on more detailed planning. These funds should be managed through some combination of the Land and Water Conservation Fund and other existing programs.

Finally, a number of existing programs that relate to coastal management are badly underfunded, and strengthening these programs would complement and support new attention to coastal storms and rising seas. Examples of existing programs needing substantially increased funding include the Coastal Zone Management Program, the Sea Grant College Program, the National Estuarine Research Reserves, and the National Estuaries Program. Additional funding in the ballpark of $150 million would roughly double current appropriations.

Adjusting Federal Spending for Flood Insurance and Disaster Assistance

As discussed earlier, the best solution to the challenges that more severe storms and rising seas pose for the flood insurance program is to manage and finance properties at risk of rising seas separately while phasing out the flood insurance policies for these properties over a long period of time (e.g., thirty years).

A first step in this process is for Congress to adopt a long-term plan to pay down the existing National Flood Insurance Program (NFIP) debt (currently about $20 billion) in small annual payments over ten or twenty years. With the debt managed, the inland (i.e., non-sea level rise) element of the flood insurance program will be able to sustain itself financially, except in the event of a catastrophic inland flood. A commitment to a long-term plan to pay off current debt is financially responsible in that it reduces the cost of carrying this huge debt as quickly as possible and conveys to policyholders a national commitment to the financial integrity of the program. What does carrying the debt cost? The NFIP paid almost $400 million in interest payments on its debt in fiscal year 2017,[1] and $3.83 billion in interest since Hurricane Katrina.[2] About $2 billion annually would pay off the debt in a dozen years.

Step two is to commit to annual appropriations from general revenues that are able to offset the annual shortfall in the operation of the coastal, sea level risk element of the program until the long-term phase-out date is reached. The reform policies described in chapter 18 will have the effect of more closely balancing premiums with payments to policyholders and administrative costs. Unfortunately, more extensive storm surge and gradual inundation driven by rising seas will result in ever higher damages and claims payments. The occasional major coastal storm could generate large one-time payments that result in a new standing debt for this coastal part of the insurance program. Given these variables, the annual shortfall in the coastal element of the program is hard to estimate, but a reasonable assumption is that shortfalls will remain at about $1.4 billion annually.

Funding these huge legacy financial obligations when starting a new effort to meet a challenge like coastal storms and rising seas is a bitter pill. The unappetizing prospect of using large sums of general revenues to slowly pay down over $20 billion in debt and cover annual shortfalls of over $1 billion with general revenues might prompt Congress to throw in the towel and simply forgive the debt. Congress already made a step in that direction when, faced with the shocking costs of Hurricanes Harvey, Irma, and Maria, it just forgave $16 billion of the NFIP debt. Although political conservatives complained, the NFIP had no way to pay the debt from its own income. Squeezing both legacy debt and annual shortfall payments from the stone that is the NFIP seems fanciful, and some debt forgiveness seems inevitable.

In the case of disaster assistance, federal general revenues are appropriated annually to the Disaster Assistance Fund, and past levels of base funding support of around $600 million seems about right. As suggested in chapter 18, expanding the commitment to predisaster planning in hazard mitigation plans will help minimize costs of future disaster assistance. Congress authorized increased funding in this area in 2018, but additional funding of about $150 million would improve plans and help link existing hazard mitigation plans with other planning efforts. Policy changes, including increasing state share of planning and response costs, and providing states with incentives to improve disaster planning would reduce recovery costs for all parties.

The big-ticket spending for disaster relief comes in the form of supplemental appropriations following major disasters. Future annual supplemental spending is hard to estimate, but the major coastal storms of 2017 resulted in supplemental appropriations of almost $120 billion and prior year appropriations have commonly been tens of billions of dollars. A key benefit of a national program responding to coastal storms and rising seas is to minimize these costs. These costs are speculative and do not fit well into annual planning but need to be recognized as part of the financial picture.

Coastal Hazards Removal Funding

Paying for the safe decommissioning of buildings and infrastructure on inundated lands is important for public safety and safe navigation. As noted in chapter 18, the US Army Corps of Engineers should manage this effort working in cooperation with states and local governments to decommission property under national standards developed by EPA.

Funds to pay for this work can come from a variety of sources. In the case of some types of infrastructure, such as electric power lines and transformers, local utilities should be responsible for safe removal of equipment. State and local governments should be responsible for decommissioning roads and other public infrastructure on public land, but costs of decommissioning private property are likely to exceed local capacity in coming years.

In the case of existing property destroyed by storm surge or inundation, the Corps should work with state and local governments to identify and rank in priority order property where decommissioning is needed and estimate the cost of this work. The Corps should seek funding from Congress annually to support decommissioning projects. This funding is likely to be minimal in the near term but grow in coming decades.

For new development projects in an area at risk of sea level rise, Congress should give the Corps new authority to require a developer to post a bond sufficient to pay the costs of decommissioning. In the case of projects approved under a wetlands permit, section 404 of the Clean Water Act, the bond condition should be in addition to other permit conditions. For new development, or substantial redevelopment, projects not requiring a 404 permit, Congress should authorize the Corps to issue a coastal hazard removal permit that would impose a permit condition requiring a developer to post a bond and pay decommissioning costs at a future date as required by the Corps. The purpose of the permit and bond is to account for the compelling interest that the federal government has in avoiding escalating costs of decommissioning projects. Precedents for this approach are a requirement for bonds from some offshore oil lessees to cover decommissioning infrastructure and a proposal in the LA SAFE report for bonding for new commercial development in high risk areas to pay future demolition costs.

Estimating the growth in new or substantially renovated properties in risky coastal areas is tricky. Population growth will drive development, but new government disincentives and regulations would discourage development. Over the decades between now and the year 2100, the proportion of coastal properties for which a decommissioning bond has been paid will increase.

Still, there are some two million existing homes at risk of 3.28 feet of sea level rise for which no bond will have been paid. Even assuming just $10,000 to decommission each property, the total bill would be as high as $20 billion, mostly coming due in the latter half of the century and could be higher. Congress should ask the Corps to report on the amount and timing of these costs as well as decommission costs on public lands that are beyond the capability of state and local government. Congress will surely pay these costs grudgingly, but the alternative is relying on costly local government legal actions against owners of abandoned property, shifting costs to state or local governments, or accepting the public safety and other risks of letting the wreckage litter the coast. Federal costs of decommissioning are likely to be modest over the next decade and funding on the order of $50 million annually.

Coastal Resilience Grants to States and Large Metropolitan Areas

The policies outlined in this part will result in long-term savings related to flood insurance, disaster response, and reduction of storm surge flooding and related sea level rise property losses and disaster relief costs. Many of these cost savings will result from development of state and local storm and sea level rise preparedness plans. But significant new funding is needed to pay for developing and implementing these plans, and the federal government should help state and local governments fund this work. Federal interests in supporting state plans include public safety (i.e., reducing loss of life), reducing federal spending (i.e., costs of flood insurance and disaster relief), and sustaining coastal communities and economies.

Congress should authorize NOAA to make grants to states and to specific large metropolitan areas to pay costs of developing plans to prepare for more severe storms and sea level rise. How much money is needed for development of plans? Annual federal funding in the range of $200 million, supported by a 40 percent match, would provide $280 million for this work. This would allow a minimum grant of several million dollars to each coastal state and the approximately one dozen large coastal metropolitan areas with additional funds for governments with the greatest land area and population at risk. Some states and metropolitan areas are already making substantial investments in this work.

Congress should also authorize grants to states and large metropolitan areas to implement response plans. The question of the amount of grant assistance the federal government should provide for implementing state and local plans is fundamental to the entire effort. It is important that the federal government set some broad guidelines on likely long-term grant support early in the planning process to help state and local governments plan accordingly.

Despite the substantial need for federal support, there are several reasons for Congress to be wary of promising too much money. If states and communities think that federal funds will largely pay for response measures, no matter how expensive, low cost but politically less attractive options might be discounted, resulting in unnecessarily expensive plans. A related problem is that starting with too large a pot of implementation money might result in some expensive projects at a scale that is found over time to be unstainable if applied to the entire coast. States quickest to Congress, or better represented on key committees, could reap the benefit of generous early funding, leaving other states to make do with less. Texas and Norfolk, Virginia, are examples of governments already appealing directly to Congress for significant coastal flood risk funding.

The amount of federal grants for implementing coastal flood risk plans will also depend on the state and local government share of these costs. Having state and local governments match federal funds increases their commitment to sound planning. Having the federal government provide a majority of funding sends a clear signal of the important national interests being addressed. Congress will debate this topic, but federal funding of a little over half of project costs (e.g., 60 percent of costs) seems a fair place to start. It is also important to keep in mind that the Coastal Property Price Stabilization Fund, proposed below, would relieve state and local governments of some costs.

Assuming a state and local match of 40 percent, and being cautious about offering too much, what would be a reasonable, initial amount of grant assistance for Congress to offer? In the property loss analysis discussed in chapter 4, implementing adaptation measures is estimated to reduce property losses by the year 2100 from roughly $3 trillion to roughly $800 billion as a result of implementing cost-effective adaptation measures (i.e., measures modeled to cost less than the value of property they protected). The cost of these measures came to roughly $160 billion, or about $2 billion per year over the next eighty years. The remaining $640 billion in losses is the value of existing property abandoned as not worth protecting. These estimates assume timely and economically strict adaptation investments, stable coastal populations, and no government costs due to property abandonment—all huge caveats suggesting this is a low estimate.

Using this $2 billion figure as a low-end point of reference, and not wanting to start too high, an initial target for implementation funding of about $4 billion, of which roughly $2.4 billion is federal funding matched by state and local governments, can serve as a placeholder for discussion purposes. This target, of course, will need to be adjusted based on approved plans and experience with implementation.

How might $2.4 billion a year be raised? One obvious option is to rely on federal appropriations, essentially taking income tax revenues from all Americans to pay for this work on the grounds that the entire country benefits, even if the coastal fringe of the country benefits more. Some other options for federal revenues are taxes linked to the release of greenhouse gases causing the problems of more severe storms and rising seas (e.g., a carbon tax or gasoline tax). A Democratic infrastructure proposal that includes spending $25 billion over ten years on "community resilience" focusing on flooding is paid for by simply "closing tax loopholes."[3]

Coastal Property Price Stabilization Fund

One last financial problem remains. As discussed in chapter 17, there is significant risk that the price of coastal property will decline over time as rising seas begin to inundate land area and supporting infrastructure. At some point in the future, prices may hit a tipping point and decline rapidly, with serious impacts on finances of property owners, on coastal communities that rely on coastal property taxes to maintain governmental services, and on the economy more generally. The value of existing coastal property that is not expected to be protected (i.e., is expected to be "abandoned" because protection is not economical) is roughly $640 billion (see ch. 4) and will surely increase as development continues in the years ahead. What could be done to reduce losses to property owners and minimize the financial instability coming with more severe storms and rising seas?

The federal government should create a voluntary program to manage the decline in value of property expected to be inundated by sea level rise and minimize the financial impacts of precipitous declines. To this end, the federal government should create a Coastal Property Price Stabilization Fund managed by the Department of Housing and Urban Development. The fund would purchase residential and commercial property from willing sellers located in areas at risk of sea level rise by 2100, as identified by NOAA.

A key feature of the fund would be to allow homeowners to sell the property to the fund but stay in the home after the sale in return for lease payments over the period between the sale and the end of safe use of the property prior to inundation. The fund would invite eligible property owners to propose a sale price and an annual lease amount. It would estimate the likely lease payments (i.e., payments times years prior to inundation), rank proposals that are most cost-effective for the fund, and accept those proposals offering the best chance of recouping the purchase price for which fund capital is available. The fund could also give priority to proposals from low-income people, including a financial subsidy, or decline a proposal not deemed cost-effective or not consistent with an approved state or local plan. The fund would pay administrative costs, including local property taxes and decommissioning, and have authority to end leases at any point that the property becomes unsafe or incurs storm damage. Lessees would pay costs of utilities, maintenance, and repairs but would not pay a sales commission or need flood insurance and could exit the agreement on an annual basis.

For the owner of risky coastal property, selling to the fund avoids having to worry about a worst-case scenario—a future quick shift in the local real estate market that results in a sudden, permanent decline in the value of a major asset. It allows the owner to take the capital value of the asset that is now depreciating and will eventually be a total loss, and invest it while giving the owner the option to continue occupancy of the property on a year-to-year basis. The only other way to avoid sudden devaluation of the asset is to sell it on the market now, or just before an elusive tipping point, but that would mean leaving a property to which many owners are strongly attached. The fund offers owners of risky coastal property a way to pull out of the downward spiral of coastal property values without immediately leaving home.

From the government's point of view, operating a Coastal Property Price Stabilization Fund helps avoid sudden coastal property price collapses. Why not just let the market sort out the price? The government has an interest in avoiding the economic and social disruption of a sudden loss in value of coastal real estate and minimizing the chances of a wave of mortgage delinquencies. Avoiding sudden loss of coastal property value also helps local governments maintain more stable property tax revenues and public services.

The Coastal Price Stabilization Fund model also has the advantage of getting risky coastal property into the hands of the government well before inundation and avoids complex issues about transfer of property ownership or takings arguments, with water lapping at the doorstep. It gives the government a measure of control to adjust the speed of stepping back from the shore to fit the degree of risk as it evolves over time. With the property out of the NFIP, that program avoids insuring a property likely to be a total loss eventually and avoids interim damage claims. And, with the property in hand, the government can manage decommissioning in the most timely and cost-effective way, enhancing public safety.

A single national fund offers nationally consistent rules and relieves state and local governments of the need to initiate their own coastal property stabilization programs. State and local governments, however, might coordinate fund investments with their plans by financially enhancing proposals from owners in specific areas or meeting specific criteria (e.g., low-income owners). For example, a state or local government might supplement homeowner proposals from a neighborhood in coordination with plans to shift the entire neighborhood to a safer location.

A major obstacle to this purchase and lease-back approach is the need for funds to make the initial purchases. One option is to look for annual appropriations, but a better approach is to seek debt authority from the Treasury in a manner similar to the debt now held by the NFIP. Just in general terms, $1 billion in annual debt authority would finance purchase of about 4,000 homes per year at $250,000 each. Assuming costs are recovered through lease-backs over an average of twenty years, the first-year government debt would be repaid in twenty years. Annual investment levels could be adjusted based on interest in the program.

Even though the program is intended to approve proposals that are likely to recoup purchase prices through lease-backs over time, setting a price that is attractive to the seller and fair to the government requires making complex assumptions about risk. With adequate demand from sellers, annual investment of $1billion in annual debt authority each year for fifty years would support purchase of 200,000 homes and, in theory, repay the government over seventy years. Five billion dollars in annual debt financing would support purchase of one million homes over that period. This debt financing approach effectively switches the government's debt investment from financing flood insurance payments in an increasingly unsustainable insurance model to financing one-time purchase of property while the property still has capacity to repay the debt through lease payments.

Finally, a Coastal Price Stabilization Fund should also have the power to make purchase agreements with coastal property owners at risk of sea level rise in the event that a property owner, insured by the NFIP, suffers significant storm damage and is willing to forgo the insurance claim as part of a purchase agreement with the fund. For example, an owner of a badly damaged $300,000 property might propose to sell the property to the fund outright for $200,000 and forgo a $150,000 NFIP claim, resulting in a cost to the fund of $200,000 but a net cost to the government of only $50,000. Although the net purchase cost is a loss to the fund, purchase by the fund avoids the insurance payment, followed by possible resale of the property, and the potential future repetitive losses. The fund should have authority to select only the most cost-effective of these "no lease" purchase proposals.

Although the fund is intended to be self-financing, there is reason to expect that it will require supplemental appropriations. Decisions to buy damaged property insured by the NFIP without a lease-back may make sense but would erode the fund, as would subsidies for low-income people. Acceleration of sea level risk or more severe storms may mean that some lease-back agreements end early and do not recoup the purchase price. Annual shortfalls in fund operating costs might need supplemental appropriations on the order of 10 percent (i.e., $500 million per year). Still, appropriations to the fund would help avoid much bigger spending for future disaster assistance and expanded shortfalls in the flood insurance program.

The scale of the problem is huge, and it is fair to ask if this approach is even in the ballpark in terms of meeting the need (e.g., a long-term target of $650 billion, assuming new development is checked in areas at risk). Five billion dollars invested annually in home price stabilization is $250 billion over 50 years and $500 billion over 100 years. Other factors to consider are that some homeowners will never sell, some states will supplement purchases, and some new homes will be built in risky areas. Major federal spending following coastal disasters might address some of the at-risk properties. So, this approach is in the ballpark, but time is a key factor and the sooner such an effort is started the better.

Unfortunately, it may take some time to build a consensus in favor of a Coastal Property Stabilization Fund. In the interim, Congress might be convinced to try other ideas. For example, Congress might promote Shoreline Adaptation Land Trusts (SALTs) that rely on property donation to a nonprofit land trust. Property owners would avoid local property taxes and could still lease back the property annually. Congress could support this concept by authorizing a national nonprofit organization to accept and manage donated coastal property and could create federal tax benefits to encourage such donations. Payment of local property taxes would vary from place to place but Congress might condition a generous federal benefit on payment of local property taxes in order to stabilize municipal finances.

A Good Investment

Senator Everett Dirksen is sometimes quoted as saying "a billion here, a billion there, pretty soon you're talking real money." A national program to improve response to coastal storms and rising seas will cost real money. It is, however, a good investment.

Just digging out of the NFIP debt and covering likely continuing coastal losses of the NFIP will cost about $3.4 billion annually for ten years unless Congress forgives the debt. Rough estimates for new federal costs for development of national response plans, along with improved disaster planning, greater support for existing coastal programs, initial investment in decommissioning flooded property, and planning grants to states and metropolitan areas come to just under $1 billion annually. Estimates of initial annual costs to the federal government of implementing response plans are about $4.6 billion annually (i.e., about $2.2 billion for national coastal infrastructure plans and ecosystem plans, and $2.4 billion for grants for state and metropolitan area plans). A coastal property stabilization fund spending $5 billion annually might be funded with debt authority or appropriations. Although these funds are expected to be paid back, operating costs and losses of about $500 million annually should be assumed.

That comes to about $9.5 billion annually, assuming stabilization fund capital is not counted, and $14.5 billion if it is. These costs, of course, would ramp up slowly at first and likely be greater in the later years of a long process of transition to a new coast. Supplemental appropriations for major storm disaster relief are a wild card. Retirement of flood insurance debt might allow resources to be shifted to other areas, such as support for state and local governments.

What can the country expect to get from a new national program responding to coastal storms and rising seas costing in the neighborhood of $10 billion a year? Benefits can be measured in monetary terms, both as property losses avoided and federal disaster and flood insurance spending minimized. Remember that the average property loss from a major storm is over $20 billion, that the largest storms can carry losses of over $100 billion, and that the three storms of 2017 generated losses of $265 billion. Federal supplemental appropriations to recover from the 2017 storms came to $120 billion, and individual major storms have cost tens of billions of dollars. As noted in chapter 1, the Congressional Budget Office estimates future annual property losses of $54 billion and federal government costs of $17 billion annually assuming just existing conditions.

Beyond the monetary savings, better preparedness for coastal storms and rising seas will save lives, help sustain coastal ecosystems, and protect coastal economies. It is true that even aggressive implementation of smart response strategies will not avoid all, or perhaps even most, costs of future storms and rising seas. The cost of implementing response strategies will increase in the decades to come. And, a full accounting of federal costs needs to include some base program spending as well as proposed new efforts. Still, over the long term, a national investment in improved response strategies has the potential to save trillions of dollars and thousands of lives.

Devising an effective national program to respond to coastal storms and rising seas is of limited value if it is not supported by adequate funding and financial mechanisms. Fortunately, the cost to the federal government of at least the initial phase of an improved response to coastal storms and rising seas is substantially less than the likely costs of federal disaster relief for future storm events. But prompt action is critical. Each year that passes without effective responses in place means that options are more limited, the margin for error gets slimmer, and the dimensions of the problem grow as millions more people move to risky coastal areas. A robust national program responding to coastal storms and rising seas is needed now. Steps to create a campaign to support adoption of such a program are discussed in the next chapter.

20

Campaign for a New Coast

Adopting and implementing a comprehensive national program responding to the risks that more severe coastal storms and rising sea level pose for the coast will require awakening policymakers, coastal residents, and the public to the scale of the potential impacts and the importance of starting now to make the hard choices that such a program will require. Ideally, the federal government would lead this effort.

Until that federal leadership emerges, however, other responsible parties need to take on this job. Key parties to this effort include Congress, state and local governments, corporations and other businesses, academic institutions, nonprofit organizations, foundations, associations, and religious organizations. Private citizens who care about the coast also need to step up to press for strategies to respond to storms and rising seas. A strategy for this engagement is preferable to improvisation and a campaign, demonstrating some energy and passion, is better than a strategy.

Even a well-planned campaign to prepare for coastal storms and rising seas, however, faces an uphill battle if its goal is just persuading government and others to take on the cold, actuarial job of keeping costs to less than they would have been. The reality of rising seas also offers a unique chance to reimagine the coast and think far enough ahead of rising seas to recast the coast to meet social, environmental, and economic values. Commenting on Boston's preparations for rising seas, Chris Reed, a professor at the Harvard Graduate School of Design, summed up the case for a positive outlook: "Planning is critical, but it needs to add up to something bigger—to inspire people, to drive public and political support, and to catalyze funding for implementation that can come from a wide variety of public, private, and philanthropic sources."[1]

Campaign for a New Coast: Cast of Key Players

A campaign to build awareness of coastal storm and sea level rise risks and a commitment to action needs to include institutions and organizations with involvement or interest in sea level rise science and adaptation. Ultimately, they all need to be enough on the same page that they can succeed in persuading Congress to provide the authority and funding needed to address the challenge. Who are these parties more specifically and what can they contribute to a campaign for a new coast?

Academic Institutions: Academic institutions are making impressive contributions to understanding the science of climate change, coastal storms, and rising seas. Less academic attention, however, is focused on laying the intellectual foundation needed to support the expensive and socially complex measures that are needed to address the problem.

One critical means of translating academic research on coastal issues into on-the-ground action is the national Sea Grant College Program. Under sponsorship of the National Oceanic and Atmospheric Administration (NOAA), the Sea Grant Program operates at thirty-three land grant colleges in coastal and Great Lakes states. The program engages undergraduate and graduate students in work with faculty and others within a state to address the full range of topics and questions related to the ocean. Sea Grant institutions have already made important contributions to sea level rise adaptation. For example, in 2016 the Georgia Sea Grant Program participated in development of a sea level rise adaptation plan for Tybee Island City, Georgia, the first such plan in the state.

In 2017, NOAA announced a series of meetings on ten Sea Grant "visioning topics,"[2] including "weather and climate" and "community response to flooding." The flood response effort looked at "vulnerabilities to both acute and chronic threats (e.g. storms, erosions, water-level changes),"[3] and the weather and climate effort was to develop "a 10 Year Weather and Climate Vision."[4]

Although the Trump administration proposed total elimination of funding for Sea Grant in the 2018 budget, Congress has not warmed to this proposed cut and approved fiscal year 2018 funding of $65 million, an increase of $2 million from the previous year. This very modest funding is divided among thirty-three universities and many topic areas, so the funding focused on community resilience or sea level rise is uncertain and is likely to be a small part of the total. And, the antipathy of the Trump administration does not bode well for future years. Still, Sea Grant programs are based in the states that they serve, bring trusted credentials to problem solving, and can play a central role in helping respond to coastal storms and rising seas.

Several major universities have invested in climate change and include sea level as part of a larger program, such as the Georgetown University Climate Center, The Earth Institute at Columbia University, and the Stanford Woods Institute for the Environment at Stanford University. In addition, federal agencies, including the US Geological Survey (USGS) and NOAA, support cooperative climate change initiatives at universities around the country, and some of this work addresses preparing for coastal storms and sea level rise.

A few universities have focused directly on sea level rise. Florida International University created a Sea Level Rise Solutions center in cooperation with local partners. Harvard University has funding from the John S. and James L. Knight Foundation to work with local partners on sea level rise in the Miami area. In Virginia, Old Dominion University is organizing forums to bring together faculty from other colleges and universities to discuss coastal resilience issues.

Business Sector Organizations: Local business groups are showing growing interest in the risks posed by more severe storms and rising seas. For example, the Fort Lauderdale Chamber of Commerce held a third *Sea-Level Rise Summit* in 2016. And, in May of 2017, the Greater Miami Chamber of Commerce convened its second annual conference on sea level rise solutions. In Massachusetts, the Climate Action Business Association developed "Business Acting on Rising Seas," an educational campaign that provides information for local businesses about the impacts of sea level rise and the implications for business growth.

National business groups have not yet shown the same degree of interest as local organizations. For example, the United States Chamber of Commerce has long-standing opposition to climate change measures, such as the Clean Power Plan, but has not adopted positions on risks associated with storms or sea level rise. The Associated General Contractors of America has environmental positions, including support for streamlining environmental project reviews and limiting federal protection of wetlands, but not on sea level rise per se. The American Insurance Association has positions on flood insurance and disasters but no position on coastal storms or sea level rise. The National Flood Association represents the interests of engineering firms, consultants, banks, and others working in the National Flood Insurance Program (NFIP) and related programs but also has not engaged sea level rise matters.

Brightening this picture somewhat is the engagement of senior business leaders with nonprofit organizations on climate change topics. The Center for Climate and Energy Solutions (C2ES) has a business leadership program focusing on reducing greenhouse gas emissions but also including attention to building resilience to climate change. In 2017, C2ES published a report on how business and communities can better work together to build climate resilience[5] and in 2013 published a report looking "at the ways multinational companies are beginning to assess and address the risks of extreme weather and other climate change impacts."[6] Both reports make useful recommendations for engaging business in building resilience to climate change generally but are less focused on the specific challenges of coastal storms or rising sea level.

The nonprofit organization CERES works with "influential investors and companies to integrate environmental, social, governance practices into core business strategies"[7] and is focused on "setting the highest bar for sustainability leadership, committing to robust sustainability goals, and improving resiliency in their operations and supply chains."[8] CERES also manages a "Disclose What Matters" campaign pressing for disclosure of sustainability issues and climate risks in financial filings. They have work groups on low carbon, disclosure, water, and forests, but not on coastal storms or sea level rise.

The good news in this story is that, although business engagement with issues related to coastal storms and sea level rise has been limited to date, some industry leaders are paying attention. The recent release of the new Climate-related Financial Disclosure (TCFD) recommendations should boost attention to climate risks generally as well as the physical risks associated to climate change, including rising seas.

Foundations, Coastal Storms, and Sea Level Rise: Foundations have been a driving force in understanding the risks of severe storms and rising seas and can continue to play a key role in helping develop and implement response plans.

The *Risky Business* report on the national costs of climate change in 2014, funded by Bloomberg Philanthropies and several other foundations, made an important contribution to understanding of the significance of coastal property loss related to storms and rising seas. The Rockefeller Foundation developed the 100 Resilient Cities program, assisting cities around the world in building more sustainable communities, including twenty-two cities in the United States. Of these, ten have identified sea level rise or coastal storms as one of the array of challenges they plan to address, including Miami, New Orleans, Norfolk, and Boston. The Pew Charitable Trusts focuses on flood risk reduction and improving the NFIP. The MacArthur Foundation and the Doris Duke Charitable Foundation support a Climate Resilience Fund making grants to improve local resilience capacity including some addressing coastal issues and rising seas. Foundations are also supporting specific communities. For example, the San Diego Foundation is supporting a sea level rise adaptation plan for San Diego Bay, and the Rockefeller Foundation is supporting the resilient design competition for San Francisco Bay.

Commendable as these efforts are, the big picture of foundation funding for climate change is underwhelming. In 2015, the presidents of the William and Flora Hewlett Foundation and the David and Lucile Packard Foundation wrote that "currently, less than 2 percent of all philanthropic dollars are being spent in the fight against climate change. That is not enough given how big of a threat we face."[9] Their call for action was largely aimed at increased funding for greenhouse gas reductions but extends as well to coping with the many consequences of climate change.

Victoria Herrmann took the bull by the horns in a 2017 paper for the Atlantic Council and, citing estimates of millions of Americans displaced by rising seas, proposed "a foundation-led effort to coordinate a program providing resources, expertise, support, and guidance to towns in need of managed retreat. Such a program could move communities forward in the absence of federal action on climate change by providing the resources to design, develop, and implement a relocation strategy with concrete milestones."[10] She outlines an initiative modeled on the 100 Resilient Cities project, noting that, "collectively, foundations and donors concerned about climate change are sitting on billions of dollars in endowments and accumulated personal wealth. The money is there to give, and there is no time to waste."[11]

Professional Associations: Professionals associated with coastal property are bound by their ethical codes to disclose material risks, including those related to floods and, arguably, sea level rise. Simple disclosure of risk, however, is a bare minimum of engagement with a problem that will have major economic, social, and environmental impacts. These professional organizations have members widely distributed around the country and can contribute insights needed to help craft informed solutions to coastal flood and inundation challenges.

The American Institute of Architects (AIA) articulates a deep concern for climate change and the role that intelligent building design can play in reducing greenhouse gas emissions. In addition, working though its associated foundation, AIA established a National Resilience Initiative (NRI) that "unites six university-led architecture studios to develop new designs, approaches, and policies that bolster resilience in the built environment."[12]

The American Society of Civil Engineers (ASCE) represents a profession that is critical to the implementation of measures to adapt to more severe storms and rising seas. In terms of advocacy, ASCE is focused on supporting rebuilding of infrastructure generally, but also advocates for "sustainability" without specifically mentioning coastal flooding or rising seas. In terms of adopted policy, the ASCE position paper on coastal development points to costly delays in coastal projects and a need for "streamlined regulatory process"[13] and support for beach nourishment projects. Noting that "among the great challenges the U.S. faces today is recognizing the magnitude of risk posed by flooding and motivating the public and decision-makers to make the investments required to reduce flood risk,"[14] ASCE calls for improved risk management policies and "use of natural processes to mitigate the consequences of flooding."[15] Although ASCE has no sea level policy specifically, the climate change policy mentions it, noting the importance of low risk of failure of critical infrastructure.

The American Planning Association has adopted a policy on climate change, speaking to both the need for planners to promote greenhouse gas reduction measures as well as "building community resilience by addressing climate-induced drought, flooding, sea-level rise, thawing permafrost, storm surge, and the many other impacts of climate change."[16] The policy does not shy away from hard choices, noting that "adaptation measures include shifting development from flood-prone areas,"[17] and that planners will also need to consider indirect impacts of climate change, including population migration "when sea-level rise inundates developed areas."[18]

The American Bar Association (ABA), guided by the 550-member ABA House of Delegates, has adopted policy on climate change and community resilience, but no policy specifically on sea level rise. The climate change policy urges the Congress to enact legislation that would, among other things, "enable the United States to adapt to existing and potential climate changes in a way that minimizes individual hardship, damage to natural resources, and economic costs"[19] and refers to sea level rise and storm damage in coastal areas. The ABA community resilience policy supports actions that "make communities more resilient to loss or damage from foreseeable hazards and also recognize property rights."[20] It refers to disasters, floods, pestilence, and climate variability, and briefly mentions climate adaptation and sea level rise.

The National Association of Realtors has adopted policy calling for extension and revision of NFIP and supporting development of private insurance options. The policy calls on the federal government to "provide pre-disaster risk mitigation options—including guaranteed loans, grants and buyouts for property owners to build stronger or relocate to higher ground."[21]

Governmental Associations: Associations representing state and local governments that are on the front lines of adapting to climate change impacts like coastal storm flooding and sea level rise inundation have policies encouraging the federal government to do more to help address the problems.

The National League of Cities has a climate change policy calling on the federal government to, "provide financial and technical assistance to support local government vulnerability assessments and climate change mitigation and adaptation implementation efforts."[22] The policy also calls for creation of a federal climate service, support for sharing best practices and climate resilient strategies, and funding of a national public service campaign to inform the public about the impacts of climate change and the need for adaptation measures.

The National Conference of Mayors has demonstrated strong leadership in helping cities define and implement greenhouse reduction strategies, including through an Alliance for a Sustainable Future, managed in cooperation with C2ES, and the Mayors Climate Protection Agreement, now signed by over 1,000 mayors. There is growing interest among mayors in disaster preparation, however, as their website notes: "When multiple impacts arrive simultaneously the resulting impacts can be catastrophic for people, property, infrastructure, and natural resources,"[23] and "what is becoming clear is that planning efforts need to evolve more quickly into action projects to mitigate disasters."[24] A network of over 400 "Climate Mayors" is committed to supporting Paris Climate Agreement goals and "building political will for effective federal and global policy action"[25] but is not focused on costal storms or sea level rise.

The National Association of Counties (NACo) represents the 3,007 county governments across the country, including 514 counties on the saltwater coast. In July 2017, NACo distinguished itself from most other governmental associations by adopting a sea level rise policy, urging Congress to "provide appropriate financial assistance and support to local governments for sea level rise and increased storm surge related initiatives and projects that aim to develop adaptive solutions to these potentially devastating events."[26]

The National Governors Association (NGA), in contrast, has steered clear of climate change to the point of drawing public protests at it 2017 meeting in Rhode Island from activists focused on sea level rise.[27] As reported by Eco RI News: "Even as the dozen or so governors in attendance at the breakout meeting avoided saying 'climate change,' they embraced the word 'resilience' as a means to address the destruction caused by more frequent extreme weather, flooding and a rising shoreline."[28] Perhaps in response to this reticence, twenty-four governors recently formed a Climate Alliance dedicated to meeting the goals of the Paris Climate Agreement and building climate resilience.

The governors' best defense from accusations of disinterest in coastal problems is their support for the Coastal States Organization (CSO). Created at a meeting of the NGA in 1970, CSO's goal is "helping states to maintain their leadership role in the development and implementation of national coastal and ocean policy,"[29] and includes engagement with Congress and providing a professional network among coastal state program managers. Much of CSO's focus is on NOAA's Coastal Zone Management Program, along with support for programs like Sea Grant and the National Marine Sanctuaries Program. Although the CSO has not addressed coastal storms or sea level rise specifically, it promotes the Community Rating System within the NFIP and encourages use of "resilient shorelines."

Several organizations focus on state and local interests in flood management programs and policy. The Association of State Floodplain Managers (ASFPM) has extensive policy positions on the NFIP and disaster response, including a 2019 report making detailed recommendations for a "holistic approach"[30] to coastal flood management, but it is less focused on rising seas. The National Association of State Wetland Managers works on wetlands science and management issues, especially those related to the Clean Water Act. The association has actively educated members on climate change issues, but not specifically addressed challenges of wetland losses due to rising seas. The National Association of Flood and Stormwater Management Agencies supports funding for water resource and stormwater management projects.

Environmental Organizations: Several national environmental organizations have focused on the problem of coastal resilience in the face of climate changes such as more severe storms and rising seas. Many of these organizations have connected to partners in government and the private sector but, in general, are working toward self-defined goals without a larger, cross-organizational strategy.

The Nature Conservancy has a legacy of land conservation and protection around the globe and, for the past ten years, has managed a group of programs advancing natural solutions to build resilience, including a "Coastal Resilience Program" that "supports practitioners around the world who are working to address coastal hazards, particularly sea level rise and storm surge, with adaptation and risk mitigation solutions."[31] Working with NOAA, the United States Geological Survey, and other partners, it has supported resilience projects in over 100 communities in eleven countries, including many in the United States. The program has published reports on the value of protecting coastal wetlands, marshes, and mangroves and financing natural infrastructure for coastal flood damage reduction. Its online tools help communities evaluate risks, assess financial impacts, and engage the Community Rating System (CRS) of the NFIP.

The Union of Concerned Scientists (UCS) is engaged in climate change issues with a focus on sea level rise impacts in the United States. UCS has published an impressive library of reports in the past several years calling attention to the impacts of rising sea level on communities, the military, national landmarks, and disadvantaged people. These reports provide extensive analysis and constructive recommendations. UCS's model of "mobilizing scientists and combining their voices with those of advocates, educators, business people, and other concerned citizens"[32] has generated major accomplishments but lacks the litigation expertise or grassroots connections found in other organizations.

The Environmental Defense Fund (EDF) is recognized as a leader in defining the economic case for environmental policies and is focused on defining the economic benefits of natural solutions to coastal protection in terms of jobs and new markets, especially along the coast of the Gulf of Mexico. Director of Coastal Resilience, Sharon Cunniff, commented in 2016 that "the truth is, we'll have to soon make choices about where, when and how we adapt to live with water, defend our coasts, and retreat. Fortunately, restoring coastal ecosystems can fit nicely with these strategies to provide human communities with benefits not only on stormy days, but year-round."[33] EDF also has conducted surveys of local government officials to identify what support they need to address flood and sea level rise risks, finding that "a pervasive belief exists that the public doesn't sufficiently understand their risks and would act differently if they did understand them."[34]

The Natural Resources Defense Council (NRDC) has an extensive program calling for global action on clean energy as well as building resilience to heat waves, drought, and floods. NRDC, known for its litigation prowess and legislative engagement, has focused on improving the NFIP, giving more attention to climate change risks in Hazard Mitigation Plans, adoption of stronger policies to reduce repetitive losses, and using funds from federally supported state revolving funds for water facilities to protect them from flooding.

Surfrider Foundation "is dedicated to the protection and enjoyment of the world's ocean, waves and beaches."[35] The organization is distinguished both by its focus on beaches and a national network of chapters and volunteers that are "the boots on the ground who collaborate on both the local and national level with regional staff and issue experts to carry out our mission."[36] Some key program areas are beach cleanups, beach access, plastic pollution, and coastal water monitoring and pollution control. Surfrider is also one of the few nonprofit organizations that have adopted a policy that explicitly "emphasizes managed retreat as an appropriate long-term strategy for dealing with coastal erosion."[37]

The National Fish and Wildlife Foundation focuses on making grants to support projects to protect fish, wildlife, and plants. The foundation does not advocate or litigate but works with fifteen federal agencies and more than forty-five corporate and foundation partners. Funds come from these partners as well as other sources, such as settlement of enforcement actions. The foundation manages the National Coastal Resilience Fund described in chapter 11.

Coastal Organizations: Several national organizations are focused on issues related to coastal environments and coastal communities. The Coastal Society is made up of private sector, academic and government professionals, and students focused on "actively addressing emerging coastal issues by fostering dialogue, forging partnerships and promoting communications and education."[38] The society has chapters at a number of universities and an active mentoring program but has not engaged storm or sea level rise policy issues.

The American Littoral Society takes a more active advocacy role in areas ranging from public beach access, to coastal land use, to opposing offshore oil and gas drilling. Climate change, including sea level rise, is recognized as a risk to coastal communities, and the society supports reducing greenhouse gases and restoring coastal habitat.

Another national organization with focus on beaches from a somewhat different perspective is the American Shore and Beach Preservation Association (ASBPA). Writing in *The Rising Sea*, sea level rise experts Orrin Pilkey and Rob Young conclude that "the organization argues that beachfront development should be encouraged and shorelines held in place. The group lobbies heavily to ensure that federal and state taxpayers fund this losing battle."[39] A 2012 ASBPA position paper states the following: "We conclude that beach nourishment remains the most effective and environmentally sound method for maintaining America's coasts in the face of rising sea level over the next century."[40]

Religious Organizations: American religious organizations have recognized climate change as a risk to people and the natural environment to varying degrees but have generally not advocated for policies to manage more severe coastal storms and rising seas. Despite limited engagement to date, churches have the potential to play an important role in supporting communities facing difficult choices, such as relocation, and in articulating the ethical case for protecting people and natural systems as sea levels rise.

Pope Francis' 2015 encyclical letter, *On Care for Our Common Home,* makes a strong argument for responding to climate change on moral grounds, noting that sea level rise can "create extremely serious situations"[41] for coastal communities worldwide, including "impoverished coastal populations who have nowhere else to go."[42] And, in polling on climate issues, American Catholics consistently expressed the highest concern for climate change and its impacts.[43] The United States Conference of Catholic Bishops supports a Catholic Climate Covenant and opposes American withdrawal from the Paris Climate Agreement but has not taken positions on the threats that more severe storms and rising seas pose to the natural environment or people.

Most major Protestant denominations have groups focused on climate change, and some have adopted climate change-related policies (e.g., the Presbyterian Mission Agency). Lutherans Restoring Creation work on a range of environmental issues including climate change but have not addressed coastal storms or rising seas. The Evangelical Climate Initiative calls for action on climate change without mentioning more severe storms and rising seas. The Coalition on the Environment and Jewish Life advocates for action on energy and climate change, among other issues, but also has not adopted positions on coastal storms or rising seas.

Organizations Focused on Sea Level Rise: One of the few nonprofit organizations to make sea level rise its primary focus is First Street Foundation with a mission "to educate citizens and elected officials on the risks, causes and solutions to sea level rise and flooding."[44] They accomplish this by "leveraging our expertise in digital marketing and communications to create tools that break down this complex subject into easy to understand topics and visualizations."[45] Key tools are *SeaLevelRise.org*, a collaboration among some twenty academic, nonprofit, government, and private sector professionals providing advice and guidance, and the FloodiQ website providing financial information about changing coastal property values. The organization offers useful information and tools but, as of now, has not focused on developing or advocating national policies or programs.

The newly organized American Flood Coalition, formerly the Seawall Coalition, is directly focused on "national solutions to flooding and sea level rise."[46] The platform includes investing in infrastructure to "ensure property values and coastal tourism remain strong,"[47] supporting standards to "build back stronger to protect communities from future flooding"[48] and increasing investments in military bases. Members of the coalition as of this writing are mostly from the Southeast and include about a dozen members of Congress, almost thirty communities, various elected officials, and six local civic organizations.

The Climigration Network has a very different take on the problem, focusing on challenges and opportunities related to relocation of coastal communities. Some fifty organizations participate in the network, which supports a newsfeed, webinars, case studies, and websites that encourage dialogue to "explore managed retreat as a part of the story of adaptation and resilience."[49] The network is one of the projects of the Consensus Building Institute (CBI) working in partnership with the Lincoln Institute of Land Policy and the Doris Duke Charitable Foundation.

Climate Central is a nonprofit organization of scientists and journalists providing

information and assessments of climate change and its impacts but "is not an advocacy organization."[50] Surging Seas is one of three focus areas for Climate Central, providing several mapping tools to help evaluate sea level rise impacts and publishing
studies, often as peer-reviewed science papers, on those impacts. A key set of reports
provides detailed assessments of risks and impacts on a state-by-state basis.

Finally, devastating flooding along the coasts and inland has resulted in a growing number of survivors of floods. Higher Ground, a project of the nonprofit Anthropocene Alliance, calls itself the "largest flood survivor organization in the country"[51]
and supports grassroots networking among local flood survivor organizations. The
organization offers practical advice on how to prepare for flooding along with suggestions for tactics to advance local flood response programs, including ideas for
public meeting slogans such as "Make America Dry Again" and "I Flood I Vote."

Politicians: Finally, although leadership from all corners of civil society is needed,
dedicated political leadership is essential to meeting the challenges of coastal storms and
rising seas. At one level, a campaign needs to persuade local, state and federal political
leaders to support actions to prepare for more severe storms and rising seas. At another
level, politicians can become advocates of the need for action and rally their colleagues
to the cause. Senator Edmund Muskie, for example, provided leadership that was essential to enactment of clean air and clean water legislation. This leadership includes the
ability to outline a vision of needed changes and the political skills to assemble a coalition that can enact new authority commensurate with the scale of the problem. Given
the scale of the coastal disruption problem, and the limited progress toward solutions to
date, this leadership seems sure to come eventually. Will it be in time?

Bricks and Mortar of a Campaign for a New Coast

The scientific community has delivered increasingly confident and increasingly dire
estimates of the impacts of more severe coastal storms and rising seas. American
civil society recognizes the significance of the science, and some governments and
organizations are constructively engaging these issues. These organizations and the
efforts to date are impressive. There is, however, no agreement on a larger, long-term
response program and no coordinated effort to create a consensus on the need for
such a program.

It is time to forge the existing science and the diverse capacity of civil society into
a campaign to build recognition of the need for a national coastal storm and sea level
rise preparedness program, frame its key elements, and promote it to the public and
ultimately Congress—it is time for a campaign for a new coast.

What are the core elements—the bricks and mortar—of such a campaign? A first step is to create organizational capacity to carry out this work. Another step is to seek funding to support and operate a campaign. Other key steps are to define a message and build the communication tools to deliver it. Specific policy and program positions need to be determined in cooperation with other interested parties. Finally, an agenda of actions needs to be agreed on and advanced.

Creating Organizational Capacity: An organizational capacity for pressing a campaign to prepare for more severe storms and rising seas might emerge in several ways. An existing organization might step forward to provide a home for this work as part of a larger advocacy agenda. Or, a new organization dedicated to defining and advancing needed policies and programs for coastal storms and rising seas might be created. In either case, a single organization speaking to these issues is likely to be less effective than the voice of a coalition or network able to represent the many organizations and interests with a stake in the topic. Coalitions of interested groups speaking to a policy topic often seek to magnify the voices of many local organizations (e.g., the Land and Water Conservation Fund Coalition) and sometimes also speak for diverse public organizations and private companies at the national level (e.g., the US Water Alliance).

Any decision to amend the work of an existing organization, to create a new organization, or to create a coalition of organizations, needs to be the result of discussions among the interested parties. The role of an organization, or network of organizations, focused on coastal storms and sea level rise would evolve over time but could start by simply sharing information among participating organizations on both scientific developments and adaptation successes. Another important initial role is to provide a forum for sorting of interests and institutional capacities of participating organizations to make the most of limited resources. Building the capacity to develop and advance national policies and programs would likely take longer.

Funding a Campaign: The present administration is not likely to endorse, much less fund, a national effort to develop a program to prepare for more severe storms and rising seas. Given the assets and investments at risk, however, finding funds to advance a strategy to reduce the costs of coastal storms and rising seas seems like an attainable goal.

One place to start in finding seed money is the foundations that have already identified a need to increase investment in climate change or have already engaged the topics of coastal storms and rising seas. One of these foundations, or a group of foundations, might provide funding to start a process of building the capacity of civil society to develop and advance the national effort to prepare for more severe storms and rising seas.

Foundation funds might be matched by small contributions from organizations that already are engaged with coastal issues and private sector organizations that want to participate in the process. The Nature Conservancy, for example, has developed a broad base of partners for its global coastal resilience initiative.

Communicating a Message: Even in the early stages of a long campaign to prepare for more severe storms and rising seas, there is a need to communicate among participating organizations, with other interested parties, with Congress, and with the public. A new campaign to prepare for these risks should be a resource for all the participating groups to communicate the key findings of new coastal storm and sea level rise science, new insights about adapting to these challenges, and success stories from governments, nonprofit organizations, and the private sector.

One key element of a communications capacity is development of a presence on social media. First Street Foundation is already building capacity in this area and sea level rise expert John Englander recently announced plans for a series of "Sea Level Rise Minutes" on science and solutions with crowdsource funding. Another communication function is to be a resource for linking experts in sea level rise science and adaptation to people, ranging from high schools and colleges to the Congress, who want credible information about more severe storms and rising seas. It is also important to have public opinion polling on coastal storm and sea level rise issues that can track changes over multiple decades.

Yet another communication function is thinking ahead to make the most of events as they unfold. For example, if a scientific journal is planning to publish a new study of relative sea level rise in the Chesapeake Bay, a good communications team could alert local press and help identify local experts to comment on the findings.

Film and print media also speak to coastal storms and rising seas, and the communications arm of a campaign should highlight new releases and otherwise promote this work. For example, the American Resilience Project released the film *Tidewater*, looking at the impact of rising seas on the Norfolk, Virginia, area, and focusing on implications for the United States Navy. The project is also developing other films on sea level impacts. In 2018, Nathan Kensinger released *"Managed Retreat,"* a short documentary that follows three New York City neighborhoods that were purchased in the aftermath of Hurricane Sandy then demolished and returned to nature. The Northern Gulf of Mexico Sentinel Site Cooperative Program has developed a series of short, state-specific videos describing storm and sea level rise risks and responses.

In addition to a number of excellent books about the impacts of Hurricane Katrina, several books have been published for the general public in the past ten years on issues related to storms and rising seas, including by Chris Mooney (*Storm*

World), Orrin Pilkey and Rob Young (*The Rising Sea*), John Englander (*High Tide on Main Street*), Chad McGuire (*Adapting to Sea Level Rise in the Coastal Zone*), Orrin Pilkey, Linda Pilkey-Jarvis, and Keith Pilkey (*Retreat from a Rising Sea*), Jeff Goodell (*The Water Will Come*), and Elizabeth Rush (*Rising*).

In addition to books, several magazines have made important contributions to the general public's understanding of sea level rise. For example, Elizabeth Colbert's 2015 article in *The New Yorker* offered a cogent summary of tensions as the city of Miami copes with both growth and rising sea level. Robinson Meyer, writing in *The Atlantic*, has covered projections of more severe storms and sea level rise along with a range of other climate change topics. Jeff Goodell writes on coastal storm and sea level rise issues for *Rolling Stone,* including a 2015 article with an interview with President Obama and others on rising sea level issues.

Define Policy and Program Positions: A campaign with the goal of preparing for more severe coastal storms and rising seas needs to go beyond coordinating information and communicating a general message about risks and success stories. A key next step is to define what needs to be done in practical terms.

These policy and program positions could initially address specific topics, such as reform of the NFIP or improved disclosure of sea level rise risks to coastal property. A campaign for a new coast might also take positions on funding of federal and state programs related to costal storms and rising seas, address federal or state legislation on these topics, and encourage citizen actions to improve preparedness for more severe storms and rising seas. A commitment to stand for specific policies requires an organizational structure that is able to resolve policy issues and press ahead with policy decisions.

Before too long, however, a campaign needs to move beyond individual policies and generate a more comprehensive vision of a national approach to preparing for more severe storms and rising seas. The nation can make progress toward preparing for rising seas without the active engagement of the federal government or the enactment of national legislation. Still, federal engagement to implement an effort authorized in national legislation and supported with national funding offers the best chance of success. To that end, a campaign for coastal storm and sea level rise preparedness should promote national legislation authorizing and funding a national program to prepare.

A Positive Vision of a New Coast

There is no avoiding the fact that the coming changes along the American coast will be expensive and painful. The work to manage these risks, however, can be undermined by an image of a Sisyphean task with high costs and small reward. Casting the challenge of coastal storms and rising seas not as just the loss of a coast but as the chance to make a new coast that can be as good as, perhaps in some ways better than, the old one is difficult but important.

Harvard professor Chris Reed recently called on Boston to "think bigger: We need to imagine a 21st-century, climate-ready version of the Emerald Necklace park system for the waterfront—a Sapphire Ring, with floodable coastal parks, hard and soft infrastructures, living shorelines and breakwaters, and active waterfront trails and destinations all around Boston Harbor that mitigate the effects of storms, protect against rising seas, provide enhanced public access, and reestablish Boston as a city built on and around its natural resources and open spaces."[52]

One way to initiate creative thinking is a "design challenge" that gives architects, planners, engineers, and local officials a chance to sketch a vision of transformation for a community that accounts for the realities of changing conditions. There are several examples of design challenges focused on coastal storm recovery and rising seas, including the "Resilient by Design/Bay Area Challenge" for San Francisco Bay, the Hurricane Sandy Design Competition, and a competition for Boston Harbor that dates to 1998. At the national level, the Department of Housing and Urban Development (HUD) offered in 2014 a one billion dollar National Disaster Resilience Grant Competition. Although HUD did not focus the competition on coastal storms or rising seas, about half the final winners proposed to address those issues to some degree. The "design competition" approach, however, has been criticized as utopian, unlikely to protect vulnerable populations, and supporting building in unsustainable places.[53]

At the far end of the spectrum of creative thinking about adapting to rising seas is the "aquatecture" movement in which homes or entire communities are designed to float as flood or sea waters rise. Although there are examples of this approach in other countries, including the Netherlands and Thailand, it has seen very limited use to date in the United States. Proponents recognize that major challenges loom, including providing public utilities and amenities needed for a fully functioning community, but the idea of living with water is seen as a way to "avoid a mass migration as water levels invade our homes and cities."[54] Aquatecture may offer some temporary benefits in some places, but it also illustrates the risk that creative ideas can also be costly, impractical diversions.

What can be said to be the benefits of preparing for more severe storms and rising seas and how might a new coast be an improvement on the old one?

A key feature of a new coast is that the shoreline will likely be more accessible to the public. Today, access to the shoreline, to beaches, and to boat launch facilities is limited by historic patterns of development. Surfrider Foundation notes that projected population increases for coastal areas make access more challenging: "As population continues to increase and more people move to coastal areas, public access opportunities can get squeezed out....As a result, the number of access points in most states is not growing fast enough to keep up with current population demand."[55] As plans are framed for stepping back from the coast, more attention will be given to public interests such as shoreline and water access. Although some coastal protection structures may necessarily have the effect of limiting shore access, other structures can promote access or include new access points at nearby locations.

In addition, a new coast can be designed with protection of natural features, such as beaches, coastal marshes, and wetlands, as a high priority. Historically, development along the coast occurred with little regard to the value of ecosystems such as wetlands. At the national scale, wetlands loss between colonial times and the 1980s is estimated at over 50 percent.[56] No estimate is available for just coastal wetland loss over this period, but it seems likely that the dense settlement along the coast resulted in significant loss of these wetlands over the years.

It is hard to estimate whether natural landward migration of coastal wetlands and beaches will increase their area or what impact affirmative efforts to promote landward migration might have. Today's understanding of the importance of coastal ecosystems for recreation, habitat, and fisheries enhancement, however, is likely to prompt a determined effort to make space for them and give these resources more care than they were accorded historically. For example, the city of Rotterdam has proposed creation of "tidal parks" in urban areas, and this idea of creating managed public spaces along shifting coastlines might be developed more broadly in the diverse settings of the American coast.

The future of coastal waterfronts is a special concern as some of these areas are iconic places of historic and cultural significance that are effectively irreplaceable. Other developed waterfronts, however, have issues including limited services, and aging infrastructure and housing stock. With an added burden of increasing storm flooding and eventual inundation, these areas will become costlier for residents and local governments. As more severe storms and rising seas prompt assessment of these waterfronts, there is a chance to reinvest in upgrading infrastructure and surrounding communities.

As people living in areas at risk of costal storm surges and rising seas become more familiar with costs and other impacts, a willingness to move may be dampened by anxiety over where to go. One way to avoid the "but where will we go" question is to focus on making the places that people relocate to be as good or better than the places they leave. Researchers at Massachusetts Institute of Technology, in their study of relocation around Boston, point out that relocation sites can be designed with a "build it better" approach that upgrades services ranging from public transportation, to open space, to shopping amenities.

With a build it better approach and relocation to less risky areas, some of the unpredictable costs and health and safety issues related to storm flooding and rising seas that come with the current coast are left behind. Yet another benefit is the opportunity to build in attention to the needs of disadvantaged populations, so that a new coast does not recreate the social inequities that evolved along the current coast.

Even if the cons of a new coast resulting from more severe storms and sea level rise stack up higher than the pros, there is something to be said for making a virtue of a necessity. John Englander observed this: "Throughout history the ocean has taught man humility. We ignore its power at our peril. Along with crisis, there is opportunity. There can be tremendous innovation and adaptation in the coming decades as we anticipate and change our coastal oriented society and economies. But getting a good return on investment requires that we see where things are headed."[57]

Citizen Action: What Grassroots Energy Can Do for a New Coast

Experts and advocates alone are not likely to get the traction in city halls, state houses, and Congress needed for a successful response to coastal storms and rising seas. Public officials need to be convinced that the experts are right, but they also need to be convinced that the public understands the problem and wants, or even demands, action to deal with it.

What can be done to convince a critical core of public officials that the time has come to prepare for more severe storms and rising seas? Millions of Americans live along the coasts and millions more think of a beach or coastal vacation spot as an important part of their lives. Eventually, the shock of seeing these treasured places slip away to coastal storms or the relentless, incremental advance of rising seas will drive a public call for action. Today, there is still time for people who care about the coast to learn about the risks and join in pressing public officials to engage in the hard work of sorting out options and implementing needed actions. The opportunities available to each person are different.

Understand Your Storm and Sea Level Risk: Most people have a general idea of the elevation of their coastal property and their exposure to storm surges or rising seas. A clearer picture of risk, however, can help inform personal decisions and prompt an interest in government response actions.

A first step in understanding this risk is to look at the current Federal Emergency Management Agency (FEMA) flood insurance maps for your community and determine if your property, or property or ecosystems you care about, are in the 100-year flood-plain or the 500-year floodplain. A second step in understanding your risk is to get a picture of the risk of major coastal storms in your area. One good way to learn this is to look at the NOAA storm impact data on the Internet showing the frequency of major storms on a county level along the coast (e.g., NOAA storm surge maps).

Finally, you might want to get a picture of how rising sea level is likely to play out on the part of the coast of most interest to you. Thanks to the development of new assessments of relative sea level rise, reflecting both localized changes in sea levels and land subsidence, it is now possible to translate global predictions about rising seas into information tailored to your place of interest. A good way to learn about relative sea level rise at a specific place and by a specific future date is to use the sea level rise calculator developed by the Army Corps of Engineers and available on the Internet. Once you have projected estimates of future changes in sea level at your location of interest, you can go to one of several mapping tools on the Internet to see how a given change in sea level (e.g., two feet or four feet) changes the shoreline, and how far inland from the current shoreline sea water would reach.

Engage Experts in Understanding Local Risk of Storms and Rising Seas: With some homework done concerning risk levels, a good next step is to seek out a general assessment of storm and sea level rise risks. For example, you might ask your community, or a local organization, to invite someone from the State Sea Grant Program at your land grant university to give a talk on this topic. Other possible sources of speakers include university faculty from other institutions, staff from the state Coastal Zone Management Program, or staff from the district office of the United States Army Corps of Engineers. If you are in an area covered by one of the twenty-eight National Estuary Programs or the National Estuarine Research Reserves, staff from these programs might also speak to these topics.

Encourage Your Community to Prepare for More Severe Storms and Rising Seas: With information about local projections of storm risk and future sea level rise in hand, a next step is to use that information to encourage your community to increase its efforts to prepare for more severe storms and rising seas. For example, you might determine whether your community participates in the Community Rating System (CRS) and, if not, encourage public officials to take the modest steps needed to join the program. If your community already participates, it is important to ask public officials what "class" your community is, what actions were implemented to earn that class rating, and what additional measures might be adopted.

If CRS measures are already in place, it is important to start engaging some of the community choices discussed in chapter 14, starting with adding storm and sea level rise information to local plans, and addressing questions such as length of planning time horizon and community tolerance for risk. In the comparatively few communities where planning for these risks is advanced, people need to push to implement plans through investments, financial incentives, or regulatory tools. Limiting development in areas at risk of storm surge and rising seas is a key first step.

Use Print and Social Media to Build Awareness of Severe Storms and Rising Seas: Most people living along the coast are generally aware that a major coastal storm could damage their property and community but have not translated national news about rising seas into an image of what rising seas will mean in their specific case. Building a broader base of public understanding of the problem can help establish a foundation of support for the plans and programs that are needed to address the problem. Social and print media can be effective tools in building this awareness.

For example, a letter to the editor of a coastal community paper or a social media post outlining what you found when you looked up the projected relative sea level rise for the area can help encourage other people to take the time to do the same thing and thus build awareness of risks. Other potential topics are the status of your community's participation in the CRS program, storm and sea level rise responses in neighboring communities, and the critical link between the long-term protection of coastal communities from rising seas and the success of global efforts to reduce greenhouse gases.

Engage Civil Society in Preparing for Storms and Rising Seas: A surprising number of people are connected to a professional or nonprofit organization with an interest in coastal property or the environment. Individuals working within these organizations can help bring more severe storm and sea level rise risks to wider attention within the organization, or prompt efforts to develop and adopt policy positions in this area.

Most environmental organizations have strong positions in favor of reducing greenhouse gases and some have policies on adapting to climate change, but only a few have spoken directly to sea level rise and storm issues. Members of these organizations should call for stronger positions to prepare for these risks.

Members of professional associations, such as the National Association of Realtors and the American Bar Association, should review existing policies relating to climate change and press to add policies specifically addressing rising seas and coastal storms. Members of the American Society of Civil Engineers and the American Institute of Architects are also well regarded in their communities as experts in the building and construction fields and can be influential voices in discussions of options related to preparing for coastal flooding and inundation. Broad-based organizations such as the Garden Clubs of America and the League of Women Voters should be encouraged to take positions on these coastal challenges.

Investor Action to Move Corporate America to Attend to Coastal Risks: Some people are not that focused on professional or other organizations but are focused on the performance of companies in which they have investments. Investors can play a role in bringing the physical risks posed to industrial infrastructure by more severe storms and rising seas to the attention of leaders of publicly traded companies.

A first step in starting this conversation is to ask whether the company has addressed climate change, including physical risks to facilities, in its annual filing with the Securities and Exchange Commission (SEC). At a minimum, the report should comply with the Climate Change Guidance developed by the SEC. If the company has major assets in coastal areas, the annual report should specifically address physical risks to these facilities from both severe storms and rising seas. Investors can also ask companies to commit to comply with the recommendations of the Financial Stability Board's Task Force on Climate-related Disclosures (TCFD). In cases of corporate denial of climate risks, investors can promote resolutions at annual stockholder meetings urging management to disclose climate risks and address them in corporate strategies.

Manage Your Coastal Land Responsibly: Many of the people most likely to suffer the impacts of severe storms and rising seas live directly on the coast. Setting a good example by responsible management of coastal land can encourage other coastal landowners to avoid damaging practices or join in response efforts.

One constructive action coastal property owners can take is to work with a local government or nonprofit organization to establish a "rolling conservation easement." As explained in chapter 15, rolling easements are agreements to allow rising seas to move inland at a natural rate without building protection structures or to prevent development on undeveloped land. Another option to consider is participating in a property buyout program if such a program is available (e.g., after a major storm or from a state).

Insist on Flood Risk Disclosure in Your Coastal Property Transactions: Over the coming decades as sea level rises there will be millions of coastal property transactions. A small percentage of these transactions will be covered by a mandatory flood and sea level rise risk disclosure requirement, and in some additional cases, coastal property professionals will insist on disclosure of at least flood risk, as an ethical matter. Another way to assure widespread disclosure of flood and sea level rise risk is for buyers of coastal property to insist that sellers make this information available as a condition of the sale. Even a gradual increase in the number of buyers asking to see this information will help to make the case that such information is "material" to the sale and thus an ethical obligation of realtors and attorneys involved in these transactions.

Engage Your Elected Officials on Storms and Rising Seas: Direct communication with public officials to press for attention to risks of coastal storms and rising sea is a basic and useful action.

At the local level, it is important that public officials hear questions about storm risk and local sea level rise projections and the adequacy of community response plans. Local officials managing public utilities, such as public water and wastewater systems, should be asked what plans they have made to prepare for these risks.

At the state level, questions might include the status of sea level rise vulnerability assessments or plans and funding for state and local response efforts. You should ask state officials about whether Coastal Zone Management Program funds allocated to the state are addressing storms and sea level rise and how these funds are being spent. State officials should also be asked to support a state law providing for mandatory disclosure of the risks at the time of the sale of coastal property. If your state has a land acquisition program, you should ask how these programs are being used to better prepare for coastal flood and inundation risks.

Members of Congress also need to hear from the public on these issues. A key topic to raise with them is support for funding of existing programs like the Coastal Zone Management Program. Another good topic is the need to support amendments to the NFIP that account for rising seas as well as storm flooding. In the event that legislation proposing a national program to prepare for rising seas and more severe storms is introduced, members of Congress need to be encouraged to review and cosponsor that legislation. Still another topic worth raising with Congress is the importance of recommitting the United States to the Paris Climate Agreement as the best hope of moderating long-term sea level rise.

No one wants to lose the existing coast, but if there must be a new one, the country should spend the time, energy, and money to make it as great as the old one. Civil society—in the form of educational institutions, foundations, advocacy organizations, and business groups—has the capacity to inform and influence government at all levels to engage this challenge. Today, different organizations are working on coastal storms and rising seas in the context of their larger mission. Focusing the message of these groups around a coordinated set of policies could attract more interest and put this topic on the national agenda. In addition, individual citizens who care about the coast can be better informed about the risks and draw on this knowledge to press elected officials to define and implement response strategies.

PART V: EPILOGUE

Campaign for a New Coast

Today, most people would consider it fanciful to suggest that policies of the last administration be resurrected, much less that the country mount a major new national effort to manage more severe storms and rising seas. The Trump administration has systematically destroyed constructive policies in this area and, when outright destruction was not possible, proposed to eliminate funding from useful programs.

Yet, the premise of part 5 is that a new national program with capacity and financial resources sufficient to deal with the problems of more severe storms and rising seas can be framed, and despite the obvious obstacles, this program, or something similar, could be enacted and funded in the foreseeable future. Given the nature of the political process, this premise can only be proven over time.

Several factors suggest success. The impacts of stronger storms and rising seas are tangible and will become harder to ignore. The benefits of preparing for these risks far outweigh the costs of failing to act. Some of the needed policies and programs are controversial but are not out of the mainstream. Building understanding of the risks can still go a long way to overcoming doubts and reluctance to deal with the problem. The diverse resources of American civil society, ranging from professional associations to nonprofit organizations to the private sector, have only begun to engage this issue. And, people who care about the coast will increasingly recognize the need for action and support new initiatives.

It seems likely that the experience of future coastal storms and relentlessly rising seas will eventually summon from the political process the will to better prepare for, and respond to, these challenges. Less clear is whether this action will come too late to avoid the most serious impacts and reap the benefits of timely action. Accelerating the engagement of the political process with the challenges posed by more severe coastal storms and rising seas will require a new, organized effort to demonstrate that effective solutions are available and to press for their timely adoption—a campaign for a new coast.

Conclusion

The story so far is that America faces serious impacts from more severe coastal storms and rising seas and should seize the opportunity to substantially reduce these risks by adopting a new national response program. The science is strong. The impacts on the American coast will be extensive, costly, and disruptive. Existing programs are not a good fit for the job. And, coastal states and communities need technical and financial support from the federal government to effectively respond to these challenges. Finally, although an effective national program to address more severe storms and rising seas will be controversial, it is not out of the mainstream and could be enacted based on a campaign to put the issue on the national agenda.

In closing the story, it is important to remember that this effort will play out in the context of the international enterprise to reduce releases of greenhouse gases and the steps other countries take to cope with more severe storms and rising seas. The progress of this work is likely to greatly influence how the American response unfolds in the decades to come.

Today, the outlook for success of the international effort under the Paris Climate Agreement to reduce releases of greenhouse gases and hold the increase in warming to less than 2°C can be described as uncertain at best. Even the emergence of a remarkably successful international effort to minimize the warming of the planet, however, will not make the problems posed by more severe coastal storms and rising sea level go away. More important, increases in global air temperatures beyond the

2°C increase will mean that sea level will rise faster and for longer, making adapting to the risk much harder everywhere.

The United States needs to be prepared for more severe storms and rising seas in whatever form they come. But, the problem will be much more manageable and far less costly if air temperatures stay below the 2°C increase cap. To that end, the United States should strongly support the Paris Climate Agreement and provide encouragement and support for other countries to do the same. The significant gap in national contributions under the agreement and the need to get on track by no later than 2030 suggest that new ideas, backed by the financial and moral authority of the world's remaining superpower, will be needed. American leadership can make a difference for the world and pay major dividends at home.

In addition, more severe coastal storms and rising seas are a risk to 75 percent of the countries in the world. For a few countries, rising seas are an existential threat. More than some other impacts of a warmer planet, they present a physical reality that will force societies around the globe to make difficult decisions to invest in protection structures or step back from the coast. Many countries have more at stake than does the United States and, due to the whims of geography, they will face significant impacts sooner. The successes and failures of other countries as they deal with costal inundation, loss of infrastructure and ecosystems, and migration of coastal populations will be evident for all to see.

Watching these developments around the world, American policymakers and people living along the American coast will have some warning of the challenges to come and a chance to avoid the mistakes of others or learn from successes. At the same time, the United States enjoys unparalleled scientific, legal, and economic resources. Marshalling these resources and stepping forward to find effective solutions to the challenges offers the United States a chance to get ahead of the risks and to demonstrate to a wider world audience effective methods, tools, and practices. Again, American leadership can make a difference for the world and pay major dividends at home.

Acknowledgments

As I started looking at the question of the future of this country's coast, I understood that I faced a steep learning curve. Still, with each passing day, my recognition of the complexities of the topic grew, as did my appreciation of the critical contributions to protecting the coast being made by scientists, academics, politicians, government employees, and advocates. I owe a major debt to this band of coastal professionals, some of whom appear in the text, and others who provided the background I needed to see the challenges that the coast faces today, and those that are to come.

I could not have conceived of this book if I had not had the opportunity to work on the problem of adapting to a changing climate at the Environmental Protection Agency. EPA colleagues from national programs and regional offices encouraged and supported this work. I was also fortunate to have a chance to work on climate adaptation issues across federal agencies while assigned to the White House Council on Environmental Quality.

As I learned about the challenges of protecting the coast, I asked dozens of people involved with these topics for suggestions and directions. I also appealed to some experts to do structured interviews to get their insights on the big picture and specific issues. These inquiries and requests were invariably met with a generous willingness to share perspectives and keep me on track. Everyone I talked with has my gratitude for their patience and thoughtfulness, but I especially want to thank Rachel Cleetus, Erika Feller, John Englander, Robert Hunter, Zoe Johnson, Jeremy Martinich, Ben McFarlane, Rob Moore, Larry Larson, Keith Rizzardi, and Erika Spangler-Siegfried. Any errors, omissions, or inaccuracies in the text, however, are strictly my own.

As helpful as the Internet can be, sometimes you need a really good library. I was fortunate to have the support of the Library of Congress, the Mary Riley Styles Public Library, and the Engineer Research and Development Center Library of the US Army Corps of Engineers.

For a first-time author, finding a first-rate editor and publisher can sometimes be an insurmountable challenge. This book could not have happened without the support of Island Press and my editor Courtney Lix. Her steady faith in the book kept it on track, focused the material, and improved the writing and presentation in countless ways.

My son James offered the boundless faith of the young that a life-long bureaucrat could change gears and write a book for a wider audience. My wife Ida patiently read numerous drafts and suggested better approaches that sent me back to the keyboard with new determination.

Appendix 1

Table A-1. Global Mean Sea Level (GMSL) rise scenarios, in feet, for 19-year averages, centered on decades through 2100

Year	Low	Intermediate-low	Intermediate	Intermediate-high	High	Extreme
2010	0.10	0.13	0.13	0.16	0.16	0.13
2020	0.20	0.26	0.33	0.33	0.36	0.36
2030	0.30	0.43	0.52	0.62	0.69	0.79
2040	0.43	0.59	0.82	0.98	1.18	1.35
2050	0.52	0.79	1.12	1.44	1.77	2.07
2060	0.62	0.95	1.48	1.97	2.53	2.95
2070	0.72	1.15	1.87	2.59	3.28	3.94
2080	0.82	1.31	2.33	3.28	4.27	5.25
2090	0.92	1.48	2.79	3.94	5.58	6.56
2100	0.98	1.64	3.28	4.92	6.56	8.20

Source: Data from William V. Sweet et al., *Global and Regional Sea Level Rise Scenarios for the United States* (Silver Spring, MD: National Oceanic and Atmospheric Administration, 2017), table 5 (meters in original converted to feet), 23.

Appendix 2

Table A-2. Projected relative sea level rise in 2100 by state/territory, by scenario

States and territories	Low (0.3 m/ 0.98 ft.)	Intermediate-low (0.5 m/ 1.64 ft.)	Intermediate (1.0 m/ 3.28 ft.)	Intermediate-high (1.5 m/ 4.92 ft.)	High (2.0 m/ 6.5 ft.)	Extreme (2.5 m/ 8.20 ft.)
Alabama	1.40	1.82	3.81	6.00	8.31	10.32
Alaska	0.03	0.35	2.00	4.06	6.46	8.34
American Samoa	1.55	2.02	3.98	6.13	8.34	10.15
California	1.01	1.47	3.33	5.60	8.16	10.09
Northern Mariana Islands	1.18	1.65	3.86	6.28	8.64	10.41
Connecticut	1.46	1.93	4.21	6.40	9.06	11.16
Delaware	1.62	2.09	4.31	6.54	9.17	11.22
District of Columbia	1.55	2.01	4.20	6.45	9.01	11.05
Florida	1.27	1.71	3.72	5.98	8.36	10.34
Georgia	1.42	1.85	3.87	6.17	8.65	10.58
Guam	1.25	1.71	3.90	6.30	8.65	10.47
Hawai'i	1.35	1.83	4.03	6.44	8.96	10.94
Louisiana	2.93	3.36	5.37	7.57	9.90	11.91
Maine	1.16	1.61	3.89	6.05	8.79	10.87
Maryland	1.63	2.09	4.30	6.55	9.14	11.19
Massachusetts	1.38	1.83	4.12	6.33	8.99	11.13
Mississippi	2.20	2.64	4.61	6.80	9.12	11.13
New Hampshire	1.13	1.56	3.84	6.03	8.72	10.83
New Jersey	1.60	2.08	4.33	6.54	9.20	11.26

continued on next page

Table A-2. continued

States and territories	Low (0.3 m/ 0.98 ft.)	Intermediate- low (0.5 m/ 1.64 ft.)	Intermediate (1.0 m/ 3.28 ft.)	Intermediate- high (1.5 m/ 4.92 ft.)	High (2.0 m/ 6.5 ft.)	Extreme (2.5 m/ 8.20 ft.)
New York	1.49	1.96	4.24	6.43	9.10	11.19
North Carolina	1.59	2.05	4.20	6.56	9.14	11.14
Oregon	0.77	1.20	2.95	5.19	7.78	9.74
Pennsylvania	1.50	1.96	4.20	6.40	9.02	11.08
Puerto Rico	1.13	1.59	3.55	5.91	8.29	10.17
Rhode Island	1.43	1.89	4.17	6.37	9.01	11.14
South Carolina	1.44	1.88	3.92	6.25	8.76	10.70
Texas	2.27	2.72	4.74	6.95	9.30	11.27
Virgin Islands	1.13	1.60	3.55	5.92	8.28	10.20
Virginia	1.66	2.13	4.32	6.61	9.19	11.23
Washington	0.51	0.91	2.61	4.80	7.39	9.38

Note: The sea-level-rise scenarios do not include vertical land movement (VLM), whereas state estimates do. Therefore, average values for states may be higher or lower than scenarios due to variations in VLM.

Source: Memorandum, "Summary of New Federal Interagency Estimates of Increased Sea Level Rise," from Environmental Protection Agency Sea Level Rise Working Group to Environmental Protection Agency Climate Change Adaptation Workgroup, (August 24, 2017), attachment 2; released under Freedom of Information Act, November 14, 2018.

Appendix 3

Table A-3. Global mean sea level (GMSL) rise scenarios, in feet, through 2200

Year	Low	Intermediate low	Intermediate	Intermediate high	High	Extreme
2100	0.98	1.64	3.28	4.92	6.56	8.20
2120	1.12	1.97	4.27	6.56	9.19	11.81
2150	1.21	2.40	5.91	10.17	14.11	18.04
2200	1.28	3.12	9.19	16.73	24.61	31.82

Note: Median values are shown.

Source: Data from William V. Sweet et al., *Global and Regional Sea Level Rise Scenarios for the United States* (Silver Spring, MD: National Oceanic and Atmospheric Administration, 2017), table 5 (meters in original converted to feet), 23.

Appendix 4

Table A-4. National assessment of shoreline change

Region	Coastal segments eroding, long and short term (%)		Average annual change rate/foot, long and short term	
New England	Long	71	Long	−1.64
	Short	70	Short	−4.92
Mid-Atlantic	Long	67	Long	−7.22
	Short	54	Short	−8.20
Southeast Atlantic	Long	45	Long	−0.23
	Short	39	Short	na
Gulf of Mexico	Long	61	Long	−5.57
	Short	55	Short	
California Cliffs	na		−1.0	
California Sandy	Long	44	Long	−0.65
	Short	66	Short	−0.65
Pacific Northwest	Long	36	Long	+2.95
	Short	44	Short	+2.95

Sources: Data from regional reports of the National Assessment of Shoreline Change; US Geological Survey (meters in original converted to feet); Cheryl Hapke et al., *National Assessment of Shoreline Change: Historical Shoreline Change along the New England and Mid-Atlantic Coast* (Washington, DC: US Geological Survey, 2011), 28; Robert Morton, Tara Miller, and Laura Moore, *National Assessment Of Shoreline Change: Part 1, Historical Shoreline Changes and Associated Coastal Land Loss along the U.S. Gulf Of Mexico* (St. Petersburg, FL: US Geological Survey, 2004), 38; Robert Morton and Tara Miller, *National Assessment Of Shoreline Change: Part 2, Historical Shoreline Changes and Associated Coastal Land Loss along the U.S. Southeast Atlantic Coast* (St. Petersburg, FL: US Geological Survey, 2004), 31; Cheryl Hapke and David Reid, *National Assessment of Shoreline Change, Part 4, Historical Coastal Cliff Retreat along the California Coast,* (Reston, VA: US Geological Survey, 2007), 1; Cheryl Hapke et al., *National Assessment of Shoreline Change, Part 3, Historical Shoreline Change and Associated Coastal Land Loss along Sandy Shorelines of the California Coast* (Reston, VA: US Geological Survey, 2007), 33, 67, 68; Peter Ruggiero et al., *National Assessment of Shoreline Change: Historical Shoreline Change along the Pacific Northwest Coast* (Reston, VA: US Geological Survey, 2013), 29–30.

Appendix 5

Representative Concentration Pathways (RCPs)

The international community of climate researchers developed Representative Concentration Pathways (RCPs) to standardize ways of describing environmental changes that might be expected in the event of specific changes in key measurements related to greenhouse gas emissions. Although the RCPs were not developed to describe the results of a given set of policies or actions, researchers have developed descriptions of future conditions associated with different RCPs. The summary below of future actions needed to attain different RCPs is derived from an RCP guide developed by the Center for International Climate Research.[1]

RCP 2.6 Low emissions: This RCP assumes ambitious greenhouse gas emissions reductions would be required over time. This future would require the following:

- declining use of oil
- low energy intensity
- a world population of 9 billion by year 2100
- use of croplands increase due to bioenergy production
- more intensive animal husbandry
- methane emissions reduced by 40 percent
- CO_2 emissions stay at 2015 levels until 2020, then decline and become negative in 2100
- CO_2 concentrations peak around 2050, followed by a modest decline to around 400 ppm by 2100

RCP 4.5 Intermediate emissions: This RCP assumes a future with relatively ambitious emissions reductions. This future is consistent with the following:

- lower energy intensity
- strong reforestation programs
- decreasing use of croplands/grasslands due to yield increases and dietary changes

- stringent climate policies
- stable methane emissions
- CO_2 emissions increase only slightly before decline commences around 2040

RCP 8.5 High emissions: This RCP is consistent with a future with no policy changes to reduce emissions. It is characterized by increasing greenhouse gas emissions that lead to high greenhouse gas concentrations over time. This future is consistent with the following:
- three times 2015 CO_2 emissions by 2100
- rapid increase in methane emissions
- increased use of croplands and grasslands driven by an increase in population
- a world population of 12 billion by 2100
- lower rate of technology development
- heavy reliance on fossil fuels
- high energy intensity
- no implementation of climate policies

Notes

Introduction

1. J. A. Church et al., "Sea Level Change," in *Climate Change 2013: The Physical Science Basis. Contribution of Working Group I to the Fifth Assessment Report of the Intergovernmental Panel on Climate Change* (Cambridge, UK and New York: Cambridge University Press, 2013), 1180, http://www.ipcc.ch/pdf/assessment-report/ar5/wg1/WG1AR5_Chapter13_FINAL.pdf.

2. James Hansen et al., "Ice Melt, Sea Level Rise, and Superstorms: Evidence from Paleoclimate Data, Climate Modeling and Modern Observations that 2 C Global Warming Could be Dangerous," *Atmospheric Chemistry and Physics* 16 (March 2016): 3799, https://www.atmos-chem-phys.net/16/3761/2016/acp-16-3761-2016.pdf.

3. Ibid.

4. William V. Sweet et al., *Global and Regional Sea Level Rise Scenarios for the United States* (Silver Spring, MD: National Ocean and Atmospheric Administration, 2017), vi, https://tidesandcurrents.noaa.gov/publications/techrpt83_Global_and_Regional_SLR_Scenarios_for_the_US_final.pdf.

5. Ibid.

6. Ibid., vii.

7. Ibid., 23.

8. United States Environmental Protection Agency, *Multi-Model Framework for Quantitative Sectoral Impacts Analysis: A Technical Report for the Fourth National Climate Assessment* (Washington, DC: Environmental Protection Agency, 2017), i, https://cfpub.epa.gov/si/si_public_record_Report.cfm?dirEntryId=335095.

9. Barbara Neumann et al., "Future Coastal Population Growth and Exposure to Sea-Level Rise and Coastal Flooding—A Global Assessment," *PloS ONE* 10.3 (March 11, 2015): table 7, https://www.ncbi.nlm.nih.gov/pmc/articles/PMC4367969/#sec001.

10. Ibid.

Chapter 1

1. "Harvey's Human Toll, Special Section: Those We Lost," *Houston Chronicle* (website) accessed July 11, 2018, https://www.houstonchronicle.com/local/hc-investigations/harvey/deaths/.

2. John P. Cangialosi, Andrew S. Latto, and Robbie Berg, *Hurricane Irma (AL112017) 30 August–12 September 2017* (Miami: National Weather Service, 2018), 13, https://www.nhc.noaa.gov/data/tcr/AL112017_Irma.pdf.

3. Mike Clary, "Hurricane Irma: Florida Keys Man Who Rode Out Storm Says it Sounded Like 'Death,'" *Sun Sentinel,* September 13, 2017, http://www.sun-sentinel.com/news/weather/hurricane/fl-hurricane-iram-keys-damage-20170912-story.html.

4. Cangialosi, *Hurricane Irma*, 7.

5. Ibid., 8.

6. Richard J. Pasch, Andrew B. Penny, and Robbie Berg, *Hurricane Maria (AL152017) 16–30*

September 2017 (Miami, FL: National Weather Service, 2018), 4, https://www.nhc.noaa.gov/data/tcr/AL152017_Maria.pdf.

7. Ibid., 5.

8. Frances Robles, Kenan Davis, Sheri Fink, and Sarah Almukhtar, "Official Toll in Puerto Rico: 64. Actual Deaths May Be 1,052," *New York Times,* December 9, 2017, https://www.nytimes.com/interactive/2017/12/08/us/puerto-rico-hurricane-maria-death-toll.html.

9. Nishant Kishore et al., "Mortality in Puerto Rico after Hurricane Maria," *New England Journal of Medicine,* special article (May 29, 2018): 1, https://www.nejm.org/doi/full/10.1056/NEJMsa1803972.

10. Central Recovery and Reconstruction Office, *Transformation and Innovation in the Wake of Devastation: An Economic and Disaster Recovery Plan for Puerto Rico* (San Juan: Government of Puerto Rico, 2018), 28, http://www.p3.pr.gov/assets/pr-draft-recovery-plan-for-comment-july-9-2018.pdf.

11. "The Great Galveston Hurricane of 1900," NOAA Celebrates 200 Years of Science, Service and Stewardship, (website) accessed July 11, 2018, https://celebrating200years.noaa.gov/magazine/galv_hurricane/welcome.html.

12. Richard D. Knabb, Jamie R. Rhome, and Daniel P. Brown, *Tropical Cyclone Report Hurricane Katrina 23–30 August 2005* (Miami, FL: National Weather Service, 2005), 9, https://www.nhc.noaa.gov/data/tcr/AL122005_Katrina.pdf.

13. Josh Keller et al., "Mapping Hurricane Sandy's Deadly Toll," *New York Times,* November 17, 2012, https://archive.nytimes.com/www.nytimes.com/interactive/2012/11/17/nyregion/hurricane-sandy-map.html?hp.

14. Eric S. Blake, Christopher W. Landsea, and Ethan J. Gibney, *The Deadliest, Costliest, and Most Intense United States Tropical Cyclones from 1851 to 2010 (and Other Frequently Requested Hurricane Facts)* (Miami, FL: National Weather Service, 2011), 7, http://www.nhc.noaa.gov/pdf/nws-nhc-6.pdf.

15. "Before and After: Coastal Change Caused by Hurricane Michael," United States Geological Survey (website) accessed December 15, 2018, https://www.usgs.gov/news/and-after-coastal-change-caused-hurricane-michael.

16. Edward N. Rappaport, "Fatalities in the United States from Atlantic Tropical Cyclones: New Data and Interpretation," *Bulletin of the American Meteorological Society* (March 2014): 342, https://journals.ametsoc.org/doi/full/10.1175/BAMS-D-12-00074.1#.

17. Ibid.

18. "Introduction," Sea, Lake, and Overland Surges from Hurricanes (SLOSH) (website) accessed July 11, 2018, National Oceanic and Atmospheric Administration, https://www.nhc.noaa.gov/surge/slosh.php.

19. "Population at Risk from Storm Surge Inundation," National Storm Surge Hazard Maps – Version 2, National Oceanic and Atmospheric Administration (website) accessed July 11, 2018, http://www.nhc.noaa.gov/nationalsurge/.

20. "CoreLogic 2018 Storm Surge Report," Core Logic (website) accessed December 5, 2018, https://www.corelogic.com/insights-download/storm-surge-report.aspx.

21. "Hurricane Frequency," Storm Surge Inundation Map; Miami–Dade County (click on Miami–Dade County) Environmental Protection Agency (website) accessed July 9, 2018, https://epa.maps.arcgis.com/apps/MapSeries/index.html?appid=852ca645500d419e8c6761b923380663.

22. S. Gopalakrishnan et al., *2016 HFIP R&D Activities Summary: Recent Result and Activities Summary* (Silver Spring, MD: National Oceanic and Atmospheric Administration, 2017), 1, http://www.hfip.org/documents/HFIP_AnnualReport_FY2016.pdf.

23. J. D. Walsh et al., "Our Changing Climate," ch. 2 in *The Third National Climate Assessment,* (Washington, DC: US Global Change Research Program, 2014), 41, http://nca2014.globalchange.gov/report/our-changing-climate/sea-level-rise.

24. J. P. Kossin et al., *"Extreme Storms,"* in *Climate Science Special Report: Fourth National Climate Assessment,* vol. 1 (Washington, DC: US Global Change Research Program, 2017), 257, https://science2017.globalchange.gov/downloads/CSSR_Ch9_Extreme_Storms.pdf.

25. "Billion Dollar Weather and Climate Disasters: Overview, National Oceanic and Atmospheric Administration (website) accessed February 18, 2019, https://www.ncdc.noaa.gov/billions/overview.

26. Ibid.

27. Ibid.

28. Ibid.

29. Ibid.

30. "Billion-Dollar Weather and Climate Disasters: Table of Events," National Oceanic and Atmospheric Administration (website) accessed February 18, 2019, https://www.ncdc.noaa.gov/billions/events/US/1980-2018.

31. Congressional Budget Office, *Expected Costs of Damage from Hurricane Winds and Storm-Related Flooding* (Washington, DC: Congressional Budget Office, 2019), 1, https://www.cbo.gov/system/files/2019-04/55019-ExpectedCostsFromWindStorm.pdf.

32. Ibid., 4.

33. Alison Slider, "Gasoline Prices Jump in Harvey's Wake," *Wall Street Journal*, August 31, 2017, https://www.wsj.com/articles/gasoline-prices-surge-as-harveys-impact-is-felt-1504202779.

34. Solomon M. Hsiang and Amir S. Jina, *The Causal Effect of Environmental Catastrophe on Long-Run Economic Growth: Evidence from 6,700 Cyclones* (Washington, DC: National Bureau of Economic Research, July 2014), 1, http://www.nber.org/papers/w20352.pdf.

35. Walsh, "Our Changing Climate," 42.

36. Kossin, *Climate Science Special Report*, 257.

37. National Oceanic and Atmospheric Administration, "Study: Climate Warming to Boost Major Hurricanes in Active Atlantic Seasons," news release, September 27, 2018, https://www.noaa.gov/study--climate-warming-to-boost-major-hurricanes-in-active-atlantic-seasons.

38. Ibid.

39. Walsh, "Our Changing Climate," 42.

40. Kevin A. Reed et al. "The Human Influence on Hurricane Florence" (website) accessed October 5, 2018, https://cpb-us-e1.wpmucdn.com/you.stonybrook.edu/dist/4/945/files/2018/09/climate_change_Florence_0911201800Z_final-262u19i.pdf.

41. James Kossin, "A Global Slowdown of Tropical Cyclone Translation Speed," *Nature* 558 (June 2018): 1, https://www.nature.com/articles/s41586-018-0158-3.epdf?referrer_access_token=dLLftJvUHt3xLu2BsdPVdtRgN0jAjWel9jnR3ZoTv0MemqNFYaQFhK1eblrz65R9wqFwJ5SEHzsxU9YMRaRIdY_gW5Vxn7OLw0iDnll3UMgjKJhhc5FjxnSlIpM2Hop0SmUfuFtd3RtpX09_ytZBPANro64yvodqFsAXCnvI_Mv3Js60tRxX9el4r0u3ypVB9ROQdVspVn62K-n4iq6MKA%3D%3D&tracking_referrer=news.nationalgeographic.com.

42. Geert Jan van Oldenborgh et al., "Attribution of Extreme Rainfall from Hurricane Harvey," *Environmental Research Letters* 12 (August 2017): abstract, https://doi.org/10.1088/1748-9326/aa9ef2.

43. Ethan D. Gutmann et al., "Changes in Hurricanes from a 13-Yr Convection-Permitting Pseudo–Global Warming Simulation," *Journal of Climate* (May 2018), https://journals.ametsoc.org/doi/10.1175/JCLI-D-17-0391.1.

44. "Hurricanes: Stronger, Slower, Wetter in the Future?" *Science News* (website) accessed May 15, 2019, https://www.sciencedaily.com/releases/2018/05/180521131532.htm.

45. Kieran T. Bhatia et al., "Recent Increases in Tropical Cyclone Intensification Rates," *Nature Communications* (February 2019): 3, https://www.nature.com/articles/s41467-019-08471-z.

46. Ibid., 7.

Chapter 2

1. D. J. Wuebbles et al., "Highlights, Executive Summary," in *Climate Science Special Report: Fourth National Climate Assessment,* vol. 1 (Washington, DC:. US Global Change Research Program, 2017), 10, https://science2017.globalchange.gov/chapter/executive-summary/.

2. Church, *Climate Change 2013,* 1139, (see introduction, n. 1).

3. Ibid., 1150.

4. W. V. Sweet et al., "Sea Level Rise," ch. 12 in *Climate Science Special Report: Fourth National Climate Assessment,* vol. 1 (Washington, DC: US Global Change Research Program, 2017), 339, https://science2017.globalchange.gov/downloads/CSSR_Ch12_Sea_Level_Rise.pdf.

5. Church, *Climate Change 2013,* 1180, (see introduction, n. 1).

6. "Projections of Future Climate Change, Table SPM 3," Climate Change 2007: Working Group I: The Physical Science Basis; Intergovernmental Panel on Climate Change (website) accessed June 29, 2018, https://www.ipcc.ch/publications_and_data/ar4/wg1/en/spmsspm-projections-of.html.

7. Walsh, "Our Changing Climate," 45, (see ch. 1, n. 23).

8. Ibid.

9. Hansen, "Ice Melt, Sea Level Rise, and Superstorms," 3762, (see introduction, n. 2).

10. Ibid.

11. Robert M. DeConto and David Pollard, "Contribution of Antarctica to Past and Future Sea-Level Rise," *Nature*531 (March 31, 2016): 1, https://www.nature.com/nature/journal/v531/n7596/full/nature17145.html.

12. University of Massachusetts Amherst, "Sea-Level Rise Could Nearly Double over Earlier Estimates in Next 100 Years," news release, March 30, 2016, http://www.umass.edu/newsoffice/article/sea-level-rise-could-nearly-double-over.

13. Sweet, *Global and Regional Sea Level Rise,* vi, (see introduction, n. 4).

14. Ibid., 21.

15. Chris Mooney, "Antarctic Ice Sheet Loss Has Tripled in a Decade. If that Continues, We Are in Serious Trouble," *Washington Post,* June 13, 2018, https://www.washingtonpost.com/news/energy-environment/wp/2018/06/13/antarctic-ice-loss-has-tripled-in-a-decade-if-that-continues-we-are-in-serious-trouble/?utm_term=.9ebea8dc4568.

16. "Greenland Ice Sheet Melt 'Off the Charts' Compared with Past Four Centuries," *Science Daily* (website) accessed December 14, 2018; https://www.sciencedaily.com/releases/2018/12/181205133942.htm.

17. "Huge Cavity in Antarctic Glacier Signals Rapid Decay," National Aeronautics and Space Administration, press release, January 30, 2019, https://www.jpl.nasa.gov/news/news.php?feature=7322.

18. Ibid.

19. M. Zemp et al., "Global Glacier Mass Changes and Their Contributions to Sea-level Rise from 1961 to 2016," *Nature* (April, 2019): 1, https://www.nature.com/articles/s41586-019-1071-0.

20. Church, *Climate Change 2013,* 1195, (see introduction, n. 1).

21. Sea-Level Change Curve Calculator (version 2017.55)," Army Corps of Engineers (website) accessed December 4, 2018, http://corpsmapu.usace.army.mil/rccinfo/slc/slcc_calc.html.

22. Walsh, "Our Changing Climate," 45, (see ch. 1, n. 23).

23. Ricarda Winkelmann et al., "Combustion of Available Fossil Fuel Resources Sufficient to Eliminate the Antarctic Ice Sheet," *Science Advances* 1, no. 8 (September 11, 2015): 1, http://advances.sciencemag.org/content/1/8/e1500589.

24. Ibid., 3.

25. Ibid.

26. Dan Reicher et al., Derisking Decarbonization: Making Green Energy Investments Blue Chip," (Palo Alto, CA: Stanford University, 2017), 3, https://energy.stanford.edu/sites/default/files/stanfordcleanenergyfinanceframingdoc10-27_final.pdf.

27. Ibid., 4.

28. United Nations Framework Convention on Climate Change, *Paris Agreement* (New York: United Nations, 2015), 4, https://unfccc.int/sites/default/files/english_paris_agreement.pdf.

29. Intergovernmental Panel on Climate Change. *Global Warming of 1.5°C: Summary for Policymakers* (New York, NY: Intergovernmental Panel on Climate Change, 2018), 8, http://report.ipcc.ch/sr15/pdf/sr15_spm_final.pdf.

30. Ibid., 21.

31. United Nations Environment Programme (UNEP), *The Emissions Gap Report 2018*, (Nairobi, Kenya: United Nations Environment Programme, 2018, xiv; http://wedocs.unep.org/bitstream /handle/20.500.11822/26895/EGR2018_FullReport_EN.pdf?isAllowed=y&sequence=1.

32. Ibid., xv.

33. Ibid.

34. Ibid., xiv.

Chapter 3

1. Sweet, *Global and Regional Sea Level Rise*, 37,(see introduction, n. 4).

2. Ibid., 38.

3. Ibid.

4. William V. Sweet et al. *Patterns and Projections of High Tide Flooding along the US Coastline Using a Common Impact Threshold,* (Silver Spring, MD: National Oceanic and Atmospheric Administration, 2018), vii, https://tidesandcurrents.noaa.gov/publications/techrpt86_PaP_of_HTFlooding.pdf.

5. Miyuki Hino, "High-tide Flooding Disrupts Local Economic Activity," *Science Advances* 5, no. 2 (February 2019): 1, http://advances.sciencemag.org/content/5/2/eaau2736.

6. Cheryl J. Hapke, et al., *National Assessment of Shoreline Change: Historical Shoreline Change along the New England and Mid-Atlantic Coast* (Washington, DC: US Geological Survey, 2011), 28, https://pubs.usgs.gov/of/2010/1118/pdf/ofr2010-1118_report_508_rev042312.pdf.

7. "Physical Therapy," Dr. Beach (website) accessed June 30, 2018, http://drbeach.org/online /physical-therapy/.

8. E. Robert Thieler and Erika S. Hammar-Klose, *National Assessment of Coastal Vulnerability to Sea-Level Rise: Preliminary Results for the U.S. Atlantic Coast,* (Woods Hole, MA: US Geological Survey, 1999), results, 1999https://pubs.usgs.gov/dds/dds68/reports/eastrep.pdf.

9. Ibid.*USUS*

10. E. Robert Thieler and Erika S. Hammar-Klose, *National Assessment of Coastal Vulnerability to Sea-Level Rise: Preliminary Results for the U.S. Pacific Coast,* (Woods Hole, MA: US Geological Survey, 1999), results, https://pubs.usgs.gov/of/2000/of00-178/pages/res.html.

11. Betsy Von Holle et al, "Effects of Future Sea Level Rise on Coastal Habitat," Journal of Wildlife Management 83 (February 3, 2019): 699, https://wildlife.onlinelibrary.wiley.com/doi/epdf /10.1002/jwmg.21633.

12. Stephen K. Gill and John R. Schultz, eds., *Tidal Datums and Their Application*, (Silver Spring, MD: National Oceanic and Atmospheric Administration, 2001), 59, https://tidesandcurrents.noaa .gov/publications/tidal_datums_and_their_applications.pdf.

Chapter 4

1. Benjamin H. Strauss et al., "Tidally Adjusted Estimates of Topographic Vulnerability to Sea Level Rise and Flooding for the Contiguous United States," *Environmental Research Letters* 7, no. 1 (March 14, 2012): 9, http://iopscience.iop.org/article/10.1088/1748-9326/7/1/014033/meta.

2. Ibid., 10.

3. Erika Spanger-Siegfried et al., *When Rising Seas Hit Home: Hard Choices Ahead for Hundreds of US Coastal Communities* (Cambridge, MA: Union of Concerned Scientists, 2017), 17, http://www .ucsusa.org/sites/default/files/attach/2017/07/when-rising-seas-hit-home-full-report.pdf.

4. Ibid.

5. Ibid.,12.

6. Kristina A. Dahl et al., "Effective Inundation of Continental United States Communities with 21st Century Sea Level Rise," *Elementa* (July 12, 2017): sect. 3.8, https://www.elementascience .org/articles/10.1525/elementa.234/.

7. Strauss, "Tidally Adjusted Estimates," 4.

8. Rhodium Group, *American Climate Prospectus Economic Risks in the United States* (New York: Rhodium Group, 2014), 90, https://gspp.berkeley.edu/assets/uploads/research/pdf/American _Climate_Prospectus.pdf.

9. Ibid., 89.

10. US Environmental Protection Agency, *Multi-Model Framework*, 113, (see introduction, n. 8).

11. Rhodium Group, *American Climate Prospectus*, 96.

12. US Environmental Protection Agency, *Multi-Model Framework*, 119, (see introduction, n. 8).

13. Ibid., 209.

14. Ibid.

15. Congressional Budget Office, *Potential Increases in Hurricane Damage in the United States: Implications for the Federal Budget* (Washington, DC: Congressional Budget Office, 2016), 2, https://www.cbo .gov/sites/default/files/114th-congress-2015-2016/reports/51518-hurricane-damage.pdf.

16. Ibid., 12.

17. US Environmental Protection Agency, *Multi-Model Framework*, 232, (see introduction, n. 8).

18. Ibid., 225.

19. Ibid., 113.

20. Ibid., 117.

21. Rhodium Group, *American Climate Prospectus*, 88.

22. US Environmental Protection Agency, *Multi-Model Framework*, 114, (see introduction, n. 8).

23. "*2014 National Population Projections Tables, table 1*," US Census Bureau (website) accessed December 12, 2018, https://www.census.gov/data/tables/2014/demo/popproj/2014-summary -tables.html.

24. United Nations, Department of Economic and Social Affairs, Population Division, *World Population Prospects: The 2015 Revision, Key Findings and Advance Tables* (New York: United Nations, 2015), 22, https://esa.un.org/unpd/wpp/Publications/Files/Key_Findings_WPP_2015.pdf.

25. National Oceanic and Atmospheric Administration, *National Coastal Population Report Population Trends from 1970 to 2020* (Washington, DC: National Oceanic and Atmospheric Administration, 2013), 4, https://aamboceanservice.blob.core.windows.net/oceanservice-prod/facts/coastal -population-report.pdf.

26. Barbara Neumann et al., "Future Coastal Population Growth and Exposure to Sea-Level Rise and Coastal Flooding - A Global Assessment," PloS ONE 10.3 (March 11, 2015): table 7, https://www .ncbi.nlm.nih.gov/pmc/articles/PMC4367969/#sec001.

Chapter 5

1. H. G. Schwartz et al., "*Transportation*," ch. 5 in *The Third National Climate Assessment* (Washington, DC: US Global Change Research Program, 2014), 131, http://nca2014.globalchange.gov /report/sectors/transportation.

2. Michael J. Savonis, Virginia R. Burkett, and Joanne R. Potter, *Impacts of Climate Change and Variability on Transportation Systems and Infrastructure: Gulf Coast Study, Phase I* (Washington, DC: US Climate Change Science Program, 2008), ES-6, http://www.iooc.us/wp-content /uploads/2010/09/Impacts-of-Climate-Change-and-Variability-on-Transportation-Systems-and -Infrastructure-Gulf-Coast-Study-Phase-I.pdf.

3. National Research Council, *Potential Impacts of Climate Change on US Transportation* (Washington, DC: National Academy of Sciences, 2008), 191, http://onlinepubs.trb.org/onlinepubs/sr /sr290.pdf.

4. Scott L. Douglass et al., *Highways in the Coastal Environment: Assessing Extreme Event* (Washington, DC: Federal Highway Administration, October 2014), 1, https://www.fhwa.dot.gov/engineering /hydraulics/pubs/nhi14006/nhi14006.pdf.

5. Jennifer M. Jacobs et al., "Recent and Future Outlooks for Nuisance Flooding Impacts on

Roadways on the US East Coast," *Transportation Research Record* (March 13, 2018): abstract, http://journals.sagepub.com/eprint/pa4UpyDkdIZ2phjmm6jR/full.

6. Ibid.

7. Ibid., 4.

8. Christopher Flavelle and Jeremy C. F. Lin, "Rising Waters Are Drowning Amtrak's Northeast Corridor," *Bloomberg Businessweek* (December 20, 2018), https://www.bloomberg.com/graphics/2018-amtrak-sea-level/.

9. Schwartz, "Transportation," 134.

10. Ibid.

11. Ibid.

12. Ibid.

13. Savonis, *Impacts of Climate Change and Variability on Transportation Systems,"* 1.

14. United States Department of Transportation, *US Department of Transportation Climate Adaptation Plan 2014 Ensuring Transportation Infrastructure and System Resilience* (Washington, DC: US Department of Transportation, 2014), 7, https://www.transportation.gov/sites/dot.gov/files/docs/DOT%20Adaptation%20Plan%202014_OCRP%20appendix_0.pdf.

15. Douglass, *Highways in the Coastal Environment,* 1.

16. Michelle A. Hummel, Matthew S. Berry, and Mark T. Stacey, "Sea Level Rise Impacts on Wastewater Treatment Systems Along the US Coasts," *Earth's Future* 6, no. 4 (March 24, 2018): 626, https://agupubs.onlinelibrary.wiley.com/doi/abs/10.1002/2017EF000805.

17. John Furlow et al., "The Vulnerability of Public Water Systems to Sea Level Rise," in *Proceedings of the Coastal Water Resource Conference,* ed. John R. Lesnik , (Middlebrook, VA: American Water Resources Association, 2002), 5.

18. Ibid.

19. Christopher Flavelle, "Miami Will Be Underwater Soon. Its Drinking Water Could Go First," *Bloomberg Businessweek* (website) accessed August 29, 2018, https://www.bloomberg.com/news/features/2018-08-29/miami-s-other-water-problem

20. Kendra-Pierre-Louis, Nadia Popovich, and Hiroko Tabuchi, "Florence's Floodwaters Breach Coal Ash Pond and Imperil Other Toxic Sites," *New York Times*, September 24, 2018, https://www.nytimes.com/interactive/2018/09/13/climate/hurricane-florence-environmental-hazards.html.

21. US Environmental Protection Agency, *Adaptation of Superfund Remediation to Climate Change* (Washington, DC: Environmental Protection Agency, 2012), 4 (author copy).

22. US Environmental Protection Agency, "Office of Water," in *Climate Change Adaptation Plan* (Washington, DC: Environmental Protection Agency, 2014), 11,https://archive.epa.gov/epa/sites/production/files/2016-08/documents/ow-climate-change-adaptation-plan.pdf.

23. Ben Bovarnick, Shiva Polefka, and Arpita Bhattacharyya, *Rising Waters, Rising Threat: How Climate Change Endangers America's Neglected Wastewater Infrastructure* (Washington, DC: Center for American Progress, 2014), 11, https://www.americanprogress.org/issues/green/reports/2014/10/31/100066/rising-waters-rising-threat/.

24. Ibid., 12.

25. America's Water Infrastructure Act of 2018, Public Law 115-270, sec. 2013, https://www.congress.gov/bill/115th-congress/senate-bill/3021text#toc-H916445DCEAB1468A89E9FCB3EF38AFC4.

26. Strategic Environmental Research and Development Program, *Assessing Impacts of Climate Change on Coastal Military Installations: Policy Implications* (Alexandria, VA: US Department of Defense, 2013), 5, http://www.dtic.mil/dtic/tr/fulltext/u2/a575273.pdf.

27. Committee on National Security Implications of Climate Change for US Naval Forces, *National Security Implications of Climate Change for US Naval Forces* (Washington, DC: National Academies Press, 2011), 74, https://www.nap.edu/catalog/12914/national-security-implications-of-climate-change-for-us-naval-forces.

28. Union of Concerned Scientists, *The US Military on the Front Lines of Rising Seas* (Cambridge,

MA: Union of Concerned Scientists, July 2016), 1, https://www.ucsusa.org/sites/default/files /attach/2016/07/us-military-on-front-lines-of-rising-seas_all-materials.pdf.

29. Ibid., 4.

30. Shepard Smith, "'Widespread, Catastrophic Damage': Every Building at Tyndall AFB Totaled by Hurricane Michael," *Fox News Insider,* October 12, 2018, http://insider.foxnews.com/2018/10/12 /hurricane-michael-trashes-tyndall-air-force-base-bay-florida.

31. Ibid.

32. Phil McKenna, "Hurricane Michael Cost This Military Base about $5 Billion, Just One of 2018's Weather Disasters," *Inside Climate News*, December 18, 2018, https://insideclimatenews.org /news/18122018/tyndall-military-hurricane-cost-2018-year-review-billion-dollar-disasters-wildfire -extreme-weather-drought-michael-florence.

33. Union of Concerned Scientists, *The US Military on the Front Lines of Rising Seas,* 8.

34. Ibid.

35. Committee on National Security Implications of Climate Change for US Naval Forces, *National Security Implications ,* 10.

36. Ibid., 74.

37. United States Department of Defense, *2014 Climate Change Adaptation Roadmap*, (Washington, DC: US Department of Defense, 2014), 4; https://www.acq.osd.mil/eie/Downloads/CCARprint _wForward_e.pdf.

38. Government Accountability Office, *Climate Change Adaptation: DOD Can Improve Infrastructure Planning*, (Washington, DC: Government Accountability Office, 2014), 26, https://www.gao .gov/assets/670/663734.pdf.

39. John A. Hall, et al., *Regional Sea Level Rise for Coastal Risk Management, Managing the Uncertainty of Future Sea Level Change and Extreme Water Levels for Department of Defense Coastal Sites Worldwide* (Washington, DC: United States Department of Defense, 2016), ES 2, https://www.serdp-estcp .org/Program-Areas/Resource-Conservation-and-Resiliency/Infrastructure-Resiliency /Regional-Sea-Level-Scenarios-for-Coastal-Risk-Management.

40. Office of the Under Secretary of Defense for Acquisition, Technology, and Logistics, *Department of Defense Climate-Related Risk to DoD Infrastructure Initial Vulnerability Assessment Survey (SLVAS) Report* (Washington, DC: United States Department of Defense, 2018), 11, https://climateand security.files.wordpress.com/2018/01/tab-b-slvas-report-1-24-2018.pdf.

41. Chris Mooney and Missy Ryan, "Pentagon Revised Obama Era Report to Remove Risks from Climate Change," *Washington Post,* May 10, 2019, https://www.washingtonpost.com/news /energy-environment/wp/2018/05/10/pentagon-revised-obama-era-report-to-remove-risks -from-climate-change/?utm_term=.43bc72c30762.

42. Rebecca Kheel, "Top Admiral Nominee: Climate Change 'Going to be a Problem' for Navy," *The Hill*, April 30, 2019, https://thehill.com/policy/defense/441361-top-admiral-nominee-climate -change-going-to-be-a-problem-for-navy.

43. Military Expert Panel, *Sea Level Rise and the US Military's Mission* (Washington, DC: The Center for Climate and Security, 2018), 16, https://climateandsecurity.files.wordpress.com/2018/02/military -expert-panel-report_sea-level-rise-and-the-us-militarys-mission_2nd-edition_02_2018.pdf.

44. Ibid., 23.

45. Ibid., 24.

46. Ibid.

47. Ibid.

48. Kheel, "Top Admiral Nominee: Climate Change 'Going to be a Problem' for Navy."

49. Ibid.

50. Megan Eckstein, "Navy Plans to Spend $21B Over 20 Years to Optimize, Modernize Public Shipyards," *USNI News,* April 17, 2018, https://news.usni.org/2018/04/17navy-plans-spend-21b -20-years-optimize-modernize-public-shipyards.

Chapter 6

1. "Learn about Coastal Wetlands," Coastal Wetlands, Environmental Protection Agency (website) accessed July 1, 2018, https://www.epa.gov/wetlands/coastal-wetlands.

2. T. E. Dahl and S. M. Stedman, *Status and Trends of Wetlands in the Coastal Watersheds of the Conterminous United States 2004 to 2009* (Washington, DC: US Fish and Wildlife Service and National Oceanic and Atmospheric Administration, National Marine Fisheries Service, 2013), 2, https://www.fws.gov/wetlands/Documents/Status-and-Trends-of-Wetlands-In-the-Coastal-Watersheds-of-the-Conterminous-US-2004-to-2009.pdf.

3. Ibid.

4. Ibid.

5. Ibid.,16.

6. Ibid.,18.

7. Ibid., 22.

8. Ibid.

9. Ibid.

10. Karen Thorne et al., "US Pacific Coastal Wetland Resilience and Vulnerability to Sea-Level Rise," *Science Advances* 4, no. 2, (February 2018): abstract, http://advances.sciencemag.org/content/4/2/eaao3270.full.

11. Mathew L. Kerwin et al., "Limits on the Adaptability of Coastal Marshes to Rising Sea Level," *Geophysical Research Letters* 37, (2010): 4, http://www.vims.edu/people/kirwan_m/pubs/Kirwan2010GRL.pdf.

12. Loraine McFadden, Tom Spencer, and Robert J. Nicholl, "Broad-scale Modelling of Coastal Wetlands: What is Required?" *Hydrobiologia,* (2007): 14, https://link.springer.com/content/pdf/10.1007%2Fs10750-006-0413-8.pdf.

13. Ibid., 12.

14. "Enhancing the Capacity of Coastal Wetlands to Adapt to Sea-Level Rise and Coastal Development," Climate Change Science Centers, Department of the Interior (website) accessed July 1, 2018, https://nccwsc.usgs.gov/projects/#/project/4f8c6557e4b0546c0c397b4c/580a8e71e4b0f497e7906c7a.

15. National Academy of Sciences, *Progress toward Restoring the Everglades: The Seventh Biennial Review–2018* (Washington, DC: National Academy of Sciences, 2018), 8, https://www.nap.edu/catalog/25198/progress-toward-restoring-the-everglades-the-seventh-biennial-review-2018.

16. Ibid.

17. National Academy of Sciences, *Progress toward Restoring the Everglades,* 9.

18. Mathew L. Kerwin et al., "Overestimation of Marsh Vulnerability to Sea Level Rise," *Nature Climate Change* 60 (February 2016): 1, https://www.researchgate.net/publication/295902318_Overestimation_of_marsh_vulnerability_to_sea_level_rise.

19. Ibid., 258.

20. M. Spalding et al., *Mangroves for Coastal Defense: Guidelines for Coastal Managers and Policy-makers* (Wetlands International and The Nature Conservancy, 2014*)*, 15, https://www.nature.org/media/oceansandcoasts/mangroves-for-coastal-defence.pdf.

21. Negative Emissions Technologies and Reliable Sequestration: A Research Agenda (Washington, DC: National Academies Press, 2018), 32, https://www.nap.edu/catalog/25259/negative-emissions-technologies-and-reliable-sequestration-a-research-agenda.

22. United States Environmental Protection Agency, *Fiscal Year 2014–2018 EPA Strategic Plan,* (Washington, DC: US Environmental Protection Agency, 2014), 65, https://www.epa.gov/sites/production/files/2014-09/documents/epa_strategic_plan_fy14-18.pdf.

23. Ariel Wittenberg and Kevin Bogardus, "EPA Falsely Claims 'No Data' on Waters in WOTUS Rule," *E&E News*, December 11, 2018, https://www.eenews.net/stories/1060109323.

24. Delaware Department of Natural Resources and Environmental Control, *Delaware Wetland Management Plan* (Dover: Delaware Department of Natural Resource and Environmental Control, 2015),

10, http://www.dnrec.delaware.gov/Admin/DelawareWetlands/Documents/2015%20Delaware%20Wetlands%20Management%20Plan.pdf.

25. National Oceanic Atmospheric Administration, *National Atlas of the United States, Coastline and Shoreline*, (Washington, DC: National Oceanic and Atmospheric Administration, 1975) table 360, https://www2.census.gov/library/publications/2010/compendia/statab/130ed/tables/11s0360.pdf.

26. "Number of BEACH Act Beaches Reported," Beach Advisory and Closing Online Notification (website) accessed July 1, 2018, https://ofmpub.epa.gov/apex/beacon2/f?p=BEACON2:12:::NO::P12_YEARS:Current.

27. Sean Vitousek et al., "A Model Integrating Longshore and Cross-shore Processes for Predicting Long-term Shoreline Response to Climate Change," *Journal of Geophysical Research: Earth Surface* 122, no. 4, (2017): 20, https://www.scribd.com/document/343229043/Vitousek-Et-Al-2017-Journal-of-Geophysical-Research-Earth-Surface.

28. Annie Sneed, "Sunken Pleasure: California Will Need Mountains of Sand to Save Its Beaches," *Scientific American* (April 18, 2017): https://www.scientificamerican.com/article/sunken-pleasure-california-will-need-mountains-of-sand-to-save-its-beaches/.

29. W. N. Heady et al., *Conserving California's Coastal Habitats: A Legacy and a Future with Sea Level Rise* (San Francisco: The Nature Conservancy and California State Coastal Conservancy, 2018), 3, https://www.conservationgateway.org/ConservationPractices/Marine/crr/library/Documents/TNC_SCC_CoastalAssessment_lo%20sngl.pdf.

30. Ibid., 4.

31. Rachel Gittman, "The Living Shoreline Approach as an Alternative to Shoreline Hardening: Implications for the Ecology and Ecosystem Delivery of Salt Marshes," (PhD diss., University of North Carolina, 2014), 37, file:///C:/Users/pijpi/Downloads/Gittman_unc_0153D_15019.pdf.

32. Ibid., 34.

33. Katie McDowell Peek et al., *Adapting to Climate Change in Coastal Parks: Estimating the Exposure of Park Assets to 1 m of Sea-Level Rise* (Fort Collins, CO: National Park Service, 2015), 22, https://www.nature.nps.gov/geology/coastal/coastal_assets_report/2015_916_NPS_NRR_Coastal_Assets_Exposed_to_1m_of_Sea_Level_Rise_Peek_et_al.pdf.

34. Ibid., 21.

35. Maria A. Caffrey, Rebecca L. Beavers, and Cat Hawkins Hoffman, *Sea Level Rise and Storm Surge Projections for the National Park Service* (Lakewood, CO: National Park Service, 2018), viii, https://www.nps.gov/subjects/climatechange/upload/2018-NPS-Sea-Level-Change-Storm-Surge-Report-508Compliant.pdf.

36. Kenneth B. Raposa et al., "Assessing Tidal Marsh Resilience to Sea Level Rise at Broad Geographic Scales with Multi-Metric Indices," *Biological Conservation* (2016): abstract, http://www.nerra.org/wp-content/uploads/2016/11/BIOC_6995_2016.pdf.

37. Ibid., 12.

Chapter 7

1. Ramakrishnan Durairajan, Carol Barford, and Paul Barford, "Lights Out: Climate Change Risk to Internet Infrastructure," Applied Networking Research Workshop (July 2018), 2, http://pages.cs.wisc.edu/~pb/anrw18_final.pdf.

2. S. C. Moser et al., "Coastal Zone Development and Ecosystems," ch. 25 in *Climate Change Impacts in the United States: The Third National Climate Assessment*, (Washington, DC: US Global Change Research Program, 2014), 589, http://nca2014.globalchange.gov/report/regions/coasts.

3. James Houston, *The Economic Value of Beaches: A 2013 Update*," *Shore & Beach* 81, no. 1 (Winter 2013): 5, http://acwc.sdp.sirsi.net/client/en_US/default/search/detailnonmodal/ent:$002f$002fSD_ILS$002f237$002fSD_ILS:237795/ada/.

4. Julie Harrington, Hongmei Chi, and Lori Pennington Gray, *Florida Tourism* (Gainesville, FL: Florida Climate Institute, 2017), 297, http://floridaclimateinstitute.org/docs/climatebook/Ch10-Harrington.pdf.

5. Ibid., 301.

6. Elizabeth Stanton and Frank Akerman, *Florida and Climate Change, The Costs of Inaction* (Medford, MA: Tufts University, 2007), vi, http://www.ase.tufts.edu/gdae/Pubs/rp/Florida_lr.pdf.

7. David L. Edgell and Carolyn E. McCormick, *Understanding Climate Change and Impacts on Tourism in the Outer Banks of North Carolina* (Whitehall, MI: Tourism Travel and Research Association, 2016), 6, https://scholarworks.umass.edu/cgi/viewcontent.cgi?article=1701&context=ttra.

8. Okmyung Bin et al., *Impacts of Global Warming on North Carolina's Coastal Economy,* Recreation and Tourism (website) accessed July 21, 2018, http://econ.appstate.edu/climate/NCSummary.pdf.

9. Ibid., 3.

10. "Disappearing Beaches: Modeling Shoreline Change in Southern California," United States Geological Survey, March 27, 2017 (website) accessed July 18, 2018, https://www.usgs.gov/news/disappearing-beaches-modeling-shoreline-change-southern-california.

11. California State Assembly Select Committee Sea Level Rise and the California Economy, *Sea-Level Rise: a Slow-Moving Emergency* (Sacramento: State of California, 2014), 29, https://lafco.smcgov.org/sites/lafco.smcgov.org/files/documents/files/Select%20Committee%20Sea-Level%20Rise%20Report.pdf.

12. "Study Predicts Sea Level Rise May Take Economic Toll on California Coast," *SF State News* (website) accessed July 1, 2018, http://www.sfsu.edu/~news/prsrelea/fy12/005.htm.

13. *Hawaii News Now,* "Experts: Rising Seas Will Wash Away Billions from Hawaii's Economy," *Hawaii News Now,* August 23, 2017, https://www.hawaiinewsnow.com/story/36025831/rising-seas-will-devour-waikiki-wash-away-billions-from-hawaiis-economy/.

14. Ibid.

15. Ibid.

16. Daniel Scott, Murray Charles Simpson, and Ryan Sima, "The Vulnerability of Caribbean Coastal Tourism to Scenarios of Climate Change Related Sea Level Rise," *Journal of Sustainable Tourism* 20, no. 6 (July 2012): 894, http://www.caribbeanhotelandtourism.com/wp-content/uploads/data_center/environmental/Journal-of-Sustainable-Toursim-2012.pdf.

17. Edgell, *Understanding Climate Change and Impacts on Tourism,* 8.

18. Harrington, *Florida Tourism,* 307.

19. Scott, "The Vulnerability of Caribbean Coastal Tourism," 894.

20. Alan Lowther and Michael Liddel, eds., *Fisheries of the United States, 2016, Current Fishery Statistics No. 2016* (Silver Spring, MD: National Marine Fisheries Service, 2016), viii, file:///C:/Users/pijpi/Downloads/FUS2016.pdf.

21. National Oceanic and Atmospheric Administration, *Fisheries Economics of the United States 2014: Economics and Sociocultural Status and Trends* Series (Silver Spring, MD: National Marine Fisheries Service, 2016), 6, https://www.st.nmfs.noaa.gov/Assets/economics/publications/FEUS/FEUS-2014/Report-and-chapters/FEUS-2014-FINAL-v5.pdf.

22. Lisa L. Colburn et al., "Indicators of Climate Change and Social Vulnerability in Fishing Dependent Communities along the Eastern and Gulf Coasts of the United States," *Marine Policy* 74 (2016): 328, https://www.st.nmfs.noaa.gov/Assets/econ-human/social/documents/Colburn%202016%20Indicators%20of%20climate%20change%20and%20social%20vulnerability%20in%20fishing%20comm%20in%20East%20US.pdf

23. Code of Federal Regulations, title 50, chap. VI, pt. 600.10, https://www.law.cornell.edu/cfr/text/50/part-600/subpart-A.

24. Ibid., pt. 600.815.4.

25. Ibid., pt. 600.915.

26. Rachael M. Gregg et al., *The State of Climate Adaptation in US Marine Fisheries Management* (Bainbridge Island, WA: EcoAdapt, 2016): 10, http://www.cakex.org/sites/default/files/documents/EcoAdapt_MarineFisheriesAdaptation_August%202016_1.pdf.

27. Terra Lederhouse, et.al., *Report from the National Essential Fish Habitat Summit* (Silver Spring, MD: National Oceanic and Atmospheric Administration, 2017), 2, https://nicholasinstitute.duke.edu/sites/default/files/publications/tm-ohc3.pdf.

28. "Essential Fish Habitat: A Smart Investment for Sustainable Fisheries," NOAA Fisheries (website) July 05, 2016, accessed July 1, 2018, https://www.fisheries.noaa.gov/feature-story/essential-fish-habitat-smart-investment-sustainable-fisheries.

29. Ibid.

30. Lederhouse, *Report from the National Essential Fish Habitat Summit*, 28.

31. *Undercurrent News*, "House of Reps Approves Magnuson Reauthorization Bill," *Undercurrent News*, (June 2, 2018), https://www.undercurrentnews.com/2015/06/02/house-of-reps-approves-magnuson-reauthorization-bill/.

32. Dr. John M. Quinn, Testimony for the Senate Commerce Committee's Subcommittee on Oceans, Atmosphere, Fisheries, and Coast Guard (Washington, DC: United States Senate, August 1, 2017), 6, https://www.commerce.senate.gov/public/_cache/files/e64e6dc9-2dfb-4f1c-8e60-424c39307102/981FC50D42378FB2FD3FCFFF08D3EE59.8.01.17-quinn-testimony.pdf.

33. Sorna Khakzadab and David Griffith, "The Role of Fishing Material Culture in Communities' Sense of Place As an Added-value in Management of Coastal Areas," *Journal of Marine and Island Cultures* 5, no. 2 (December 2016): 111, https://ac.els-cdn.com/S2212682116300336/1-s2.0-S2212682116300336-main.pdf?_tid=da204f70-3612-4940-a519-e5e91d9a5759&acdnat=1521316594_5af91d464cd76e602c478ba1b76ad8df.

34. J. S. Dell et al., *"Energy Supply and Use Climate Change Impacts in the United States,"* ch. 4 in *The Third National Climate Assessment* (Washington, DC: US Global Change Research Program, 2014), 119, http://nca2014.globalchange.gov/report/sectors/energy.

35. Ibid.

36. James Bradbury et al., *Climate Change and Energy Infrastructure Exposure to Storm Surge and Sea-Level Rise* (Oak Ridge, TN: United States Department of Energy, 2015), 3, https://energy.gov/sites/prod/files/2015/07/f24/QER%20Analysis%20-%20Climate%20Change%20and%20Energy%20Infrastructure%20Exposure%20to%20Storm%20Surge%20and%20Sea-Level%20Rise_0.pdf.

37. Ibid., 15–16.

38. Ben Strauss and Remik Ziemlinski, *Sea Level Rise Threats to Energy Infrastructure: A Surging Seas Brief Report* (Climate Central, 2012), 2, http://slr.s3.amazonaws.com/SLR-Threats-to-Energy-Infrastructure.pdf.

39. R. Bierkandt et al., "US Power Plant Sites at Risk of Future Sea-Level Rise," *Environmental Research Letters,* 10 (December 2015): 1, http://iopscience.iop.org/article/10.1088/1748-9326/10/12/124022/pdf.

40. Dave Lochbaum, "Fission Stories #48: Hurricane Andrew vs. Turkey Point," *All Things Nuclear* (blog), Union of Concerned Scientists, July 12, 2011, http://allthingsnuclear.org/dlochbaum/fission-stories-48-hurricane-andrew-vs-turkey-point.

41. Jerry Iannelli, "State Senator Says FPL Isn't Preparing Miami's Nuclear Plant for Sea-Level Rise," *Miami New Times,* October 5, 2017, http://www.miaminewtimes.com/news/turkey-point-miami-nuclear-plant-sea-level-rise-plan-inadequate-miami-lawmaker-warns-9722390.

42. Ibid.

43. Jeffrey Baran, "Separate Views of Commissioner Baran, Nuclear Regulatory Commission" (website) accessed April 11, 2019, 1, https://www.nrc.gov/docs/ML1902/ML19023A353.pdf.

44. Bradbury, *Climate Change and Energy Infrastructure Exposure*, 9.

45. Ibid., 10.

46. Christina Carlson, Gretchen Goldman, and Kristina Dahl, *Stormy Seas, Rising Risks: What Investors Should Know about Climate Change Impacts at Oil Refineries* (Cambridge, MA: Union of Concerned Scientists, February 2015), 3, https://www.ucsusa.org/sites/default/files/attach/2015/02/stormy-seas-rising-risks-ucs-2015.pdf.

47. Mike Lee, "Beyond what #Exxonknew. Refineries Face Flooding Dangers," *E&E News Reporter, Energywire,* November 13, 2015, https://www.eenews.net/stories/1060027921/.

48. Associated Press, "Big Oil Asks Government to Protect its Facilities from Climate Change," *CBS News,* August 22, 2018, https://www.cbsnews.com/news/texas-protect-oil-facilities-from-climate-change-coastal-spine/.

49. "Hurricane Katrina and Rita," Bureau of Safety and Environmental Enforcement, United States Department of Interior (website) accessed July 2, 2018, https://www.bsee.gov/research-record /category/hurricane-katrina-and-rita.

50. Bureau of Safety and Environmental Enforcement, US Department of Interior, "Impact Assessment of Offshore Facilities from Hurricanes Katrina and Rita," Release: #3418, January 19, 2006, https://www.boem.gov/boem-newsroom/press-releases/2006/press0119.aspx.

51. "Hurricanes and the Oil and Natural Gas Industry Preparations," American Petroleum Institute (website) assessed July 2, 2018, http://www.api.org/news-policy-and-issues/hurricane -information/hurricane-preparation.

52. United States Department of Energy, *DOE Climate Change Adaptation Plan* (Washington, DC: Department of Energy, 2014), 15, https://energy.gov/sites/prod/files/2014/10/f18/doe_ccap_2014.pdf.

53. United States Department of Energy, *US Energy Sector Vulnerabilities to Climate Change and Extreme Weather* (Washington, DC: Department of Energy, 2013), 44, https://energy.gov/sites/prod /files/2013/07/f2/20130716-Energy%20Sector%20Vulnerabilities%20Report.pdf.

Chapter 8

1. Anthony Leiserowitz et al., *Climate Change in the American Mind: December 2018* (New Haven, CT: Yale University and George Mason University, 2018), 3, http://climatecommunication.yale.edu /wp-content/uploads/2019/01/Climate-Change-American-Mind-December-2018.pdf.

2. Ibid.

3. "Executive Summary: Survey Results: US Views on Climate Adaptation," Stanford Woods Institute for the Environment and the Center for Ocean Solutions (website) accessed July 2, 2018, https://woods.stanford.edu/sites/default/files/documents/Climate_Survey_Exec_Summ_US.pdf.

4. GfK Custom Research North America, *Stanford University Climate Adaptation National Poll* (Palo Alto, CA: Stanford Woods Institute for the Environment and the Center for Ocean Solutions, 2013), 3, https://woods.stanford.edu/sites/default/files/documents/Climate-Adaptation-Results -TOPLINE.pdf.

5. Tom Horton, "On the Chesapeake, A Precarious Future of Rising Seas and High Tides," *Yale Environment 360*, Yale School of Forestry and Environmental Studies (website) accessed July 2, 2018, https:// e360.yale.edu/features/on-the-chesapeake-a-precarious-future-of-rising-seas-and-high-tides.

6. North Carolina Coastal Resources Commission's Science Panel on Coastal Hazards, *North Carolina Sea-Level Rise Assessment Report* (Raleigh, NC: North Carolina Department of Environment and Natural Resources, 2010), 12, https://files.nc.gov/ncdeq/Coastal%20Management/documents /PDF/Coastal%20Hazards%20Storm%20Information/NC_Sea_Level_Rise_Assessment_Report _2010_CRC_Science_Panel.pdf.

7. Wade Rawlins, "North Carolina Lawmakers Reject Sea Level Rise Predictions," *Reuters, Environment*, July 3, 2012, https://www.reuters.com/article/us-usa-northcarolina/north-carolina-lawmakers -reject-sea-level-rise-predictions-idUSBRE86217I20120703.

8. Dave Dewitt, "The State That 'Outlawed Climate Change' Accepts Latest Sea-Level Rise Report," *North Carolina Public Radio*, May 4, 2015, http://wunc.org/post/state-outlawed-climate-change -accepts-latest-sea-level-rise-report#stream/0.

9. Ibid.

10. Orrin H. Pilkey, "A Disaster is Going to Happen on the N.C. Coast. Here's How to Avoid It," *Charlotte Observer*, March 02, 2017, http://www.charlotteobserver.com/opinion/article135926383.html.

11. "Elon Poll: After Hurricane Florence, North Carolinians Believe Climate Change Impacting Coast," Elon University (website) October 11, 2018, https://www.elon.edu/E-Net/Article/167006.

12. Tristram Koten, "In Florida, Officials Ban Term 'Climate Change'," *Florida Center for Investigative Reporting*, March 8, 2015, https://fcir.org/2015/03/08/in-florida-officials-ban-term -climate-change/.

13. Ibid.

14. Brady Dennis and Darryl Fears, "Florida Governor Has Ignored Climate Change Risks, Critics Say,"

Washington, Post, September 8, 2017, https://www.washingtonpost.com/national/health-science/florida-governor-has-ignored-climate-change-risks-critics-say/2017/09/08/04a8c60a-94a0-11e7-aace-04b862b2b3f3_story.html?utm_term=.751e372e5f9d.

15. James Bruggers, "In Florida, a New Governor Shifts Gears on Environment, and Maybe Climate Change," *Inside Climate News,* January 28, 2019, https://insideclimatenews.org/news/28012019/florida-governor-desantis-environment-directives-sea-level-rise-climate-change-republican.

16. Cameron Wake, "How One State Bridged the Cultural Divide on Climate Change to Prepare for a Stormier Future," *The Conversation,* https://theconversation.com/how-one-state-bridged-the-cultural-divide-on-climate-change-to-prepare-for-a-stormier-future-88898.

17. Chris Mooney and Brady Dennis, "On Climate Change, Scott Pruitt Causes an Uproar—and Contradicts the EPA's Own Website," *Washington Post,* March 9, 2017, https://www.washingtonpost.com/news/energy-environment/wp/2017/03/09/on-climate-change-scott-pruitt-contradicts-the-epas-own-website/?utm_term=.6d725d4233aa.

18. Chris Cillizza, "Donald Trump Buried a Climate Change Report Because 'I Don't Believe It,'" *CNN,* November 27, 2018, https://www.usatoday.com/story/news/politics/2018/11/26/trump-dire-economic-forecast-climate-change-dont-believe/2118152002/.

19. Editorial Board, "Antarctic Ice Is Melting Faster. Coastal Cities Need to Prepare—Now," *Washington Post,* June 22, 2018, https://www.washingtonpost.com/opinions/antarctic-ice-is-melting-faster-coastal-cities-need-to-prepare--now/2018/06/22/9e1e83b6-74af-11e8-805c-4b67019fcfe4_story.html?utm_term=.0aec3576faec.

20. Carol Vaughn, "President Trump Chats On Phone with Tangier Mayor," *delmarva now,* June 13, 2017, https://www.delmarvanow.com/story/news/2017/06/13/president-trump-chats-phone-tangier-mayor/391375001/.

21. Executive Order 13653 (Revoked), "Preparing the United States for the Impacts of Climate Change," Section 1, https://sftool.gov/learn/annotation/427/executive-order-13653-preparing-united-states-impacts-climate-change-archived.

22. United States Army Corps of Engineers, "Procedures to Evaluate Sea Level Changes: Impacts, Responses, and Adaptation," Technical Letter No. 1100-2-1 (Washington, DC: Army Corps of Engineers, 2014), https://www.publications.usace.army.mil/Portals/76/Publications/EngineerTechnicalLetters/ETL_1100-2-1.pdf.

23. John Englander, "Testimony to House Subcommittee on Energy and Mineral Resources," July 28, 2015, 5, http://docs.house.gov/meetings/II/II06/20150728/103852/HHRG-114-II06-Wstate-EnglanderJ-20150728.pdf.

24. Congressman Frank Pallone Jr., "Pallone, LoBiondo Launch Bipartisan Caucus to Advocate for Coastal Communities," press release, February 24, 2015, https://pallone.house.gov/press-release/pallone-lobiondo-launch-bipartisan-caucus-advocate-coastal-communities.

25. Ibid.

26. Tony Doris, "'We're Scared': Sea-level Rise Issue Prompts Senate Hearing in West Palm," *Palm Beach Post,* April 10, 2017, https://www.palmbeachpost.com/weather/scared-sea-level-rise-issue-prompts-senate-hearing-west-palm/pLjIv5luRoyFPcPm9N39KK/.

27. "Senate Oceans Caucus Adds Seven New Members: Bipartisan Group's Membership Grows to 30," Press Release, Senator Sheldon Whitehouse, January 16, 2016, https://www.whitehouse.senate.gov/news/release/senate-oceans-caucus-adds-seven-new-members.

28. International Conservation Caucus Foundation, "Oceans Caucus Sets Priorities for 115th Congress," News Release, January 21, 2017, http://www.internationalconservation.org/oceans-caucus-leadership-sets-priorities-for-115th-congess.

29. Joint Ocean Commission Initiative, *Ocean Action Agenda: Supporting Regional Ocean Economics and Ecosystems,* (Washington, DC: Joint Ocean Commission Initiative, 2017), https://oceanactionagenda.org/wp-content/uploads/2017/04/OceanActionAgenda.pdf.

30. 162 Cong. Rec. S6723 (December 6, 2016) (statement of Senator Whitehouse), https://congress.gov/crec/2016/12/06/CREC-2016-12-06.pdf.

31. 16 US Code Section 7504(a), https://www.law.cornell.edu/uscode/text/16/7504.

32. Coastal State Climate Preparedness Act of 2017, H. R. 3533, 115th Cong. (2017), https://www.congress .gov/115/bills/hr3533/BILLS-115hr3533ih.pdf.

33. "Amy's Plan to Build America's Infrastructure," *Medium* (website) accessed April 14, 2019, https:// medium.com/@AmyforAmerica/amys-plan-to-build-america-s-infrastructure-671b08a10751.

Chapter 9

1. James M. Wright, *The Nation's Responses to Flood Disasters: A Historical Account* (Madison, WI: The Association of State Floodplain Managers, 2000), 9, http://www.floods.org/PDF/hist_fpm.pdf.

2. "Galveston Seawall," *Wikipedia*, accessed July 9, 2018, https://en.wikipedia.org/wiki/Galveston _Seawall.

3. Wright, *The Nation's Responses to Flood Disasters*, 12.

4. Ibid., 16.

5. Ibid., 31.

6. Ibid., 34.

7. Ibid., 35.

8. Inspector General, *FEMA Needs to Improve Management of Its Flood Mapping Programs* (Washington, DC: Department of Homeland Security, 2017), 3, https://www.oig.dhs.gov/sites/default /files/assets/2017/OIG-17-110-Sep17.pdf.

9. Sarah Pralle, "Drawing Lines: FEMA and the Politics of Mapping Flood Zones," Paper for Conference of American Political Science Association, 2017, 16, https://www.maxwell.syr.edu/uploaded Files/faculty/psc/Pralle_Drawing%20Lines_APSA2017.pdf.

10. "Unit 5: The NFIP Floodplain Management Requirements," Federal Emergency Management Agency (website) accessed July 2, 2018, 5-1, https://www.fema.gov/pdf/floodplain/nfip_sg_unit_5.pdf.

11. Ibid., 5-51.

12. Ibid., 5–54.

13. "Summary of the NFIP Program Changes Effective April 1, 2018 and January 1, 2019, "National Flood Insurance Program, Federal Emergency Management Agency (website) accessed July 21, 2018, https://nfip-iservice.com/Stakeholder/pdf/bulletin/Attachment%20A%20-%20Summary %20of%20the%20NFIP%20April%202017%20Program%20Changes%20Final.pdf.

14. Property Casualty Insurers of America, *White Paper: True Market-Risk Rates for Flood Insurance* (Chicago, IL: Property Casualty Insurers of America, 2011), 6, https://www.pciaa.net/pciwebsite /common/page/attachment/13821.

15. "Community Rating System," Federal Emergency Management Agency (website) June 2017, accessed July 2, 2018, https://www.fema.gov/media-library-data/1507029324530 -082938e6607d4d9eba4004890dbad39c/NFIP_CRS_Fact_Sheet_2017_508OK.pdf.

16. Federal Emergency Management Agency, *National Flood Insurance Program Community Rating System: Coordinator's Manual* (Washington, DC: Federal Emergency Management Agency, 2017), 400–414, https://www.fema.gov/media-library-data/1493905477815-794671adeed5beab6a6304 d8ba0b207/633300_2017_CRS_Coordinators_Manual_508.pdf.

17. United States Fish and Wildlife Service, *The Coastal Barrier Resources Act: Harnessing the Power of the Market to Conserve the Coast and Save Taxpayers Money* (Washington, DC: Division of Federal Program Activities, 2002), 27, https://www.heartland.org/_template-assets/documents/publications /10683.pdf.

18. Andrew S. Coburn and John C. Whitehead, "An Analysis of Federal Expenditures Related to the Coastal Barrier Resources Act (CBRA) of 1982," *Journal of Coastal Research* (March 2019): 4, https://shoreline.wcu.edu/Andy/Coburn&Whitehead_2019_JCR.pdf.

19. Wright, *The Nation's Responses to Flood Disasters: A Historical Account*, 41.

20. Ibid., 39.

21. Property Casualty Insurers of America, *White Paper: True Market-Risk Rates for Flood Insurance*, 5.

22. Caitlin Bronson, "10 Years After Katrina: What an Unstable NFIP Means for Flood Insurance,"

Insurance Business America, August 26, 2015, http://www.insurancebusinessmag.com/us/news /catastrophe/10-years-after-katrina-what-an-unstable-nfip-means-for-flood-insurance-24137.aspx.

23. United States Federal Emergency Management Agency, *The Watermark* 1 (1st quarter, 2018): 5, https:// www.fema.gov/media-library-data/1522167351921-a5e457454262dd100e2f15a7210d21c5 /Watermark_FY18_Q1_v6_508.pdf.

24. Congressional Budget Office, *The National Flood Insurance Program: Financial Soundness and Affordability*, (Washington, DC: Congressional Budget Office, 2017), 1, https://www.cbo.gov/system /files/115th-congress-2017-2018/reports/53028-nfipreport2.pdf.

25. Ibid., 13.

26. Ibid., 15.

27. Government Accountability Office, *High-risk Series: Progress on Many High-risk Areas, While Substantial Efforts Needed on Others* (Washington, DC: Government Accountability Office, 2017), 619, https://www.gao.gov/assets/690/682765.pdf.

28. United States Federal Emergency Management Agency, *The Watermark*, 5.

29. Ibid., 4,

30. Pete Kasperowicz, "House Retreats from 2012 Flood Reforms," *The Hill*, March 14, 2014, http:// thehill.com/blogs/floor-action/199900-house-votes-to-retreat-from-2012-flood-reforms.

31. Ibid.

32. 162 Cong. Rec. H 9235 (November 14, 2017) (statement of Rep. Duffy), https://www.congress .gov/congressional-record/2017/11/14/house-section/article/H9209-1.

33. United States Congress, "The Cassidy-Gillibrand Flood Insurance Affordability and Sustainability Act of 2017, section summary," (Washington, DC: United States Congress, 2017), 2, https://www.cassidy.senate.gov/imo/media/doc/Cassidy-Gillibrand%20Section%20Summary %204_25_17_%20FINAL.pdf.

34. Ray Lehmann, "Private Flood Insurance Market Is Getting Bigger, More Competitive, Less Profitable," *Insurance Journal* (March 18, 2018), https://www.insurancejournal.com/blogs/right -street/2018/03/18/483689.htm.

35. M. Mick Mulvaney, director, Office of Management and Budget to Mike Pence, President of the Senate, October 4, 2017, encl. 1, https://www.whitehouse.gov/sites/whitehouse.gov/files /omb/Letters/Letter%20regarding%20additional%20funding%20and%20reforms%20to %20address%20impacts%20of%20recent%20natural%20disasters.pdf.

36. Technical Mapping Advisory Council, *Future Conditions Risk Assessment and Modeling* (Washington, DC: Federal Emergency Management Agency, 2015), 4, https://www.fema.gov/media-library -data/1454954261186-c348aa9b1768298c9eb66f84366f836e/TMAC_2015_Future_Conditions _Risk_Assessment_and_Modeling_Report.pdf.

37. Ibid., 10.

38. Ibid., 12.

39. Ibid., 13.

40. Mulvaney to Pence, encl. 2.

41. Government Accountability Office, *National Flood Insurance Program: Actions to Address Repetitive Loss Properties* (Washington, DC: Government Accountability Office, 2004), 2, https://www.gao .gov/assets/120/110626.pdf.

42. Becky Hayat and Robert Moore, "Addressing Affordability and Long-Term Resiliency Through the National Flood Insurance Program," *Environment Law Reporter*, 4 (2015): 4–2015, https://www. nrdc.org/sites/default/files/blog-national-flood-insurance-program-report.pdf.

Chapter 10

1. Government Accountability Office, *Federal Disaster Assistance: Federal Departments and Agencies Obligated at Least $277.6 Billion during Fiscal Years 2005 through 2016* (Washington, DC: Government Accountability Office, 2016), 14, https://www.gao.gov/assets/680/679977.pdf.

2. Government Accountability Office: *Federal Disaster Assistance: Improved Criteria Needed to Assess a Jurisdiction's Capability to Respond and Recover on Its Own* (Washington, DC: Government Accountability Office, 2012), summary page, https://www.gao.gov/assets/650/648162.pdf.

3. Multi-Hazard Mitigation Council, *Natural Hazard Mitigation Saves: An Independent Study to Assess the Future Savings from Mitigation Activities, vol. 1: Findings, Conclusions and Recommendations* (Washington, DC: National Institute of Building Sciences, 2005), iii, http://www.floods.org/PDF/MMC_Volume1_FindingsConclusionsRecommendations.pdf.

4. Federal Emergency Management Agency, *Hazard Mitigation Assistance Guidance, Hazard Mitigation Grant Program, Pre-Disaster Mitigation Program, and Flood Mitigation Assistance Program* (Washington, DC: Federal Emergency Management Agency, 2015), 4, https://www.fema.gov/media-library-data/1424983165449-38f5dfc69c0bd4ea8a161e8bb7b79553/HMA_Guidance_022715_508.pdf.

5. "Program Overview," FEMA Hazard Mitigation Grant Program (website) accessed July 3, 2018, https://www.fema.gov/hazard-mitigation-grant-program.

6. Federal Emergency Management Agency, *Hazard Mitigation Assistance Guidance,* 16.

7. Missy Stults, "Integrating Climate Change into Hazard Mitigation Planning: Opportunities and Examples in Practice," *Climate Risk Management*17 (2017): 30, https://ac.els-cdn.com/S2212096316300869/1-s2.0-S2212096316300869-main.pdf?_tid=795cd444-10f3-11e8-8dd8-00000aab0f6b&acdnat=1518550033_7e69b91011c02da05c11b79e97f8227d.

8. Ibid., 30.

9. Government Accountability Office, *Hurricane Sandy: An Investment Strategy Could Help the Federal Government Enhance National Resilience for Future Disasters* (Washington, DC: Government Accountability Office, 2015), highlights, http://www.gao.gov/assets/680/671796.pdf.

10. Association of State Floodplain Managers, *Meeting the Challenge of Change* (Madison, WI: Association of State Floodplain Managers Foundation, 2016), 4, http://www.asfpmfoundation.org/ace-images/forum/Meeting_the_Challenge_of_Change.pdf.

11. Federal Emergency Management Agency, *National Pre-Disaster Mitigation Fund: September 1, 2017 Fiscal Year 2017 Report to Congress* (Washington, DC: Federal Emergency Management Agency, 2017), 3, https://www.dhs.gov/sites/default/files/publications/FEMA%20-%20National%20Pre-Disaster%20Mitigation%20Fund.pdf.

12. Ibid., 4.

13. Ibid., 2.

14. "General Program Information," FEMA Flood Mitigation Assistance Grant Program (website) accessed July 3, 2018, https://www.fema.gov/flood-mitigation-assistance-grant-program.

15. "FY 2017 Flood Mitigation Assistance (FMA) Grant Program, Fact Sheet," Federal Emergency Management Agency (website) accessed July 3, 2017, https://www.fema.gov/media-library-data/1499793315357-c31fef3839ece1533d9fccfe5caee71d/FMA_FactSheet_FY2017_508.pdf.

16. Government Accountability Office, *"Hurricane Sandy,"* highlights.

17. Federal Emergency Management Agency, *2017 Hurricane Season FEMA after Action Report* (Washington: DC: Federal Emergency Management Agency, 2017), ii, https://www.fema.gov/media-library-data/1531743865541-d16794d43d3082544435e1471da07880/2017FEMAHurricaneAAR.pdf.

18. Government Accountability Office, *2017 Hurricanes and Wildfires Initial Observations on the Federal Response and Key Recovery Challenges* (Washington, DC: Government Accountability Office, 2018), 3, https://www.gao.gov/assets/700/694231.pdf.

19. Chris Edwards, "The Federal Emergency Management Agency Floods, Failures, and Federalism," *Policy Analysis* 764 (November 18, 2014): 9, https://object.cato.org/sites/cato.org/files/pubs/pdf/pa764_1.pdf.

20. David Inserra, "FEMA Reform Needed: Congress Must Act," *Heritage Foundation Issue Brief 4342,* (Heritage Foundation, 2015): 1 https://www.heritage.org/homeland-security/report/fema-reform-needed-congress-must-act.

21. National Infrastructure Advisory Council: *Water Sector Resilience: Final Report and Recommendations* (Washington, DC: United States Department of Homeland Security, 2016), 41, https://www.dhs.gov/sites/default/files/publications/niac-water-resilience-final-report-508.pdf.

Chapter 11

1. Executive Order No. 13547Stewardship of the Ocean, Our Coasts, and the Great Lakes, 2010, 1 https://www.gpo.gov/fdsys/pkg/FR-2010-07-22/pdf/2010-18169.pdf.

2. Ibid.

3. "President Donald J. Trump is Promoting America's Ocean Economy," Council on Environmental Quality (website) accessed February 21, 2019, https://www.whitehouse.gov/briefings-statements /president-donald-j-trump-promoting-americas-ocean-economy/.

4. National Ocean Council, *National Ocean Policy 2016 Annual Work Plan* (Washington, DC: National Ocean Council, 2016), iii, https://obamawhitehouse.archives.gov/sites/default/files /microsites/ostp/2016_annual_work_plan_final_-_160105.pdf.

5. National Ocean Council, *National Ocean Policy Implementation Plan* (Washington, DC: National Ocean Council, 2013), 15, https://obamawhitehouse.archives.gov/sites/default/files/national _ocean_policy_implementation_plan.pdf.

6. National Ocean Council, *National Ocean Policy 2016 Annual Work Plan*, 6.

7. Ibid.

8. "Disaster Resilient Communities," Governors South Atlantic Alliance (website) accessed July 3, 2018, http://southatlanticalliance.org/disaster-resilient-communities/.

9. Coastal Zone Management Act, 16 USC, Chapter 33, Section 1452; Declaration of Purpose, https://www.law.cornell.edu/uscode/text/16/1452.

10. Coastal Zone Management Act, 16 USC, Chapter 33, Section 1453, Definitions, https://www.law .cornell.edu/uscode/text/16/1453.

11. Coastal Zone Management Act, 16 USC, Chapter 33, Section 1453, Definitions.

12. Harold F. Upton, *Coastal Zone Management: Background and Reauthorization Issues* (Washington, DC: Congressional Research Service, 2008), 7, https://digital.library.unt.edu/ark:/67531 /metadc94133/m1/1/high_res_d/RL34339_2008Jun20.pdf.

13. Code of Federal Regulations, 923.25 Shoreline Erosion and Mitigation Planning, subsection (a), https://www.law.cornell.edu/cfr/text/15/923.25.

14. Code of Federal Regulations, 923.25 Shoreline Erosion and Mitigation Planning, subsection (c).

15. Ibid.

16. "Funding Summary 2017," NOAA Coastal Zone Management Program (website) accessed July 3, 2018, https://coast.noaa.gov/czm/media/funding-summary.pdf

17. Ibid.

18. Ibid.

19. Coastal Zone Management Act, 16 USC, Chapter 33, Section 456(a)(2), Coastal Zone Enhancement Grants, https://www.law.cornell.edu/uscode/text/16/1453.

20. National Ocean and Atmospheric Administration, *Coastal Zone Management Act Section 309 Program Guidance 2016 to 2020 Enhancement Cycle* (Silver Spring, MD: National Oceanic and Atmospheric Administration, 2014), 5, https://coast.noaa.gov/czm/media/Sect-309_Guidance_ June2014.pdf.

21. James B. London, "Sea Level Rise and Coastal Management," *Oxford Research Encyclopedia, Environmental Science,* 2017, 4, http://environmentalscience.oxfordre.com/view/10.1093/ acrefore/9780199389414.001.0001/acrefore-9780199389414-e-377?print=pdf.

22. Ibid.

23. "Coastal Resilience Grants for Coastal Communities," National Oceanic and Atmospheric Administration (website) accessed July 3, 2018, https://www.coast.noaa.gov/data/resilience /factsheet-resilience-grants.pdf.

24. Government Accountability Office, *Climate Change: Information on NOAA's Support for States' Marine Coastal Ecosystem Resilience Effort* (Washington, DC: Government Accountability Office, 2016), 13, https://www.gao.gov/assets/690/680099.pdf.

25. "Coastal Resilience Grants for Coastal Communities," National Oceanic and Atmospheric Administration.

26. "Coastal Resilience Grant Program Project Summaries," National Oceanic and Atmospheric Administration (website) accessed July 3, 2018, https://www.coast.noaa.gov/resilience-grant/projects/.

27. "Funding Summary 2017," NOAA's National Estuarine Research Reserve (website) accessed July 23, 2018, https://coast.noaa.gov/data/docs/nerrs/funding-summary.pdf.

28. NOAA Office of Coastal Management, *The National Estuarine Research Reserve System Strategic Plan 2017-2022* (Silver Spring, MD: National Oceanic and Atmospheric Administration, undated), 4, https://coast.noaa.gov/data/docs/nerrs/StrategicPlan.pdf.

29. "Overview of the National Estuary Program," US Environmental Protection Agency (website) accessed July 3, 2018, https://www.epa.gov/nep/overview-national-estuary-program.

Chapter 12

1. Joel B. Smith et al. *Climate Change: A Call for Federal Leadership* (Arlington, VA: Pew Center on Global Climate Change, 2010), 21, https://www.c2es.org/site/assets/uploads/2010/04/adapting-to-climate-change-call-for-federal-leadership.pdf.

2. Ibid., 2.

3. Executive Order No. 13653: Preparing the United States for the Impacts of Climate Change, Fed. Reg. vol. 78, no. 215 (November 6, 2013), 66819. https://sftool.gov/learn/annotation/427/executive-order-13653-preparing-united-states-impacts-climate-change-archived.

4. Executive Office of the President, *President's State, Local and Tribal Leaders Task Force on Climate Preparedness and Resilience: Recommendations to the President* (Washington, DC: Executive Office of the President, 2014), 13, https://obamawhitehouse.archives.gov/sites/default/files/docs/task_force_report_0.pdf.

5. Ibid., 33.

6. Ibid., 35.

7. Ibid., 37.

8. Exec. Order No. 13653, 66821.

9. Ibid.

10. Jane A. Leggett, *Change Adaptation by Federal Agencies: An Analysis of Plans and Issues for Congress* (Washington, DC: Congressional Research Service, 2015), summary 2, https://fas.org/sgp/crs/misc/R43915.pdf.

11. Ibid., 18.

12. Ibid., 20.

13. Hannah Conners, Kathleen D. White, and Jeffrey R. Arnold, *Comparison of 2014 Adaptation Plans: Report Providing Comparison of Adaptation Plans Submitted to the White House in 2014* (Washington, DC: United States Army Corps of Engineers, 2015), https://www.iwr.usace.army.mil/Portals/70/docs/pubs/Comparison_of_2014_Adaptation_Plans_JUNE%202015.pdf.

14. Patrick Woolsey, *Consideration of Climate Change in Federal EISs, 2009–2011* (New York: Columbia Law School, 2012), 7, https://web.law.columbia.edu/sites/default/files/microsites/climate-change/files/Publications/Students/Woolsey%20NEPA%20report.pdf.

15. Council on Environmental Quality, *Final Guidance for Federal Departments and Agencies on Consideration of Greenhouse Gas Emissions and the Effects of Climate Change in National Environmental Policy Act Reviews* (Washington, DC: Executive Office of the President, 2016), 24, https://ceq.doe.gov/docs/ceq-regulations-and-guidance/nepa_final_ghg_guidance.pdf.

16. Government Accountability Office, *High-risk Series: Progress on Many High-risk Areas, While Substantial Efforts Needed on Others* (Washington, DC: Government Accountability Office, 2017), 153, https://www.gao.gov/assets/690/682765.pdf.

17. Ibid., 154.

18. Ibid.

19. Government Accountability Office, "Limiting the Federal Government's Fiscal Exposure by Better Managing Climate Change Risks" (website) accessed April 11, 2019, https://www.gao.gov /highrisk/limiting_federal_government_fiscal_exposure/why_did_study.

20. Committee on Engineering Implications of Changes in Relative Mean Sea Level, *Responding to Changes in Sea Level: Engineering Implications* (Washington, DC: National Academy Press, 1987), 125, https://www.nap.edu/read/1006/chapter/11#125.

21. United States Army Corps of Engineers, *Technical Letter No. 1100-2-1, Procedures to Evaluate Sea Level Change: Impacts, Response, and Adaptation* (Washington, DC: United States Army Corps of Engineers, 2014), 1–3, http://www.publications.usace.army.mil/Portals/76/Publications/Engineer TechnicalLetters/ETL_1100-2-1.pdf.

22. Ibid., 1–4.

23. United States Army Corps of Engineers, *North Atlantic Coast Comprehensive Study: Resilient Adaptation to Increasing Risk, Main Report, Final Report* (Washington, DC: United States Army Corps of Engineers, 2015), preface, http://www.nad.usace.army.mil/Portals/40/docs/NACCS/NACCS_main _report.pdf.

24. Ibid.

25. Ibid.

26. US Environmental Protection Agency, "Coastal Wetland Reviews: Highlights" (Washington, DC: Environmental Protection Agency, n.d.), 4, https://www.epa.gov/sites/production/files/2015-04 /documents/cwr_highlights-updated.pdf.

27. Lisa Friedman, "Trump Signs Order Rolling Back Environmental Rules on Infrastructure," *New York Times,* August 15, 2017, https://www.nytimes.com/2017/08/15/climate/flooding-infrastructure -climate-change-trump-obama.html.

Chapter 13

1. Peter Byrne, "The Cathedral Engulfed: Sea-Level Rise, Property Rights, and Time," *Louisiana Law Review* 73, no. 1 (2012): 99–100, https://scholarship.law.georgetown.edu/cgi/viewcontent.cgi ?article=2136&context=facpub.

2. Michael A. Hiatt, "Come Hell or High Water: Reexamining the Takings Clause in a Climate Changed Future," *Duke Environmental Law and Policy Forum* 18 no. 371 (2008): 397, https://schol-arship.law.duke.edu/cgi/viewcontent.cgi?article=1072&context=delpf.

3. Ibid.

4. Christopher Flavelle, "The Fighting Has Begun over Who Owns Land Drowned by Climate Change," *Bloomberg Business Week*, April 25, 2018, https://www.bloomberg.com/news /features/2018-04-25/fight-grows-over-who-owns-real-estate-drowned-by-climate-change.

5. Devon Applegate, "The Intersection of the Takings Clause and Rising Sea Levels: Justice O'Connor's Concurrence in Palazzolo Could Prevent Climate Change Chaos," *Boston College Environmental Affairs Law Review* 43, no. 2, article 11 (2016): 512, http://lawdigitalcommons.bc.edu/ealr/vol43/iss2/11.

6. Ibid.

7. Christopher Flavelle, "The Nightmare Scenario for Florida's Coastal Homeowners: Demand and Financing Could Collapse Before the Sea Consumes a Single House," *Bloom*berg News, April 19, 2017, https://www.bloomberg.com/news/features/2017-04-19/the-nightmare-scenario -for-florida-s-coastal-homeowners.

8. Ibid.

9. Kristina A. Dahl et al., *Underwater* (Cambridge, MA: Union of Concerned Scientists, 2018), 21, https://www.ucsusa.org/sites/default/files/attach/2018/06/underwater-analysis-full-report.pdf.

10. Elizabeth Fleming et al., "Coastal Effects," ch. 8 in The *Fourth National Climate Assessment* (website) accessed December 5, 2018, https://nca2018.globalchange.gov/chapter/8/#fn:62.

Chapter 14

1. "Lesson Plan: Sea Level Rise: Climate Change, Everglades National Park," National Park Service (website) accessed July 5, 2018, https://www.nps.gov/teachers/classrooms/sealev.htm.

2. Hawai'i Climate Change Mitigation and Adaptation Commission, *Hawai i Sea Level Rise Vulnerability and Adaptation Report* (Honolulu, HI: State of Hawaii Department of Land and Natural Resources, 2017), 253, https://climateadaptation.hawaii.gov/wp-content/uploads/2017/12/SLR -Report_Dec2017.pdf.

3. "Gov. Henry McMaster Creates Statewide Commission to Address Flooding," Press Release, State of South Carolina, October 15, 2018, https://governor.sc.gov/Newsroom/Pages/GovHenryMcMaster CreatesStatewideCommissiontoAddressFlooding.aspx.

4. Ibid.

5. "FEMA and Flood Zones: Real Estate Agent Frequently Asked Questions Answered," CRES Insurance Services (website) accessed July 5, 2018, https://www.cresinsurance.com/fema-and -flood-zones-real-estate-agent-frequently-asked-questions-answered/.

6. "NAR Legal Guidance Disclosure of Flood Insurance Requirements, Rates, and Rate Increases by Brokers and Agents," National Association of Realtors (website) April 2, 2014, accessed July 5, 2018, http://narfocus.com/billdatabase/clientfiles/172/4/1816.pdf.

7. Ian Urbina, "Perils of Climate Change Could Swamp Real Estate," *New York Times*, November 24, 2016, https://www.nytimes.com/2016/11/24/science/global-warming-coastal-real-estate.html.

8. Dena Adler, "State Disclosure Laws Leave Homebuyers in the Dark about Flood Risks," Climate Law Blog (blog), Sabine Center for Climate Change Law, August 16, 2018, http://blogs.law .columbia.edu/climatechange/2018/08/16/state-disclosure-laws-leave-homebuyers-in-the-dark -about-flood-risks/.

9. Urbina, "Perils of Climate Change."

10. Aaron Applegate, "Buyer Beware: Check If the Property is in a Flood Zone," *Virginia Pilot Online*, January 24, 2015, https://pilotonline.com/news/local/environment/buyer-beware-check-if-the -property-is-in-a-flood/article_ced5780a-aadd-5db8-88aa-313017bea9aa.html.

11. Maryland Commission on Climate Change Adaptation and Response Working Group, *Comprehensive Strategy for Reducing Maryland's Vulnerability to Climate Change Phase I: Sea-level Rise and Coastal Storms* (Annapolis, MD: State of Maryland, 2008), 16, http://dnr.maryland.gov/ccs /Publication/Comprehensive_Strategy.pdf.

12. 2016 Florida Statutes, Title XI, chapter 161, part III, 161.57, Coastal properties disclosure statement, https://law.justia.com/codes/florida/2016/title-xi/chapter-161/part-iii/section-161.57.

13. Federal Emergency Management Agency, *National Flood Insurance Program Community Rating System: Coordinator's Manual,* 110–16.

14. Governor's Commission to Rebuild Texas, Eye of the Storm: Report of the Governor's Commission to Rebuild Texas (College Station, TX: Governor's Commission to Rebuild Texas, 2018), 154, https:// gov.texas.gov/uploads/files/press/RebuildTexasHurricaneHarveyEyeOfTheStorm_12132018.pdf.

15. Ocean Protection Council, *State of California Sea Level Rise Guidance: 2018 Update* (Sacramento, CA: State of California, 2018), 18, http://www.opc.ca.gov/webmaster/ftp/pdf/agenda_items/20180314 /Item3_Exhibit-A_OPC_SLR_Guidance-rd3.pdf.

16. Stults, "Integrating Climate Change into Hazard Mitigation Planning," 30, (see ch. 10, n. 7).

17. Ibid., 24.

18. Ibid.

19. "Section 145 Climate Change and Sea Level," Rhode Island Coastal Resources Management Program (website) accessed July 5, 2018, 6, http://sos.ri.gov/documents/archives/regdocs/released /pdf/CRMC/7264.pdf.

20. "Management Plan," RI Shoreline Change Special Area Management Plan (website) accessed July 5, 2018, http://www.beachsamp.org/management-plan/.

21. Ibid.

22. State of New York, *State Environmental Quality Review Act Findings Statement for Amendments to 6*

NYCRR part 167 (2018) (Albany, NY: State of New York, 2018), 24, http://www.dec.ny.gov/docs/permits_ej_operations_pdf/617fnlfindings.pdf.

23. Hawaii Governor's Office, "Governor David Ige Signs Bills to Set Carbon-neutral Goal and Combat Climate Change," News Release, June 4, 2018, https://governor.hawaii.gov/newsroom/latest-news/governors-office-news-release-governor-david-ige-signs-bills-to-set-carbon-neutral-goal-and-combat-climate-change/.

24. Jessica Grannis, *Adaptation Tool Kit: Sea-Level Rise and Coastal Land Use: How Governments Can Use Land-Use Practices to Adapt to Sea-Level Rise* (Washington, DC: Georgetown University, 2011), 17, http://www.georgetownclimate.org/files/report/Adaptation_Tool_Kit_SLR.pdf.

25. City of Virginia Beach, *City of Virginia Beach Comprehensive Plan—It's Our Future: A Choice City* (Virginia Beach, VA: City of Virginia Beach, 2016), 2–55, https://www.vbgov.com/government/departments/planning/2016ComprehensivePlan/Documents/Section%202.2_%20Environmental%20Stewardship_Final_5.17.16.pdf.

26. Ibid., 2–56.

27. Maryland Commission on Climate Change Adaptation and Response Working Group, *Comprehensive Strategy,* 2.

28. Ibid.

29. New Hampshire Coastal Risk and Hazards Commission, *Preparing New Hampshire for Projected Storm Surge, Sea-Level Rise, and Extreme Precipitation* (Concord, NH: State of New Hampshire, 2016), v, http://www.nhcrhc.org/wp-content/uploads/2016-CRHC-final-report.pdf.

30. City of Charleston, *Sea Level Rise Strategy* (Charleston, SC: City of Charleston, 2015), 3, https://www.charleston-sc.gov/DocumentCenter/View/10089.

31. Ibid., 12.

32. Ibid., 11.

33. Ibid., 13.

34. Mark Lubell, *The Governance Gap: Sea Level Rise in the San Francisco Bay Area* (Davis, CA: University of California, 2018), 6, http://www.bcdc.ca.gov/cm/2018/UCDavisGovernanceGapSeaLevelRiseFinalReport.pdf.

35. "Living Laboratory," RISE Resilience Innovations (website) accessed December 13, 2018, https://riseresilience.org/living-laboratory/.

36. "Project Overview," Regional AdaptLA: Coastal Impacts Planning in the Los Angeles Region (website) accessed December 12, 2018, https://dornsife.usc.edu/uscseagrant/adaptla/.

37. Los Angeles Collaborative for Climate Action and Sustainability, *A Greater LA Climate Action Framework* (Los Angeles, CA: Los Angeles Collaborative for Climate Action and Sustainability, 2016), 90, http://climateaction.la/wp-content/themes/larc/report/AGreaterLA_ClimateActionFramework_Dec-19-2016.pdf.

38. "What is the Compact?" South Florida Regional Compact: Climate Change (website) accessed July 5, 2018, http://www.southeastfloridaclimatecompact.org/about-us/what-is-the-compact/.

39. Southeast Florida Regional Climate Change Compact Counties, *A Region Responds to a Changing Climate: Regional Climate Action Plan* (Miami, FL: Southeast Florida Regional Climate Change Compact Counties, October 2012), v, http://www.southeastfloridaclimatecompact.org/wp-content/uploads/2014/09/regional-climate-action-plan-final-ada-compliant.pdf.

40. Ibid., vi.

41. Robert Behre, "Raising Charleston's Low Battery Expected to Cost $100 Million and Take Decade or More to Complete," *Post and Courier,* July 14, 2017, https://www.postandcourier.com/news/raising-charleston-s-low-battery-expected-to-cost-million-and/article_e6b95248-6747-11e7-8ee7-1f03feedbb40.html.

42. Governor's Commission to Rebuild Texas, *Request for Federal Assistance Critical Infrastructure Projects* (College Station, TX: Governor's Commission to Rebuild Texas, 2017), 1, https://www.documentcloud.org/documents/4164748-Rebuild-Texas-REQUEST-FOR-FEDERAL-ASSISTANCE.html.

43. Texas General Land Office, *Texas Coastal Resiliency Master Plan* (Austin, TX: General Land Office, 2019), vi, http://coastalstudy.texas.gov/resources/files/2019-coastal-master-plan.pdf.

44. Army Corps of Engineers, "Notice of Intent to Prepare a Draft Environmental Impact Statement for the Lake Pontchartrain and Vicinity General Re-Evaluation Report, Louisiana," 84 Fed. Reg. 12598 (April 2, 2019), https://www.govinfo.gov/content/pkg/FR-2019-04-02/pdf/2019-06354.pdf.

45. Committee on Mitigating Shore Erosion Along Sheltered Coasts, *Mitigation Shore Erosion Along Sheltered Coasts* (Washington, DC: National Academies Press, 2007), 4, https://www.nap.edu /catalog/11764/mitigating-shore-erosion-along-sheltered-coasts.

46. Ibid., 2.

47. "Beach Nourishment Viewer," Program for the Study of Developed Shorelines at Western Carolina University (website) accessed March 8, 2019, http://beachnourishment.wcu.edu/.

48. "National Beach Nourishment Database," Digital Coast, National Oceanic and Atmospheric Administration, (website) accessed July 5, 2018, https://coast.noaa.gov/digitalcoast/tools/beach -nourishment.html.

49. North Carolina Division of Coastal Management, *Coastal Erosion Study* (Morehead City, NC: North Carolina Department of Environmental Quality, 2016), 30, https://ncdenr.s3.amazonaws .com/s3fs-public/Coastal%20Management/documents/PDF/North%20Carolina%20Beach %20Erosion%20Study%20DRAFTvMASTER%2020150211.pdf.

50. Nicholas Kusnetz, "In the Outer Banks, Officials and Property Owners Battle to Keep the Ocean at Bay," *Inside Climate News*, November 28, 2017, https://insideclimatenews.org/news/28112017 /nags-head-north-carolina-beach-erosion-climate-change-sea-level-rise.

51. Casey Hedrick, *State, Territory, and Commonwealth Beach Nourishment Programs* (Silver Spring, MD: National Oceanic and Atmospheric Administration, 2000), 4, https://coast.noaa.gov/czm /media/finalbeach.pdf.

52. Joey Flechas, "Miami Beach to Begin New $100 Million Flood Prevention Project in Face of Sea Level Rise," *Miami Herald*, January 28, 2017, http://www.miamiherald.com/news/local/community /miami-dade/miami-beach/article129284119.html.

53. Philip Marcello "Mayor Offers Waterfront Plan to Save Boston from Rising Seas," *Associated Press*, October 17, 2018, https://apnews.com/4f7375861a7a4a4094c314bf07cb1f6e.

54. Chelsie Papiez, *Coastal Land Conservation in Maryland: Targeting Tools and Techniques for Sea Level Rise Adaptation and Response* (Annapolis, MD: Department of Natural Resources, 2012), 21, http:// dnr.maryland.gov/ccs/publication/coastalland_conserv_md.pdf.

55. Connecticut Department of Energy and Environmental Protection, *Connecticut Coastal and Estuarine Land Conservation Plan* (Hartford, CN: Department of Energy and Environmental Protection, 2015), 23, http://www.ct.gov/deep/lib/deep/long_island_sound/coastal_management /connecticut_coastal_and_estuarine_land_conservation_program_plan_october_2015.pdf.

56. "Full Text: Measure AA: The San Francisco Bay Clean Water, Pollution Protection and Habitat Restoration Fund," *Yes on AA* (website) accessed July 5, 2018, https://www.yesonaaforthebay .com/moreinfo/file/SFBRA_AA_Text_SCC.pdf.

57. Melody Gutierrez, "SF Bay Protection: Measure AA Passes," *SFGate*, June 8, 2016, https://www .sfgate.com/politics/article/SF-Bay-protection-Measure-AA-passes-7970365.php.

58. Florida Department of Environmental Protection, *Florida Forever Five Year Plan 2017* (Tallahassee, FL: Department of Environmental Protection, 2017), introduction, http://publicfiles.dep.state.fl.us/DSL/ OES/FloridaForeverAnnualRpts/FLDEP_DSL_OES_FloridaForeverAnnualReport2017_20170920.pdf.

59. Kristen L. Miller, Janet L. Kaminski Leduc, and Kevin E. McCarthy, *Sea Level Rise Adaptation Policy in Various States* (Hartford, CN: Office of Legislative Research, 2012), https://www.cga.ct .gov/2012/rpt/2012-R-0418.htm.

60. Gulf Coast Ecosystem Restoration Council, *Comprehensive Plan Update, 2016: Restoring the Gulf's Ecosystem and Economy* (New Orleans, LA: Gulf Coast Ecosystem Restoration Council, 2016), 9, https:// www.restorethegulf.gov/sites/default/files/CO-PL_20161208_CompPlanUpdate_English.pdf.

61. Ibid., 12.

62. State Income Tax Credit for Fortification Measures, South Carolina Department of Insurance (website) accessed July 5, 2018, http://www.doi.sc.gov/593/State-Income-Tax-Credit-for -Fortificatio.

63. Judd Schechtman and Michael Brady, *Cost-Efficient Climate Change Adaptation in the North Atlantic* (Groton, CN: Connecticut Sea Grant College Program, 2013), 177, http://media.ctseagrant.uconn .edu/publications/CEANA/OceanCity.pdf.

64. City of Oakland v. BP PLC et al., Case 3:17-cv-06012-WHA, Document 236, 15, https://www.gpo .gov/fdsys/pkg/USCOURTS-cand-3_17-cv-06012/pdf/USCOURTS-cand-3_17-cv-06012-20.pdf.

65. State of Rhode Island v. Chevron et al., PC-2018-4716 (2018), 4, http://www.riag.ri.gov/documents /KilmartinVChevronEtAl.pdf.

66. Nicholas Kusnetz, "Faced with Costs of Climate Change, Rhode Island Sues Fossil Fuel Companies," July 3, 2018, https://www.bostonglobe.com/metro/2018/07/02/faced-with-costs-climate-change -sues-fossil-fuel-companies/mGflVecLITflLHCmBHYzAM/story.html.

67. Michael Pawlukiewicz, Prema Katari Gupta, and Carl Koelbel, *Ten Principles of Coastal Development*, (Washington, DC: Urban Land Institute, 2007), 2, https://uli.org/wp-content/uploads /ULI-Documents/Ten-Principles-for-Coastal-Development.pdf.

68. Ibid.

69. Grannis, *Adaptation Tool Kit: Sea-Level Rise and Coastal Land Use,* 19.

70. South Florida Regional Planning Council, *Adaptation Action Areas: A Planning Guide for Florida's Local Governments* (Hollywood, FL: South Florida Regional Planning Council, 2015), 8, http:// www.floridajobs.org/docs/default-source/2015-community-development/community-planning /crdp/aaaguidebook2015.pdf?sfvrsn=2.

71. Greenwich Municipal Code, §6-94, DIVISION 9, USE REGULATIONS, 8, 9–41, http://www .greenwichct.org/upload/medialibrary/23f/pzRegsDivision09.pdf.

72. Federal Emergency Management Agency, *National Flood Insurance Program Community Rating System: Coordinator's Manual,* 42021, (see ch. 9, n. 16).

73. Woods Hole Sea Grant, University of Hawaii Sea Grant, and Barnstable County, *Model Bylaw For Effectively Managing Coastal Floodplain Development* (Barnstable, MA: Cape Cod Commission, 2009), 15, http://www.capecodcommission.org/resources/bylaws/Coastal_Floodplain _Bylaw_Dec2009.pdf.

74. Ibid., 35.

75. James Buggers, "Not Trusting FEMA's Flood Maps, More Storm-Ravaged Cities Set Tougher Rules," *Inside Climate News*, March 19, 2019,Not Trusting FEMA's Flood Maps, More Storm-Ravaged Cities Set Tougher Rules," https://insideclimatenews.org/news/19032019/fema -flood-maps-risk-zones-cities-climate-change-mexico-beach-houston-outer-banks?utm _source=InsideClimate+News&utm_campaign=414f174ba8-&utm_medium=email&utm _term=0_29c928ffb5-414f174ba8-327889757.

76. Molly Loughney Melius and Margaret R. Caldwell, *2015 California Coastal Armoring Report: Managing Coastal Armoring and Climate Change Adaptation in the 21st Century* (Stanford, CA: Stanford University Environment and Natural Resources Law & Policy Program, 2015), 5, http://law.stanford.edu/wp-content/uploads/2015/07/CalCoastArmor-FULL-REPORT-6.17 .15.pdf.

77. John Englander, pers. comm., January 6, 2019.

78. A. Dan Tarlock and Deborah M. Chizewer, "Living with Water in a Climate Changed World: Will Flood Policy Sink or Swim," *Environmental Law* 46, no. 491 (September 2016): 535, https://law .lclark.edu/live/files/22558-46-3tarlockpdf.

79. Surfrider Foundation, *State of the Beach Report Card: 2018* (Surfrider Foundation, 2018), 8, https:// s3-us-west-2.amazonaws.com/publicfiles.surfrider.org/Surfrider-Foundation-State-Of-The-Beach -Report-2018.pdf.

80. Fleming, "Coastal Effects," ch. 8 in the *Fourth National Climate Assessment*, (see ch. 13, n. 10).

Chapter 15

1. R. Lempert et al., "Beyond Incremental Change," ch. 28 in *Impacts, Risks, and Adaptation in the United States: Fourth National Climate Assessment, Volume II* (Washington, DC: US Global Change Research Program, 2018), 16, https://nca2018.globalchange.gov/chapter/28/.

2. Ibid.

3. Ben Walker, "An Island Nation Turns Away from Climate Migration, Despite Rising Seas," *Inside Climate News,* November 20, 2017, https://insideclimatenews.org/news/20112017/kiribati-climate -change-refugees-migration-pacific-islands-sea-level-rise-coconuts-tourism.

4. Richard Marles, "Australia Must Not Be Afraid of its Obligations to Pacific Climate Migrants," *Guardian,* November 10, 2015, https://www.theguardian.com/commentisfree/2015/nov/11 /australia-must-not-be-afraid-of-its-obligations-to-pacific-climate-migrants.

5. Blaine Friedlander, "Rising Seas Could Result in 2 Billion Refugees by 2100," *Cornell Chronicle* (June 19, 2017): 1, http://news.cornell.edu/stories/2017/06/rising-seas-could-result-2-billion-refugees-2100.

6. David López-Carr and Jessica Marter-Kenyon, "Human Adaptation: Manage Climate-Induced Resettlement," *Nature* 517 (January 14, 2015): 265, https://www.nature.com/news /human-adaptation-manage-climate-induced-resettlement-1.16697#/ref-link-4.

7. Ibid., 267.

8. António Guterres, "Migration, Displacement and Planned Relocation," The UN Refugee Agency (December, 31 2012), conclusion, http://www.unhcr.org/en-us/news/editorial/2012/12/55535d6a9 /migration-displacement-planned-relocation.html.

9. "Shoreline Management Plans," Programmes, BBC (website) accessed July 20, 2018, http://www .bbc.co.uk/programmes/articles/2tQfptZpznVzGzCW1k2ZYtq/shoreline-management-plans.

10. "Welsh Village to Sue Government Over 'Alarmist' Rising Sea Level Claim," *Telegraph* (website) February 11, 2016, accessed July 20, 2018, https://www.telegraph.co.uk/news/earth/environment /climatechange/12152240/Welsh-village-to-sue-government-over-alarmist-rising-sea-level -claim.html.

11. Ibid.

12. "Shoreline Management Plan Policies—What Do They Mean?" Shoreline Management Plans, Natural Resources, Wales (website) accessed July 21, 2018, https://www.naturalresources .wales/flooding/managing-flood-risk/flood-risk-map-guidance/shoreline-management-plan /?lang=en.

13. Committee on Climate Change, *Managing the Coast in a Changing Climate* (London, UK: Committee on Climate Change, 2018), 3, https://www.theccc.org.uk/wp-content/uploads/2018/10 /Managing-the-coast-in-a-changing-climate-October-2018.pdf.

14. Ibid., 11.

15. Andy Walker, "Environment Agency Calls for New Approach to Flood and Coastal Resilience," *Infrastructure Intelligence*, May 9, 2019, http://www.infrastructure-intelligence.com/article/may-2019 /environment-agency-calls-new-approach-flood-and-coastal-resilience.

16. Environment Agency, *Draft National Flood and Coastal Erosion Risk Management Strategy for England* (London, UK: Environment Agency, 2019), 20, https://consult.environment-agency.gov.uk /fcrm/national-strategy-public/user_uploads/fcrm-strategy-draft-final-1-may-v0.13-as-accessible -as-possible.pdf.

17. Victoria Herrmann, *The United States' Climate Change Relocation Plan* (Washington, DC: The Atlantic Council, 2017), 1, http://eleep.eu/sites/default/files/publications/Herrmann%2C%20Victoria .%20%282017%29.%20The%20United%20States%20Climate%20Change%20Relocation %20Plan.pdf.

18. Executive Office of the President, *President's State, Local and Tribal Leaders Task Force.* 30, (see ch. 12, n. 4).

19. Ibid.

20. Christopher Flavelle, "Obama's Final Push to Adapt to Climate Change: Saving the Toughest Question for Last: Which Towns Should the Government Help Move?" *Bloomberg News,*

December 16, 2016, https://www.bloombergquint.com/opinion/2016/12/16/obama-s-final -push-to-adapt-to-climate-change.

21. Ibid.

22. Roger-Mark De Souza et al., *Building Coastal Resilience for Greater US Security* (Stanford, CA: Hoover Institution Press, 2018), 25, https://www.hoover.org/sites/default/files/research/docs/costal resilience_web_final.pdf.

23. James B. London, "Sea Level Rise and Coastal Management," *Oxford Research Encyclopedia, Environmental Science* (USA, Oxford University Press, 2017), 4, http://environmentalscience.oxfordre.com /view/10.1093/acrefore/9780199389414.001.0001/acrefore-9780199389414-e-377?print=pdf.

24. Blue Ribbon Committee on Shoreline Management, *Final Report: Recommendations for Improved Beachfront Management in South Carolina* (Columbia, SC: South Carolina Department of Health and Environmental Control, 2013), 6, http://www.scdhec.gov/library/CR-010631.pdf.

25. Scott Dance, "At Smith Island, a Chance to Shore Up the Future," *Baltimore Sun,* July 8, 2017, http://www.baltimoresun.com/news/maryland/bs-md-smith-island-jetty-20170707-story.html.

26. Ibid.

27. State of Virginia Executive Order Twenty Four—Increasing Virginia's Resilience to Sea Level Rise and Natural Hazards, 2018, 3, https://www.governor.virginia.gov/media/governorvirginiagov/executive -actions/ED-24-Increasing-Virginias-Resilience-To-Sea-Level-Rise-And-Natural-Hazards.pdf.

28. Hawaii Coastal Zone Management Program, *Assessing the Feasibility and Implications of Managed Retreat Strategies for Vulnerable Coastal Areas in Hawaii: Final Report* (Honolulu, HI: Hawaii Department of Business Economic Development & Tourism, 2019), 57, http://files.hawaii.gov/dbedt /op/czm/ormp/assessing_the_feasibility_and_implications_of_managed_retreat_strategies_for _vulnerable_coastal_areas_in_hawaii.pdf.

29. Ibid.

30. "Quinault Indian Nation Plans for Relocation," EPA Climate Change Adaptation Resource Center (website) accessed July 6, 2018, https://www.epa.gov/arc-x/quinault-indian-nation-plans-relocation.

31. Adriana Brasilero, "In Miami, Battling Sea Level Rise May Mean Surrendering Land," *Reuters*, July 20, 2017, https://www.reuters.com/article/us-miami-sealevelrise/in-miami-battling-sea-level-rise -may-mean-surrendering-land-idUSKBN1A601L.

32. Amy Green, "With Governor and Legislators in Denial, This Tiny Florida Town Tries to Adapt to Climate Change," WMFE Radio (website) assessed July 20, 2018, http://www.wmfeindepth .org/with-governor-and-legislators-in-denial-this-tiny-florida-town-tries-to-adapt-to-climate -change/406.

33. Ibid.

34. Brent D. Ryan, David Vega-Barachowitz, and Lily Perkins-High, *Rising Tides: Relocation and Sea Level Rise in Metropolitan Boston* (Cambridge, MA: MIT Norman B. Leventhal Center for Advanced Urbanism, 2015), 10, http://lcau.mit.edu/sites/lcau.mit.edu/files/attachments/project/Rising Tides_ProjectPrimer.pdf.

35. Regional Plan Association, *The Fourth Regional Plan: Making the Region Work for All of Us: Executive Summary* (New York: Regional Plan Association, 2017), 17, http://library.rpa.org/pdf/RPA-4RP -Executive-Summary.pdf.

36. A. Bukvic and G. Owen, "Attitudes Towards Relocation Following Hurricane Sandy: Should We Stay or Should We Go?" *Disasters* 41, no. 1 (2016): 119, https://onlinelibrary.wiley.com/doi /full/10.1111/disa.12186.

37. Tim Henderson, "Rising Tide of State Buyouts Fights Flooding," *Stateline*, December 3, 2015, http://www.pewtrusts.org/en/research-and-analysis/blogs/stateline/2015/12/03/rising-tide -of-state-buyouts-fights-flooding.

38. "Blue Acres Buyout Program," ReNewJerseyStronger (website) accessed July 6, 2018, http://www .renewjerseystronger.org/homeowners/blue-acres-buyout-program/.

39. "3 Years Long, 3 Years Strong: New Jersey's Successful Approach to Purchasing Homes Along Sandy's Flooded Path," FEMA Mitigation Best Practices (website) accessed July 6, 2018, https://

www.fema.gov/media-library-data/1447357426269-e95b4a3f2805922a9b8b0fde5f91789f/01
_3years-long_3years-strong_web-rev.pdf.

40. "Blue Acres Buyout Program," ReNewJerseyStronger.

41. Resettlement, ReNewJerseyStronger (website) accessed July 6, 2018, http://www.renewjersey
stronger.org/homeowners/resettlement/.

42. Robert Freudenberg et al., *Buy-In for Buyouts: The Case for Managed Retreat from Flood Zones*
(Cambridge, MA: Lincoln Institute for Land Policy, 2016), 4, https://www.lincolninst.edu/sites
/default/files/pubfiles/buy-in-for-buyouts-full.pdf.

43. David Hammer, "Examining Post-Katrina Road Home Program: 'It's More Than the Money. It's the
Hoops We Had to Jump Through to Do It'," *The Advocate,* April 23, 2015, http://www.theadvocate
.com/baton_rouge/news/article_f9763ca5-42ba-5a62-9935-c5f7ca94a7c4.html.

44. Tegan Wendland, "Louisiana Says Thousands Should Move From Vulnerable Coast, But Can't
Pay Them," *National Public Radio,* January 4, 2018, https://www.npr.org/2018/01/04/572721503
/louisiana-says-thousands-should-move-from-vulnerable-coast-but-cant-pay-them.

45. Coastal Protection and Restoration Authority of Louisiana, *Louisiana's Comprehensive Master Plan
for a Sustainable Coast* (Baton Rouge, LA: OST-State Printing, June 2, 2017), 104, http://coastal.
la.gov/wp-content/uploads/2017/04/2017-Coastal-Master-Plan_Web-Book_CFinal-with-Effective
-Date-06092017.pdf.

46. Tegan Wendland, "Louisiana Says Thousands Should Move."

47. Freudenberg, *"Buy-In for Buyouts,"* 5.

48. Ibid.

49. Ibid.

50. James Titus, *Rolling Easements* (Washington, DC: Environmental Protection Agency, 2011), 8,
https://www.epa.gov/sites/production/files/documents/rollingeasementsprimer.pdf.

51. Ibid.

52. Institute on Science for Global Policy (ISGP), *ISGP Climate Change Program (ICCP): Sea Level Rise:
What's Our Next Move?* (Washington, DC: Institute on Science for Global Policy, 2015), 37, http://
scienceforglobalpolicy.org/wp-content/uploads/56e30928039ff-ISGP%20Sea%20Level%20Rise.pdf.

53. Richard Turner Henderson, "Sink or Sell: Using Real Estate Purchase Options to Facilitate Coastal Retreat," *Vanderbilt Law Review* (March 2018), 664, https://wp0.vanderbilt.edu
/lawreview/2018/03/sink-or-sell-using-real-estate-purchase-options-to-facilitate-coastal-retreat/.

54. Joseph L. Sax, "The Fate of Wetlands in the Face of Rising Sea Levels: A Strategic Proposal," *UCLA
Journal of Environmental Law and Policy* 9, no.143 (1990): 154, https://scholarship.law.berkeley
.edu/cgi/viewcontent.cgi?article=2783&context=facpubs.

55. US Const. amend. V.

56. Byrne, "The Cathedral Engulfed," 98, (see ch. 13, n. 1).

57. Ibid.

58. *Penn Central Transportation Co. v. New York City*, Supreme Court, Legal Information Institute, Cornell Law School (website) accessed July 6, 2018, 124, https://www.law.cornell.edu/supremecourt
/text/438/104.

59. Ibid.

60. Byrne, "The Cathedral Engulfed," 109, (see ch. 13, n. 1).

61. James G. Titus, "Rising Seas, Coastal Erosion, and the Takings Clause: How to Save Wetlands and
Beaches Without Hurting Property Owners," *Maryland Law Review* 57, no. 4 (1998): 1358, http://
digitalcommons.law.umaryland.edu/mlr/vol57/iss4/3/.

62. Thomas Ruppert, "Reasonable Investment Back Expectations: Should Notice of Rising Seas Lead to
Falling Expectations for Coastal Property Purchasers?" *Journal of Land Use,* (Spring, 2011): 276, https://
www.flseagrant.org/wp-content/uploads/2012/01/Ruppert-239-277-FINAL-_-5-10-11_1.pdf.

63. Applegate, "The Intersection of the Takings Clause," 524, (see ch. 13, n. 5).

64. Ibid., 513.

65. Matthew Scarano, "Withholding Municipal Services to Facilitate Coastal Retreat: Legal Risks and Possibilities," *Environmental Claims Journal (*March, 2018): 28, https://www.law.columbia.edu /sites/default/files/news/files/matthew_scarano_witholding_municipal_services.pdf.

66. Mathew E. Hauer, Jason M. Evans, and Deepak R. Mishra, "Millions Projected to Be At Risk of Sea Level Rise in Continental United States," *Nature Climate Change* 6 (March 14, 2016): 1, https://www.nature.com/articles/nclimate2961.

67. Mathew E. Hauer, "Migration Induced by Sea-level Rise Could Reshape the US Population Landscape," *Nature Climate Change* 7 (April 17, 2017): 1, https://www.nature.com/articles /nclimate3271.

68. Ibid., 3.

69. Ibid.

70. Rebecca Mordechai, "'We Will Miss the Warm Winters.' Retirees Are Fleeing Florida as Climate Change Threatens Their Financial Future," *Money*, March 19, 2019, http://money.com /money/5638871/we-will-miss-the-warm-winters-retirees-are-fleeing-florida-as-climate-change -threatens-their-financial-future/.

71. Charles Geisler and Ben Currens, "Impediments to Inland Resettlement Under Conditions of Accelerated Sea Level Rise," *Land Use Policy* 66, (July 2017): 328, https://www.sciencedirect.com /science/article/pii/S0264837715301812?via%3Dihub.

72. Ibid., 327.

73. Ibid.

74. Ibid.

75. State of California, *State of California Fourth Climate Change Assessment: Statewide Summary* (Sacramento, CA: State of California, 2018), 66, http://www.climateassessment.ca.gov/state /docs/20180827-StatewideSummary.pdf.

76. Orrin H. Pilkey, Linda Pilkey-Jarvis, and Keith C. Pilkey, *Retreat from a Rising Sea: Hard Choices in an Age of Climate Change* (New York, NY: Columbia University Press, 2016), 9.

Chapter 16

1. MDC, *Before the Storm*, (Chapel Hill, MDC, 2009), 1, https://www.mdcinc.org/wp-content /uploads/2017/11/Before-the-Storm_0.pdf.

2. "Chapter 8: Coastal Effects," *Fourth National Climate Assessment* (website) accessed December 5, 2018, https://nca2018.globalchange.gov/chapter/8/#fn:62.

3. Martinich et al., *"Risks of Sea Level Rise to Disadvantaged Communities in the United States,"* *Mitigation and Adaptation Strategies for Global Change* 18, no. 2, (February 2012): 177, https://link .springer.com/article/10.1007/s11027-011-9356-0#Sec6.

4. Ibid.

5. Ibid., 178.

6. Ibid., 179.

7. Ibid., 180.

8. Junia Howell and James R Elliott, "Damages Done: The Longitudinal Impacts of Natural Hazards on Wealth Inequality in the United States," *Social Problems* (August 2018): 1, https://academic .oup.com/socpro/advance-article/doi/10.1093/socpro/spy016/5074453.

9. Rebeca Hersher and Robert Benincasa, "How Federal Disaster Money Favors the Rich," National Public Radio, March 5, 2019, https://www.npr.org/2019/03/05/688786177/how-federal-disaster -money-favors-the-rich.

10. Erika Bolstad, "High Ground Is Becoming Hot Property as Sea Level Rises: Climate Change May Now Be a Part of the Gentrification Story in Miami Real Estate," *Scientific American (*May 1, 2017), https:// www.scientificamerican.com/article/high-ground-is-becoming-hot-property-as-sea-level-rises/.

11. Caitlin Dewey, "Seven Weeks After Hurricane Maria, Puerto Ricans Still Can't Access Programs That Fed Millions in Texas and Florida," *Washington Post,* November 7, 2017, https://www

.washingtonpost.com/news/wonk/wp/2017/11/07/seven-weeks-after-hurricane-maria-puerto-ricans-still-cant-access-programs-that-fed-millions-in-texas-and-florida/?noredirect=on&utm_term=.e59a72c40616.

12. Hilda Lloréns et al., "Hurricane Maria: Puerto Rico's Unnatural Disaster," *Social Justice Journal* (January 22, 2018), http://www.socialjusticejournal.org/hurricane-maria-puerto-ricos-unnatural-disaster/.

13. Ryan, *Rising Tides: Relocation and Sea Level Rise in Metropolitan Boston,* 8, (see ch.15, n. 34).

14. Alice Kaswan, "Domestic Climate Change Adaptation and Equity," *Environmental Law Reporter* 42 (December 2012): 11134, https://papers.ssrn.com/sol3/papers.cfm?abstract_id=2184551.

15. Ibid.

16. Ibid.

17. Ted Jackson, "On the Louisiana Coast, A Native Community Sinks Slowly into the Sea," *YaleEnvironment* 360, Yale School of Forestry and Environmental Studies (website) March 15, 2018, accessed July 2, 2018, https://e360.yale.edu/features/on-louisiana-coast-a-native-community-sinks-slowly-into-the-sea-isle-de-jean-charles.

18. Chris Sellers, "Storms Hit Poorer People Harder, from Superstorm Sandy to Hurricane Maria," *The Conversation,* November 19, 2017, https://theconversation.com/storms-hit-poorer-people-harder-from-superstorm-sandy-to-hurricane-maria-87658.

19. Ibid.

20. Kaswan, "Domestic Climate Change Adaptation and Equity," 11139.

21. Susan Clayton, Christie Manning, and Caroline Hodge, *Beyond Storms and Droughts: The Psychological Impacts of Climate Change,* (Washington, DC: American Psychological Association and ecoAmerica, 2014), 18, https://elpnet.org/sites/default/files/portfolio/eA_Beyond_Storms_and_Droughts_Psych_Impacts_of_Climate_Change.pdf.

22. Clayton, *Beyond Storms and Droughts,* 18.

23. A. Crimmins et al., *The Impacts of Climate Change on Human Health in the United States: A Scientific Assessment* (Washington, DC: US Global Change Research Program, 2016), 220, https://health2016.globalchange.gov/.

24. Herrmann, *The United States' Climate Change Relocation Plan,* 5, (see ch. 15, n. 17).

25. Ryan Kailath, "In Coastal Louisiana, Home Buyouts Raise Questions and Fears," *National Public Radio,* January 18, 2017, http://wwno.org/post/coastal-louisiana-home-buyouts-raise-questions-and-fears.

26. Ryan, *Rising Tides: Relocation and Sea Level Rise in Metropolitan Boston*, 5, (see ch. 15, no. 34).

27. Phil Diehl, "No 'Retreat' from Rising Sea Level for Homes in Del Mar," *San Diego Union Tribune,* December 4, 2017, http://www.sandiegouniontribune.com/communities/north-county/sd-no-sea-level-20171129-story.html.

28. Diehl, "No 'Retreat' from Rising Sea Level for Homes in Del Mar."

29. Sherri Brokopp Binder, *Resilience and Postdisaster Relocation: A Study of New York's Home Buyout Plan in the Wake of Hurricane Sandy* (Boulder, CO: University of Colorado, 2013), 4, https://hazards.colorado.edu/uploads/quick_report/binder_2013.pdf.

30. Ibid., 9.

31. Mary Beth Griggs, "These Staten Islanders Lost their Neighborhood to Sandy: Here's Why They're Not Getting it Back," *Popular Science,* October 13, 2017, https://www.popsci.com/storm-sandy-staten-island-five-years-later#page-4.

32. Bukvic, "Attitudes Towards Relocation Following Hurricane Sandy," 110, (see ch. 15, n. 36).

33. Miyuki Hino, Christopher B. Field, and Katharine J. Mach, "Managed Retreat as a Response to Natural Hazard Risk," *Nature Climate Change* (March 27, 2017): 1, http://www.theurbanclimatologist.com/uploads/4/4/2/5/44250401/hinoetal2017managedretreatoverview.pdf.

34. Ibid., 4.

35. Ibid., 5.

36. Ibid., 4.

37. Christopher Flavelle, "Louisiana, Sinking Fast, Prepares to Empty Out Its Coastal Plain," *Bloomberg News,* December 22, 2017, https://www.bloomberg.com/news/articles/2017-12-22/louisiana-sinking-fast-prepares-to-empty-out-its-coastal-plain.

38. Mark T. Gibbs, "Why is Coastal Retreat Is So Hard to Implement? Understanding the Political Risk to Coastal Adaptation Pathways," *Ocean and Coastal Management*, 130 (2016): 111, https://www.sciencedirect.com/science/article/pii/S0964569116301119.

39. Ibid., 110.

40. Ibid., 109.

41. Bukvic, "Attitudes Towards Relocation Following Hurricane Sandy," 119, (see ch. 15, n. 36).

42. Ibid.

Chapter 17

1. Keith W. Rizzardi, "Rising Seas, Receding Ethics? Why Real Estate Professionals Should Seek the Moral High Ground," *Washington and Lee Journal of Energy, Climate and the Environment* 6, no. 2 (2015): 449, https://scholarlycommons.law.wlu.edu/jece/vol6/iss2/4/.

2. Sean Becketti, "Life's a Beach," *Insight,* Freddie Mac, April 26, 2016 (website) accessed July 7, 2018, http://www.freddiemac.com/research/insight/20160426_lifes_a_beach.html.

3. Bryan Walsh, "The Coastal Mortgage Time Bomb: Experts Worry That if Insurers Start to Pull Out of Flood-prone Seaside Communities it Could Cause a Crisis Worse Than 2008," *Inman*, December 18, 2017, https://www.inman.com/2017/12/18/coastal-mortgage-time-bomb/.

4. Ibid.

5. Ibid.

6. Ibid.

7. Christopher Flavelle, "Florida's Real-estate Reckoning Could Be Closer Than You Think," *Seattle Times,* December 29, 2017, https://www.seattletimes.com/nation-world/floridas-real-estate-reckoning-could-be-closer-than-you-think/.

8. Jeanna Simeone, "The Use of Credit Ratings for Mortgage-Backed Securities," *Developments in Banking Law*, Boston University School of Law 31, (Fall 2017): 108, https://www.bu.edu/rbfl/files/2013/09/UseOfCreditRatingsMBS.pdf.

9. Becketti, "Life's a Beach."

10. Ibid.

11. Ibid.

12. Asaf Bernstein, Matthew Gustafson, and Ryan Lewis, "Disaster on the Horizon: The Price Effect of Sea Level Rise," *Journal of Financial Economics* (November 2017): 23, https://papers.ssrn.com/sol3/papers.cfm?abstract_id=3073842.

13. First Street Foundation, "Rising Seas Erode $15.8 Billion in Home Value from Maine to Mississippi," news release, February 27, 2019, https://assets.floodiq.com/2019/02/9ddfda5c3f7295fd97d60332bb14c042-firststreet-floodiq-mid-atlantc-release.pdf.

14. Karen Doyle, "Rising Sea Levels Are Hurting Home Values in These 40 Cities," *GOBankingRates,* May 7, 2019, https://finance.yahoo.com/news/rising-sea-levels-hurting-home-090000627.html.

15. "Ocean at the Door," *Climate Central and Zillow* (website) accessed December 5, 2018, 3, http://assets.climatecentral.org/pdfs/Nov2018_Report_OceanAtTheDoor.pdf?pdf=OceanAtTheDoor-Report.

16. Mike Maciag, "Analysis: Areas of the US With Most Floodplain Population Growth," *Governing*, August 2018, http://www.governing.com/gov-data/census/flood-plains-zone-local-population-growth-data.html#methodology.

17. Brooke Jarvis, "When Rising Seas Transform Risk Into Certainty Along Parts of the East Coast, the Entire System of Insuring Coastal Property is Beginning to Break Down," *New York Times Magazine*, April, 18, 2017, https://www.nytimes.com/2017/04/18/magazine/when-rising-seas-transform-risk-into-certainty.html.

18. Becketti, "Life's a Beach."

19. BlackRock Investment Institute, *Getting Physical: Scenario Analysis for Assessing Climate-related Risks* (New York: BlackRock Investment Institute, 2019), 14; https://www.blackrock.com/us /individual/literature/whitepaper/bii-physical-climate-risks-april-2019.pdf.

20. Ibid.

21. Ibid.

22. Walsh, "The Coastal Mortgage Time Bomb."

23. Ibid.

24. BlackRock, *Getting Physical*, 10.

25. Ibid.

26. Ibid.

27. Moody's Investor Service, "Climate Change is Forecast to Heighten US Exposure to Economic Loss Placing Short- and Long-term Credit Pressure on US States and Local Governments," news release, November 28, 2017, https://www.moodys.com/research/Moodys-Climate-change-is-forecast -to-heighten-US-exposure-to--PR_376056.

28. Michael Bonanno and Andrew Teras, "Rating Agencies and Municipal Climate Risk," *ESG Newsletter*, January 3, 2018, https://www.breckinridge.com/insights/details/rating-agencies-and -municipal-climate-risk/.

29. Jarvis, "When Rising Seas Transform Risk Into Certainty." https://www.nytimes.com/2017/04/18 /magazine/when-rising-seas-transform-risk-into-certainty.html.

30. Glenn D. Rudebusch, "Climate Change and the Federal Reserve," *FRBSF Economic Letter* (March 25, 2019): 2, https://www.frbsf.org/economic-research/files/el2019-09.pdf.

31. Ibid. 3.

32. Ibid.

33. Securities and Exchange Commission, "Commission Guidance Regarding Disclosure Related to Climate Change," 17 CFR parts 211, 231 and 241, February 20, 2010, 1, https://www.sec.gov/rules /interp/2010/33-9106.pdf.

34. Ibid. 12.

35. Ibid., 26.

36. Ibid., 7.

37. David Gelles, "S.E.C. Is Criticized for Lax Enforcement of Climate Risk Disclosure," *New York Times*, January 23, 2016, https://www.nytimes.com/2016/01/24/business/energy-environment /sec-is-criticized-for-lax-enforcement-of-climate-risk-disclosure.html.

38. Ibid.

39. CERES, *Turning Point: Corporate Progress on the Ceres Roadmap for Sustainability* (Boston, MA: CERES, 2018), 56, https://www.ceres.org/node/2275.

40. Securities and Exchange Commission, "Business and Financial Disclosure Required by Regulation S-K," 17 CFR parts 210, 229, 230, 232, 239, 240 and 249, April 22, 2016, 23973, https://www.federalregister.gov/documents/2016/04/22/2016-09056/business-and-financial -disclosure-required-by-regulation-s-k.

41. Linda Lowson, "Global Climate Change and Sustainability Financial Reporting: An Unstoppable Force with or without Trump," Harvard Law School Forum on Corporate Governance and Financial Regulation, April 30, 2017 (website) assessed July 18, 2018, https://corpgov.law.harvard .edu/2017/04/30/global-climate-change-and-sustainability-financial-reporting-an-unstoppable -force-with-or-without-trump/.

42. Task Force on Climate-Related Financial Disclosures, *Final Report: Recommendations of the Task Force on Climate-related Financial Disclosures* (Basel, SZ: Financial Stability Board, 2017), i, https:// www.fsb-tcfd.org/wp-content/uploads/2017/06/FINAL-TCFD-Report-062817.pdf.

43. Ibid., 6.

44. Ibid., i.

45. Ibid., 41.

46. Ibid., 42.

47. Task Force on Climate-Related Financial Disclosures, *Task Force on Climate-Related Financial Disclosures: Status Report,* (Basel, SZ: Financial Stability Board, 2018), iv, https://www.fsb-tcfd.org/wp-content/uploads/2018/08/FINAL-2018-TCFD-Status-Report-092518.pdf.

48. Jose Luis Blasco and Adrian King, *The Road Ahead,* (Amstelveen, NL: KPMG, 2017), 34.

49. Ibid., 31.

50. Anya Khalamayzer, "Why Voluntary Climate Risk Disclosure is Going Mainstream," *GreenBiz,* January 9, 2018, https://www.greenbiz.com/article/why-voluntary-climate-risk-disclosure-going-mainstream.

51. "Investor-Driven Climate-Risk Transparency Is Taking Hold," Institute for Energy Economics and Financial Analysis (website) February 8, 2018, accessed July 7, 2018, http://ieefa.org/investor-driven-climate-risk-transparency-taking-hold/.

52. Nick Dawson, "Breaking the Ice: Investors Warm to Climate Change," *Harvard Law School Forum on Corporate Governance and Financial Regulation*, June 9, 2017, https://corpgov.law.harvard.edu/2017/06/09/breaking-the-ice-investors-warm-to-climate-change/.

53. Marianne Lavelle, "Exxon Shareholders Approve Climate Resolution: 62% Vote for Disclosure," *Inside Climate News,* May 31, 2017, https://insideclimatenews.org/news/31052017/exxon-shareholder-climate-change-disclosure-resolution-approved.

54. Ibid.

55. CERES, "New Exxon Report Is a Step Forward for Investor Disclosure on Climate Change, but Falls Short on Detail", news release, February 5, 2018, https://www.3blmedia.com/News/New-Exxon-Report-Step-Forward-Investor-Disclosure-Climate-Change-Falls-Short-Detail.

56. Rizzardi, "Rising Seas, Receding Ethics," 403.

57. Executive Office of the President, *President's State, Local and Tribal Leaders Task Force,* 39, (see ch. 12, n. 4).

58. Ibid.

59. National Association of Realtors, *Code of Ethics and Standards of Practice of the National Association of Realtors,* article 2, https://www.nar.realtor/about-nar/governing-documents/code-of-ethics/2018-code-of-ethics-standards-of-practice.

60. American Bar Association, *Model Rules of Professional Conduct,* rule 4.1, https://www.americanbar.org/groups/professional_responsibility/publications/model_rules_of_professional_conduct/rule_4_1_truthfulness_in_statements_to_others.html.

61. Rizzardi, "Rising Seas, Receding Ethics," 406.

62. Ibid., 437.

63. Patrice Taddonio, "How Much Do Insurance Companies Profit After a Natural Disaster?" *Frontline, National Public Radio* (May 24, 2016), https://www.pbs.org/wgbh/frontline/article/how-much-do-insurance-companies-profit-after-a-natural-disaster/.

64. Robert Hunter, "Profiting Off Disaster: Frontline—All Things Considered, 5/24/16," *Consumer Federation of America* (blog) May 31, 2016, accessed July 9, 2018, https://consumerfed.org/profiting-off-disaster/.

65. Ibid.

Part 4 Epilogue

1. Orrin Pilkey and Rob Young, *The Rising Sea* (Washington, DC: Island Press, 2009), 181.

2. Ibid.

Chapter 18

1. Sea Level Rise: Hearing Before the Committee on Energy and Natural Resources, 112th Cong.

34 (2012) (remarks of Senator Maria Cantwell), https://www.gpo.gov/fdsys/pkg/CHRG-112 shrg76897/pdf/CHRG-112shrg76897.pdf.

2. Mitchell Chester, *Palm Beach Post,* July 10, 2018, accessed July 18, 2018, https://www-palm beachpost-com.cdn.ampproject.org/c/s/www.palmbeachpost.com/news/opinion/point-view -coastal-residents-need-set-aside-money-for-flooding/3eN5Fl3IRUo5XA3VXaMsPI/amp.html.

3. Independent Advisory Committee on Applied Climate Assessment, "Evaluating Knowledge to Support Climate Action: A Framework for Sustained Assessment," Weather Climate and Society (April 2019): online release line 198; file:///C:/Users/pijpi/AppData/Local/Packages/microsoft .windowscommunicationsapps_8wekyb3d8bbwe/LocalState/Files/S0/8352/Attachments/wcas -d-18-0134.1[9568].pdf.

4. "Regional Sea-level Change and Coastal Impacts," World Climate Research Programme, (website) accessed December 3, 2018, https://www.wcrp-climate.org/component/content/article /687-gc-sealevel-regional-impacts-overview?catid=28&Itemid=264.

5. Judith Curry, *Special Report: Sea Level and Climate Change* (Reno, NV: Climate Forecast Applications Network, 2018), 72, https://curryja.files.wordpress.com/2018/11/special-report-sea-level -rise3.pdf.

6. National Infrastructure Advisory Council: *Water Sector Resilience: Final Report and Recommendations* (Washington, DC: United States Department of Homeland Security, 2016), 41, https://www .dhs.gov/sites/default/files/publications/niac-water-resilience-final-report-508.pdf.

7. "Regulatory Program Frequently Asked Questions," *Army Corps of Engineers* (website) accessed August 18, 2018, https://www.usace.army.mil/Missions/Civil-Works/Regulatory-Program-and-Permits /Frequently-Asked-Questions/.

8. Military Expert Panel, *Sea Level Rise and the US Military's Mission,* 48, (see ch. 5, n. 43).

9. Thorne, "US Pacific Coastal Wetland Resilience and Vulnerability to Sea-level Rise," 5, (see ch. 6, n. 10).

10. Oliver Milman, "Americans, The Next Climate Migrants," *The Guardian,* September 24, 2018, https:// www.theguardian.com/environment/2018/sep/24/americas-era-of-climate-mass-migration-is-here.

11. David Ford and Doug Plasencia, *Meeting the Challenge of Change: Implementing the Federal Flood Risk Management Standard and a Climate-Informed Science Approach* (Washington, DC: Association of State Floodplain Managers, 2016, 4, http://www.asfpmfoundation.org/ace-images/forum /Meeting_the_Challenge_of_Change.pdf.

Chapter 19

1. United States Federal Emergency Management Agency, *The Watermark,* 1, (see ch. 9, n. 23).

2. Ibid., 3.

3. "A Blueprint to Rebuild America's Infrastructure," *Democratic Policy and Communications Committee* (website) accessed August 22, 2018, 1, https://www.democrats.senate.gov/imo/media/doc /ABlueprinttoRebuildAmericasInfrastructure1.24.17.pdf.

Chapter 20

1. Chris Reed, "Climate Readiness: Think Big, Act Fast," *Boston Globe,* March 8, 2018, https://www .bostonglobe.com/opinion/2018/03/08/climate-readiness-think-big-act-fast/WDKRVHmW KlocucHWodjGZJ/story.html.

2. "Network Visioning," *Sea Grant: Inside Sea Grant* (website) accessed July 9, 2018, https://seagrant .noaa.gov/insideseagrant/Implementation/Network-Visioning.

3. "Community Response to Flooding Proposal," *Sea Grant: Inside Sea Grant* (website) accessed July 9, 2018, https://seagrant.noaa.gov/Portals/1/Network%20Visioing/CommunityResponse toFlooding_NetVis_2017_Proposal.pdf.

4. "Climate and Weather Proposal," *Sea Grant: Inside Sea Grant* (website) 2017, accessed July 9, 2018, https://seagrant.noaa.gov/Portals/1/Network%20Visioing/ClimateandWeather_NetVis_2017 _Proposal.pdf.

5. Ashley Lawson, Katy Maher, and Janet Peace, *Guide to Public-Private Collaboration on City Climate Resilience Planning* (Arlington, VA: Center for Climate and Energy Solutions, 2017), https://www .c2es.org/site/assets/uploads/2017/05/guide-public-private-collaboration-city-climate-resilience -planning.pdf.

6. Meg Crawford and Stephen Seidel, *Weathering the Storm: Building Resilience to Climate Change* (Arlington, VA: Center for Climate and Energy Solutions, 2013), ix, https://www.c2es.org /document/weathering-the-storm-building-business-resilience-to-climate-change-2/.

7. "Networks," CERES (website) accessed July 9, 2018, https://www.ceres.org/networks.

8. "CERES Company Network," CERES (website) accessed July 9, 2018, https://www.ceres.org /networks/ceres-company-network.

9. Larry Kramer and Carol Larson, "Foundations Must Move Fast to Fight Climate Change," *Chronicle of Philanthropy* (April 2015): 1, https://www.philanthropy.com/article/Foundations -Must-Move-Fast-to/229509?cid=pt&utm_source=pt&utm_medium=en.

10. Herrmann, *The United States' Climate Change Relocation Plan*, 2, (see ch. 15, n. 17).

11. Ibid., 12.

12. Architects Foundation, *Forging Connections: The American Institute of Architects National Resilience Initiative Annual Report* (Washington, DC: American Institute of Architects, 2016), 1, http://aiad8 .prod.acquia-sites.com/sites/default/files/2017-05/NRI%20Annual%20Report%20FINAL %205-3-17%20%281%29_1.pdf.

13. "Policy Statement 162–Coastal Development," American Society of Civil Engineers (website) accessed July 9, 2018, https://www.asce.org/issues-and-advocacy/public-policy/policy-statement -162---coastal-development/.

14. "Policy Statement 545–Flood Risk Management," American Society of Civil Engineers (website) accessed July 9, 2018, https://www.asce.org/issues-and-advocacy/public-policy/policy -statement-545--flood-risk-management/.

15. Ibid.

16. "APA Policy Guide on Climate Change, Executive Summary," American Planning Association (website) 2011, accessed July 9, 2018, https://www.planning.org/policy/guides/adopted/climatechange.htm.

17. "APA Policy Guide on Climate Change," American Planning Association.

18. Ibid.,

19. Section of Environment, Energy, and Resources, *Policy* 109, American Bar Association, 2008), 1, https://www.americanbar.org/content/dam/aba/directories/policy/2008_my_109.pdf.

20. Standing Committee on Disaster Response and Preparedness, Section of State and Local Government Law; *Policy 108*, (American Bar Association, 2017), 1, https://www.americanbar.org /content/dam/aba/images/disaster/Resolution%20108%20FINAL.pdf.

21. "National Flood Insurance Program: Overview," National Association of Realtors (website) accessed July 9, 2018, https://www.nar.realtor/national-flood-insurance-program.

22. National League of Cities, *2018 National Municipal Policy and Resolutions* (Washington, DC: National League of Cities, 2017), 25, https://www.nlc.org/sites/default/files/users/user167/2018 -National-Municipal-Policy-Book.pdf.

23. "City Resiliency Planning Takes Data Analytic Turn. Will It Make a Difference?" United State Conference of Mayors (website) accessed July 9, 2018, https://www.usmayors.org/2018/02/05 /city-resiliency-planning-takes-data-analytic-turn-will-it-make-a-difference/.

24. Ibid.

25. "US Mayors Demonstrate Ambitious, Collective Climate Leadership," Climate Mayors (website) accessed July 13, 2018, http://climatemayors.org/.

26. National Association of Counties, *NACo American County Platform and Resolutions 2017–2018: Environment, Energy and Land Use* (Washington, DC: National Association of Counties, 2017), 22, http://www.naco.org/sites/default/files/documents/2017-2018%20EELU.pdf.

27. Avory Brookins, "RI Environmental Activists Rally For More Aggressive Climate Action During NGA Meeting," *Rhode Island Public Radio*, July 16, 2017, http://ripr.org/post/ri-environmental

-activists-rally-more-aggressive-climate-action-during-nga-meeting#stream/0.

28. Tim Faulkner, "Climate Change: Governors Talk Resilience, Protesters Demand Action," *ecoRI news,* July 17, 2017, https://www.ecori.org/climate-change/2017/7/17/ieg809naq5dt0k6g6otu50rpcukyz1.

29. "Serving the States," Coastal States Organization (website) accessed July 9, 2018, http://www.coastalstates.org/about-cso/.

30. Jake Thickman, *Beyond Elevation: Exploring a Holistic Approach to Coastal Flood Risk Management* (Madison, WI: Association of State Floodplain Managers, 2019) 1, https://www.floodsciencecenter.org/products/holistic-approach-coastal-flood-risk-management/.

31. "Natural Solutions Toolkit, Coastal Resilience, The Nature Conservancy (website) accessed July 9, 2018, http://coastalresilience.org/natural-solutions/toolkit/.

32. "Our History and Accomplishments," Union of Concerned Scientists (website) accessed July 9, 2018, https://www.ucsusa.org/about/history-of-accomplishments.html#.WrgIDOjwZPY.

33. Shannon Cunniff, "Amid Dramatic Sea Level Rise, Nature Itself Can Provide a Much-needed Solution," *Environmental Defense Fund* (blog) April 7, 2016, https://www.edf.org/blog/2016/04/07/amid-dramatic-sea-level-rise-nature-itself-can-provide-much-needed-solution.

34. Shannon E. Cunniff, "A Tale of Two Surveys: What Coastal Communities Need to Meet the Challenges Posed by Climate Change," *Shore & Beach* 84, no. 4, (Fall 2016): 42, https://www.edf.org/sites/default/files/Cunniff_Shore_and_Beach_Vol_84_No_4_Fall_2016.pdf.

35. "Champions of Surf & Sand," Surfrider Foundation (website) accessed July 9, 2018, https://www.surfrider.org/our-approach.

36. "Chapters: Our Network Grows Stronger with Every Chapter We Add," Surfrider Foundation (website) accessed July 9, 2018, http://www.surfrider.org/chapters.

37. "Managed Retreat," Beachapedia, Surfrider Foundation (website) accessed July 18, 2018, http://www.beachapedia.org/Managed_Retreat.

38. "The Coastal Society Mission," The Coastal Society (website) accessed December 20, 2018, https://new.thecoastalsociety.org/?page_id=10.

39. Pilkey, *The Rising Sea,* 163–4 ,(see part 4 epilogue, n. 1).

40. Science and Technology Committee, *Managing Sea Level Rise on Shore and Beaches* (Fort Myers, FL: American Shore and Beach Preservation Association, 2012), 1, http://asbpa.org/wpv2/wp-content/uploads/2016/04/Managing-Sea-Level-Rise-FINAL.pdf.

41. Pope Francis, *Encyclical Letter Laudato Si of the Holy Father Francis On Care of our Common Home* (Vatican, 2015), 20, http://w2.vatican.va/content/dam/francesco/pdf/encyclicals/documents/papa-francesco_20150524_enciclica-laudato-si_en.pdf.

42. Ibid., 34.

43. Anthony Leiserowitz et al., *Climate Change in the American Christian Mind* (New Haven: Yale Project on Climate Change Communication/George Mason Center for Climate Change Communication, 2015), 10, http://climatecommunication.yale.edu/wp-content/uploads/2015/04/Global-Warming-Religion-March-2015.pdf.

44. "Why We Exist," First Street Foundation (website) accessed August 2, 2018, https://firststreet.org/why-we-exist.

45. Ibid.

46. "Homepage," American Flood Coalition (website) accessed October 8, 2018, https://medium.com/american-flood-coalition.

47. "Our Policy Platform," American Flood Coalition (website) accessed December 20, 2018, https://floodcoalition.org/platform/#Economy.

48. Ibid.

49. "Climigration Platform Fosters Dialogue on Coastal Adaptation," Consensus Building Institute (website) accessed April 10, 2019, https://www.cbi.org/case/climigration-network-fosters-dialogue/.

50. "What We Do," Climate Central (website) accessed July 9, 2018, http://www.climatecentral.org/what-we-do#wwd.

51. "Higher Ground," Anthropocene Alliance (website) accessed May 14, 2019, https://anthropocene alliance.org/higherground.

52. Reed, "Climate Readiness: Think Big, Act Fast."

53. Ashley Dawson, *Extreme Cities: The Peril and Promise of Urban Life in the Age of Climate Change* (Brooklyn, NY: Verso, 2017), https://www.versobooks.com/books/2558-extreme-cities.

54. Erica Williams, "Aquatecture: Architectural Adaptation to Rising Sea Levels, (Tampa, FL: University of South Florida, 2009), vii, http://scholarcommons.usf.edu/cgi/viewcontent.cgi?article=1084&context=etd.

55. Katie Tannenbaum, "Surfrider Foundation's Stance on Beach Access," Beachapedia (website) accessed July 9, 2018, http://www.beachapedia.org/Beach_Access.

56. Thomas E. Dahl, *Wetlands Loss Since the Revolution* (St. Petersburg, FL: US Fish and Wildlife Service, 1990), 1, https://www.fws.gov/wetlands/Documents%5CWetlands-Loss-Since-the-Revolution.pdf.

57. Englander, "Testimony to House Subcommittee on Energy and Mineral Resources," July 28, 2015, 5, (see ch. 8, n. 23).

Appendix 5

1. Christian Bjørnæs, "A Guide to Representative Concentration Pathways," Center for International Climate Research (website) 2015 accessed July 9, 2018, http://www.cicero.oslo.no/en/posts/news/a-guide-to-representative-concentration-pathways.

Index